Chemical Diversity of Plant Specialized Metabolites
A Biosynthetic Approach

Chemical Diversity of Plant Specialized Metabolites

A Biosynthetic Approach

By

Bratati De

ROYAL SOCIETY OF CHEMISTRY

Print ISBN: 978-1-83767-045-1
PDF ISBN: 978-1-83767-147-2
EPUB ISBN: 978-1-83767-148-9

A catalogue record for this book is available from the British Library

The Royal Society of Chemistry is a charity, registered in England and Wales, Number 207890, and a company incorporated in England by Royal Charter (Registered No. RC000524), registered office: Burlington House, Piccadilly, London W1J 0BA, UK, Telephone: +44 (0) 20 7437 8656.

Visit our website at www.rsc.org/books

Printed in the United Kingdom by CPI Group (UK) Ltd, Croydon, CR0 4YY, UK

Preface

Plant specialized metabolites play diverse important roles in sessile plants to resist and withstand different biotic (pathogens, herbivores, insects, and other pests) and abiotic (salinity, drought, water logging, extreme high or low temperatures, UV-radiation, pollutants including heavy metals *etc.*) stresses. Such metabolites also play important roles in pollination, breeding, and plant propagation. In addition, plants are sources of different bioactive secondary metabolites (specialized metabolites) for food and medicine including modern medicine. So, these plant specialized metabolites are extensively studied and researched in different fields of the biological sciences (plant sciences, biochemistry, biotechnology, food, nutrition *etc.*), pharmaceutical sciences, medicinal chemistry, pharmacology, phytochemistry and pharmacognosy. This book provides the readers with a comprehensive knowledge of different classes of plant specialized metabolites, their diverse structures, biosynthesis, and functions. Compartmentations regarding the availability of metabolites and biosynthetic enzymes have also been discussed for some metabolites.

With more than 40 years of research and 32 years of teaching experience, particularly in phytochemistry, pharmacognosy, and plant stress biology, I felt the need for a book with updated information on structures, biosynthesis, and functions of plant specialized metabolites. So, after retirement, as Professor at University of Calcutta, India, I wished to publish a book supplementing some of these areas. Different reviews and research papers (up to 2021) were consulted to present updated information.

This book, dedicated to the Almighty, was written with tremendous support from my husband, my son Bratiraj, and daughter-in-law Smriti. I pay my gratitude to my late parents and grandparents who encouraged me always to contribute to the advancement of knowledge in the field of science. I am very much indebted to my husband, Chanchal, for tirelessly helping and

Chemical Diversity of Plant Specialized Metabolites: A Biosynthetic Approach
By Bratati De

Published by the Royal Society of Chemistry, www.rsc.org

supporting me throughout the process of writing this book. Marvin was used for drawing, displaying, and characterizing chemical structures, substructures and reactions, Marvin 17.21.0, ChemAxon (https://www.chemaxon.com). Kyoto Encyclopaedia of Genes and Genomics (KEGG), developed by Kanehisa Laboratories,[1–3] was consulted wherever necessary for drawing structures and biosynthetic pathways. Pubchem was also consulted occasionally for drawing structures. Finally, I wish to convey my acknowledgement to Royal Society of Chemistry for the publication of the book.

Bratati De
Formerly Professor of Botany
University of Calcutta, India
bratatide@hotmail.com

References

1. M. Kanehisa and S. Goto, KEGG: Kyoto Encyclopedia of genes and genomics, *Nucleic Acids Res.*, 2000, **28**, 27–30.
2. M. Kanehisa, Toward understanding the origin and evolution of cellular organisms, *Protein Sci.*, 2019, **28**, 1947–1951.
3. M. Kanehisa, M. Furumichi, Y. Sato, M. Ishiguro-Watanabe and M. Tanabe, KEGG: integrating viruses and cellular organisms, *Nucleic Acids Res.*, 2021, **49**, D545–D551.

Contents

Chemical Diversity of Plant Specialized Metabolites: A Biosynthetic Approach
By Bratati De

Published by the Royal Society of Chemistry, www.rsc.org

CHAPTER 1

Introduction

Plants have been used as sources of food and medicine throughout history. Physicians were initially interested in studying plants as these were the only sources of medicine available. Gradually chemists became involved in isolation and identification of plant metabolites. Many plant metabolites were isolated and identified by phytochemists. Plants are metabolically diverse because of their sessile mode of life which means they need to adapt to different adverse conditions. Two classes of plant metabolites are identified. These are primary metabolites and plant specialized metabolites. **Primary metabolites** are essential for cellular functions and are either substrates or products of primary metabolism. Primary metabolites include nucleotides in nucleic acids, amino acids in proteins, sugars as energy sources, polysaccharides, and phospholipids as constituents of cell membranes. In fact, the core of primary metabolism of plant species is mostly like that of non-plant species.[1] **Plant specialized metabolites** is a recently proposed name for what are traditionally called secondary metabolites[2] because exact biochemical boundaries between primary and secondary metabolites have not been established.[3] Plant specialized metabolites are products of metabolism by different plants specialized for specific environmental interactions.[4] Moreover, these metabolites have applications in humans for alleviating human ailments as pharmaceutical drugs, as nutrients, and for many other purposes in human daily life such as flavourings, dyes, natural preservatives *etc.* Plant specialized metabolites belong to different classes such as different types of phenolic compounds, terpenoids, alkaloids and nitrogen-containing compounds, sulfur containing compounds, and fatty acid derivatives.[1] Glycosylation also plays a major role in diversifying the structures of plant specialized metabolites. Thus, chemically diverse plant specialized metabolites are produced by different plants. Glycosylation has an impact on the properties of a metabolite affecting solubility, cellular localization, and bioactivity. Glycosides of

Chemical Diversity of Plant Specialized Metabolites: A Biosynthetic Approach
By Bratati De

Published by the Royal Society of Chemistry, www.rsc.org

plant specialized metabolites help in attracting pollinators, and many other biological functions.[5] Usually, mono-, di- or oligosaccharides are involved in the glycosylation process. Diversifications also arise due to prenylation, methoxylation, acylation *etc.*

During coevolution with animals and microorganisms, sessile plants developed some adaptive strategies by producing metabolites having specialized functions such as to defend against these organisms, and to allow continuing reproduction in different harsh environments.[6,7] Moreover, there is sufficient evidence to suggest that individual specialized metabolites are not always secondary at all. Specialized metabolites can also be produced from different precursors in primary metabolism. There are approximately 4 000 000 extant species of vascular plants.[8] These plants are sources of between 200 000 and 1 million plant specialized metabolites.[1,9]

Plant specialized metabolites have important roles in the global ecosystem and plant life in different ways such as providing signals to pollinators through volatiles or colour, serving as allelochemicals and herbivore deterrents, inducing chemical defences called phytoalexins, alleviating damage to abiotic stress conditions, establishing symbiotic interactions, providing phytohormonal responses, protecting plants from UV light and desiccation, systematic movement of water, storage of compounds, providing nutritional value and aroma to fruits and mediating phototropism.[1,2,6] Plant specialized metabolites can donate or accept electrons and often possess both pro- and antioxidant activities. Different taxa produce different specialized metabolites by different environmental interactions.[4]

The interest in these plant specialized metabolites is fast expanding and drawing the attention of scientists from different areas of expertise. Apart from drug discovery, many other areas of research on these metabolites including the biosynthesis of specialized metabolites and the role and mechanism of plant environment interactions are being unravelled connecting chemistry to function.[10] This has been possible due to innovations in the fields of genomics, transcriptomics, proteomics and metabolomics using very high-end instrumental facilities along with computational tools. Thus, a vast amount of knowledge and information has been generated in the highly popular subject area of plant specialized metabolites. It is not possible to accommodate all areas of this subject here, not just because of its vastness but also due to the multidisciplinary aspects of the subject.

Many plant metabolites are used as drugs for treating different human ailments either without modification or as precursors to produce semisynthetic compounds. The bottlenecks for utilizing specialized plant products as sources of drugs are low yield, low purification efficiency, endangered or slow growing plant sources, infeasible total synthesis *etc.*[7] Knowledge about biosynthetic processes would help in overcoming some of the bottlenecks to boost production in organisms by metabolic engineering.

Recent discoveries have revealed that the plant metabolites regularly travel between two or more compartments for completion of biosynthetic processes. Such compartments may occur at the cellular, intracellular (plastids, vacuoles, peroxisomes, endoplasmic reticulum, phenyloplasts, tannosomes *etc.*), tissue or interorgan level. For production or accumulation of plant specialized metabolites, some dedicated structures are often found as compartments *e.g.*, laticifers, resin ducts, nectaries or glandular hairs. Interspecies interactions between plants and fungi/bacteria are also reported for production of plant specialized metabolites. Often a single plant produces a different profile of specialized metabolites in different tissues. Physiological states, developmental stage, and environmental conditions also affect formation of specialized metabolites.[11]

There are four major pathways for specialized metabolism. These are (1) the acetate malonate pathway for biosynthesis of fatty acid derivatives and polyketides, (2) the acetate mevalonate and methylerythritol phosphate pathway for biosynthesis of terpenoids, (3) the shikimic acid pathway for phenols, and (4) the alkaloid metabolism from amino acids. All these metabolic pathways are dependent on primary metabolic pathways and form a complex metabolic grid. However, other simple molecules like sugars often are tools to produce specialized plant metabolites by glycosylation to bring diversity in structure. Furthermore, many carbohydrates elicit defense responses, accumulate Reactive Oxygen Species (ROS) and pathogenesis-related proteins, protect plants against pathogen infections and play an important role in plant immunity.[12] As such, carbohydrates and their role in plant protection and human utilization are also discussed in this book, apart from the other four metabolic pathways.

References

1. C. Garagounis, N. Delkis, and K. K. Papadopoulou, Unravelling the roles of plant specialized metabolites: using synthetic biology to design molecular biosensors, *New Phytol.*, 2021, **231**, 1338–1352.
2. L. Yuan, and E. Grotewold, Plant specialized metabolism, *Plant Sci.*, 2020, **298**, 110579.
3. M. Erb, and D. J. Kliebenstein, Plant secondary metabolites as defenses, regulators, and primary metabolites: the blurred functional trichotomy, *Plant Physiol.*, 2020, **184**, 39–52.
4. L. Chae, T. Kim, R. Nilo-Poyanco, and S. Y. Rhee, Genomic signatures of specialized metabolism in plants, *Science*, 2014, **344**, 510–513.
5. T. Louveau, and A. Osbourn, The sweet side of plant-specialized metabolism, in *Additional Perspectives on Engineering Plants for Agriculture*, ed. P. C. Ronald, Cold Spring Harbor Perspectives in Biology, 2019, vol. 11, p. a034744.
6. E. Lachhini, and A. Goossens, Combinatorial control of plant specialized metabolism: mechanisms, functions, and consequences, *Annu. Rev. Cell Dev. Biol.*, 2020, **36**, 1.1–1.23.
7. J. R. Jacobowitz, and J. Weng, Exploring uncharted territories of plant specialized metabolism in the postgenomic era, *Annu. Rev. Plant Biol.*, 2020, **71**, 27.1–27.28.

8. *State of the World's Plants 2017. Report (London: Royal Botanic Gardens, Kew)*, ed. K. J. Willis, 2017.
9. S. Wang, S. Alseekh, A. R. Fernie, and J. Luo, The structure and function of major plant metabolite modifications, *Mol. Plant*, 2019, **12**, 899–919.
10. J. Weng, J. H. Lynch, J. O. Matos, and N. Dudareva, Adaptive mechanisms of plant specialized metabolism connecting chemistry to function, *Nat. Chem. Biol.*, 2021, **17**, 1037–1045.
11. A. Tissier, J. Zeigler, and T. Vogt, Specialized plant metabolites: Diversity and biosynthesis, in *Ecological Biochemistry: Environmental and Interspecies Interactions*, ed. K. Gerd-Joachim, and H. N. Dietrich, Wiley-VCH Verlag GmbH & Co. KGaA, 2015.
12. S. Trouvelot, M. Heloir, B. Poinssot, A. Gauthier, F. Paris, C. Guillier, M. Combier, L. Trda, X. Daire, and M. Adrian, Carbohydrates in plant Immunity and Plant protection: roles and potential application as foliar sprays, *Front. Plant Sci.*, 2014, **5**, 592.

CHAPTER 2

Carbohydrates and Glycosides

2.1 Carbohydrates

Carbohydrates are polyhydroxy aldehydes or ketones and their derivatives with the general formula $C_x(H_2O)_y$.[1,2] Carbohydrates may be categorized into three classes. These are monosaccharides, di- and oligosaccharides, and polysaccharides.

2.1.1 Monosaccharides

Monosaccharides may be of different types depending on the number of carbon atoms such as trioses (C = 3), tetroses (C = 4), pentoses (C = 5), hexoses (C = 6), *etc.* Glyceraldehyde and dihydroxyacetones are the simplest monosaccharide trioses. Monosaccharides occur in two forms D and L. According to the linear form as shown in Fischer projection, the aldehyde group (–CHO) of glyceraldehyde is at the top. Glyceraldehyde with the aldehyde group leads to a series of aldoses (Figure 2.1). Dihydroxyacetone with the keto group leads to ketoses[3] (Figure 2.2).

The monosaccharides with ≥ five carbons may also exist in cyclic hemiacetal ring form (*e.g.*, the six membered ring pyranose) or hemiketal ring form (*e.g.*, the five membered ring furanose) (Figure 2.3). The cyclic hemiacetal has one more asymmetric carbon than that of the linear form. For example, glucose in the linear form has four asymmetric carbons, whereas the cyclic form of glucose has five asymmetric carbons. The newly formed asymmetric carbon (C1) has two anomers indicated as α and β[2,3] (Figure 2.4). The cyclic monosaccharides may be drawn in several ways as shown in Figure 2.4.

Some **deoxy sugars** such as the deoxy hexoses *e.g.*, rhamnose (6-deoxy-L-mannose) and fucose (6-deoxy-galactose) (Figure 2.5) are found in plants as part of polysaccharides or glycosides. Formation of acids (known as sugar

Chemical Diversity of Plant Specialized Metabolites: A Biosynthetic Approach
By Bratati De

Published by the Royal Society of Chemistry, www.rsc.org

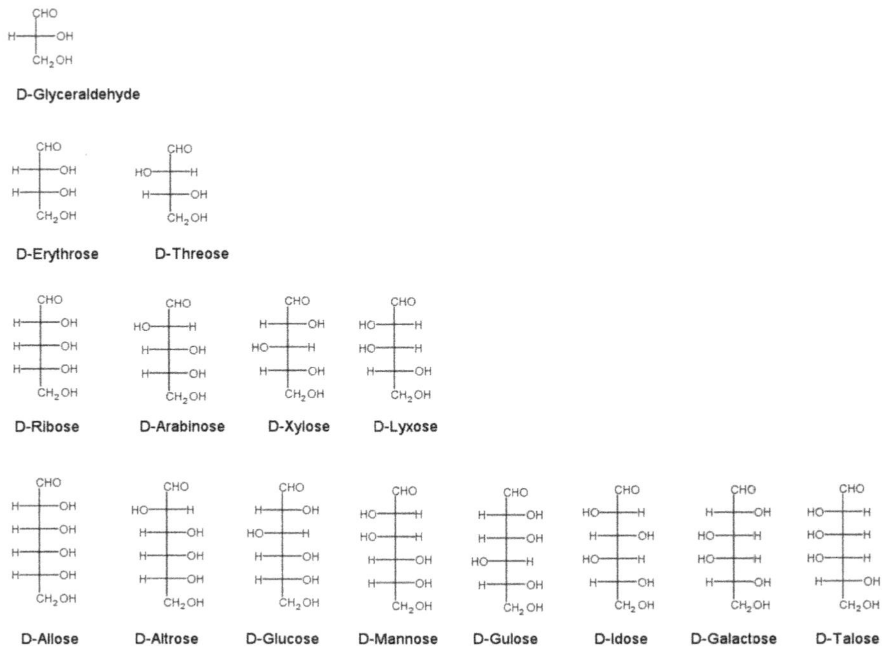

Figure 2.1 Linear D- series aldoses.

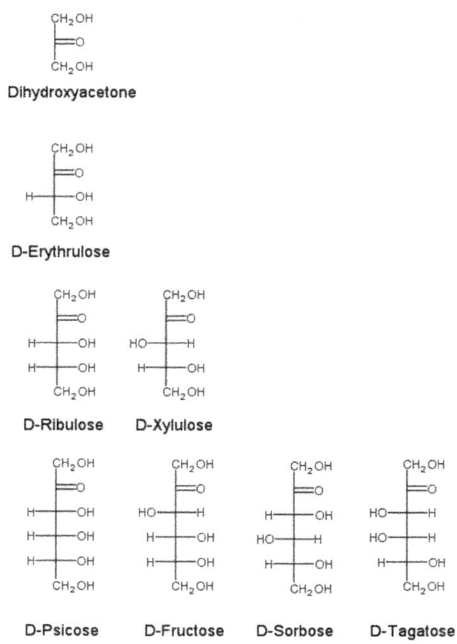

Figure 2.2 Linear ketoses.

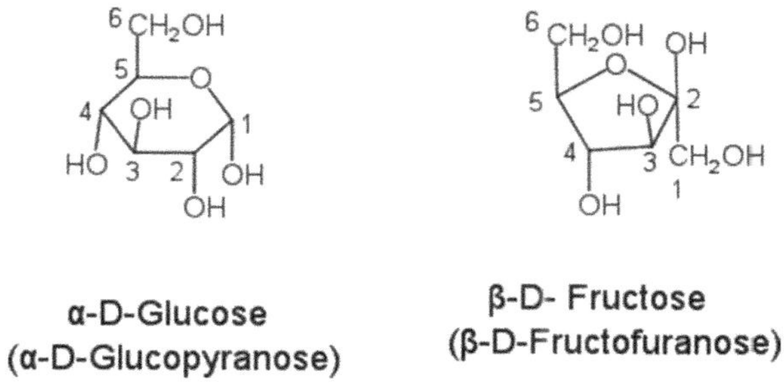

Figure 2.3 Pyranose and furanose forms of monosaccharides.

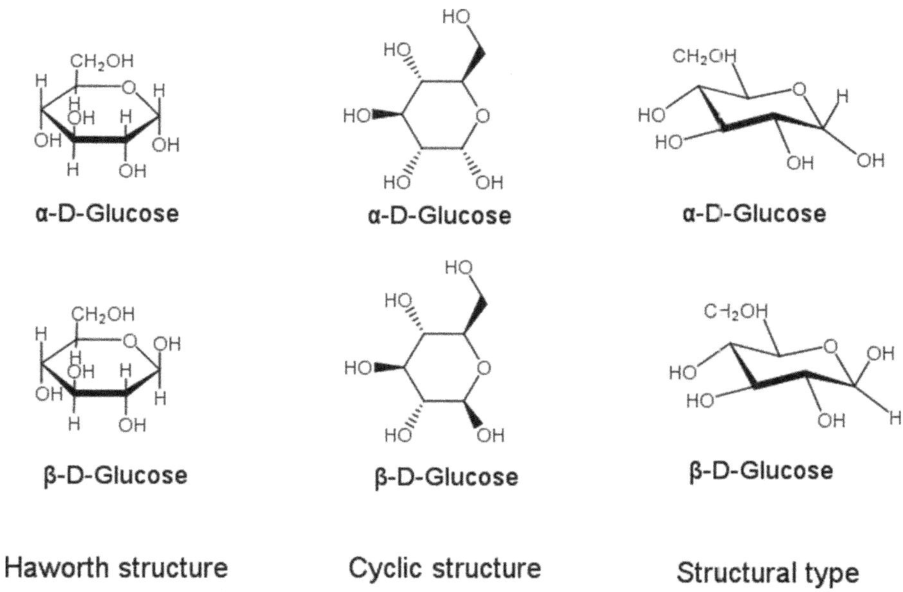

Figure 2.4 Various structural representations using D-glucose as an example.

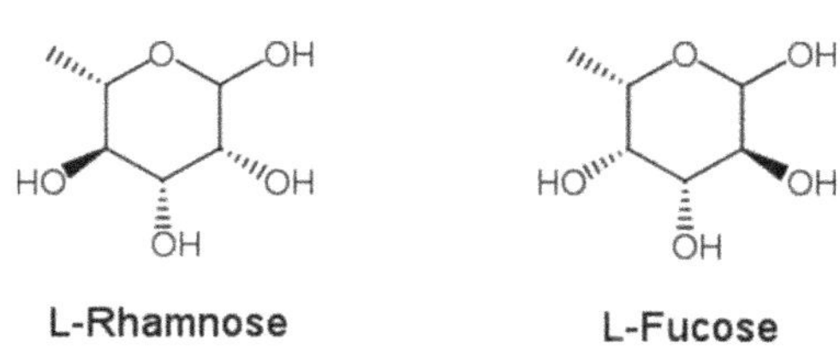

Figure 2.5 Deoxy hexoses.

acids) is often detected in polysaccharides due to oxidation of the substrate. For example, **aldonic acids** (D-gluconic acid, D-xylonic acid, L-arabonic acid, D-galactonic acid, D-mannonic acid) are formed by the oxidation of the terminal aldehyde group of a monosaccharide to form a carboxylic acid. **Uronic acids** (D-glucuronic acid, D-xyluronic acid, L-arabinuronic acid, D-galacturonic acid, D-mannuronic acid) are generated by the replacement of the $-CH_2OH$ group of monosaccharides by a carboxyl group. **Aldaric acids** (D-glucaric acid, D-xylaric acid, L-arabinaric acid, *meso*-galactaric acid, D-mannaric acid) are formed by oxidizing both the aldehyde and the terminal hydroxyl groups of an aldose, generating the dicarboxylic acid.[3,4] Saccharic acid is generated by simultaneous oxidation and replacement.[3]

Polyols (sugar alcohols) are reduced forms of aldose and ketose sugars and are found naturally in fruits, vegetables, and mushrooms. The frequently found polyols in plants are mannitol, sorbitol, glycerol, arabitol and erythritol.[5] Cyclitol and myo-inositol are ubiquitous. Mannitol and sorbitol are primary photosynthetic products in mature leaves.

2.1.2 Oligosaccharides

An oligosaccharide is a carbohydrate consisting of 3–10 monosaccharide units linked by *O*-glycosidic bonds. However, many researchers consider that the carbohydrates consisting of 2–10 monosaccharide units are oligosaccharides and carbohydrates with a degree of polymerization of up to 20–25 units are often considered as oligosaccharides.[6,7] Oligosaccharides may be of different types such as primary oligosaccharides and secondary oligosaccharides.[8]

Primary oligosaccharides are synthesized in the plant from a mono or oligosaccharide and a glucosyl donor by the action of glycosyl transferase. Sucrose, a disaccharide, is the most common primary oligosaccharide. In plants, primary oligosaccharides are mainly composed of glucose oligosaccharides, oligosaccharides based on sucrose, and oligosaccharides containing polyhydric alcohols[6,8] (Table 2.1, Figure 2.6). **Secondary oligosaccharides** are formed from the hydrolysis of primary oligosaccharides or higher oligosaccharides and polysaccharides[8] (Table 2.2).

Table 2.1 Primary oligosaccharides.

Oligosaccharides	Examples	Sources
Sucrose (Disaccharide)		
Glucose oligosaccharides	α,α-**Trehalose:**O-α-D-glucopyranosyl-(1→1)-α-D-glucopyranoside	
	Selaginose:O-α-D-glucopyranosyl-(1→2)-O-α-D-glucopyranosyl-(1→1)-O-α-D-glucopyranoside	Soluble reserve carbohydrate in pteridophytes

Table 2.1 (*Continued*)

Oligosaccharides	Examples	Sources
Oligosaccharides based on sucrose	**Raffinose family oligosaccharides**: Formed by continuous transfer of D-galactopyranosyl, D-glucopyranosyl or D-fructofuranosyl units to sucrose. The process can continue up to nonasaccharide.	
	Raffinose (Trisaccharide): α-D-galactopyranosyl-(1→6)-α-D-glucopyranosyl-(1→2)-α-D-fructofuranoside	Fabaceae
	Stachyose: α-D-galactopyranosyl-(1→6)-α-D-galactopyranosyl-(1→6)-α-D-glucopyranosyl-(1→2)-α-D-fructofuranoside	Fabaceae
	Verbascose: α-D-galactopyranosyl-(1→6)-[α-D-galactopyranosyl-(1→6)]$_2$-α-D-glucopyranosyl-(1→2)-α-D-fructofuranoside	Fabaceae
	Fructan (also known as **fructooligosaccharides** or **FOS**): this group is classified into four subgroups	
	1-Kestose series: [β-D-fructofuranosyl-(2→1)-β-D-fructofuranosyl]$_n$-α-D-glucopyranoside; (n = 1,2 to 8) *e.g.*, **1-Kestotriose (isokestose), 1,1-kestotetraose (nystose), 1,1,1-kestopentaose (fructosylnystose)**	Liliaceae, Asteraceae
	6-Kestose series: [β-D-fructofuranosyl-(2→6)-β-D-fructofuranosyl]$_n$-α-D-glucopyranoside; (n = 1,2 to 8): Fructosyl–fructose linkage is β-(2→6) and the fructosyl–glucose linkage is β-(2→1). Examples are: **Graminan** (for branched structures) and **phleins** (for linear structures).	Poaceae
	Mixed type: [β-D-fructofuranosyl-(2→6)-β-D-fructofuranosyl]$_n$-[β-D-fructofuranosyl-(2→1)-β-D-fructofuranosyl]*m*-α-D-glucopyranoside ($n + m$ = 2 to 8) *e.g.*, **bifurcose**	Poaceae, Agavaceae
	Neoseries: [*O*-β-D-fructofuranosyl-(1→6)-β-D-fructofuranosyl]$_n$-α-D-glucopyranoside (n = 1, 2 to 10). example: **neokestose**	Asparagaceae, Liliaceae
Oligosaccharide containing polyhydric alcohol	**Galactinol:***O*-α-D-galactopyranosyl-(1→1)-myo-inositol	

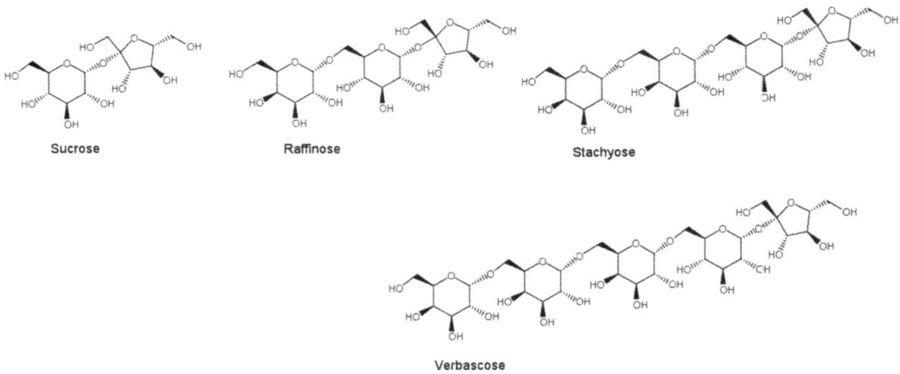

Figure 2.6 Raffinose family oligosaccharides.

Table 2.2 Secondary oligosaccharides.

Formed by hydrolysis of primary oligosaccharides	**Melibiose** [*O*-α-D-galactopyranosyl-(1→6)-D-glucose] is produced from raffinose [α-D-galactopyranosyl-(1→6)-α-D-glucopyranosyl-(1→2)-α-D-fructofuranoside] after removal of D-fructose unit.
	Gentiobiose [*O*-β-D-glucopyranosyl-(1→6)-D-glucose] is produced from gentianose [*O*-β-D-glucopyranosyl-(1→6)-*O*-α-D-glucopyranosyl-(1→2)-β-D-fructofuranoside]
Formed by hydrolysis of higher oligosaccharides or polysaccharides	**Cellobiose** [*O*-β-D-glucopyranosyl-(1→4)-D-glucose] is produced from cellulose.
	Isomaltose [*O*-α-D-glucopyranosyl-(1→6)-D-glucose] is produced from branching unit of amylopectin.
	Amylose and **amylopectin** from starch

Non-digestible oligosaccharides: Based on digestibility, some carbohydrates are digested easily by human digestive enzymes, and some are not. Non-digestible oligosaccharides (NDOs) are carbohydrates where the anomeric C atom (C1 or C2) of the monosaccharide units of the oligosaccharides has a configuration that makes their osidic bonds resist hydrolysis by human salivary and intestinal digestive enzymes. NDOs occur naturally in many plants such as whole grains, fruits, and vegetables. The NDOs of the main food ingredients contain monosaccharide units *e.g.*, fructose, galactose, glucose, and/or xylose.[9,10] Different plants are natural sources of NDOs.[11,12] Examples of some NDOs along with their sources are given in Table 2.3.

2.1.3 Polysaccharides

Polysaccharides are complex carbohydrates with more than 10 repeating units of monosaccharides bound by glycosidic linkages. Some examples are starches, cellulose, hemicelluloses, pectin, gums and mucilages.

Table 2.3 Non-digestible oligosaccharides.

Non-digestible oligosaccharides	Sources
Galactooligosaccharides include raffinose, stachyose, verbascose	Soybean, pulses, and other seeds of Fabaceae
Fructooligosaccharides consisting of a linear chain of fructose units linked by β-(2→1)-linkage	Asparagus, sugar beet, garlic, onion, wheat, artichoke, banana, tomato, honey
Xylooligosaccharides (the mixture of xylobiose, xylotriose, xylotetrose)	Bamboo shoots, fruits, vegetables, honey, milk
Isomaltulose (or palatinose)	Honey, sugarcane juice
Inulin type fructans with linear chain of fructose polymer (β2→1 linkages)	Onion, garlic, asparagus, chicory roots, artichoke
Soybean oligosaccharides that include raffinose, sucrose, xylose	Soybean seeds and other legumes

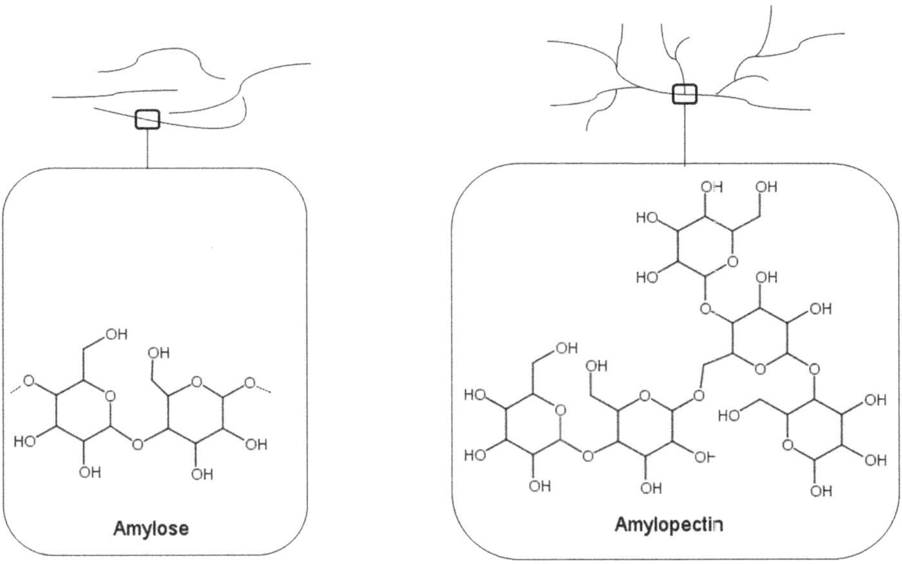

Figure 2.7 Amylose and amylopectin structures.

2.1.3.1 *Starches*

Starches are the most common digestible polysaccharides found in plants. Starch has two structurally different glucan polymers: amylose and amylopectin (Figure 2.7). Normally, starches contain 20–30% amylose and 70–80% amylopectin. Amylose contains 250 to 300 D-glucopyranose units linked by α-(1→4) glucosidic bonds forming a linear, unbranched chain. Amylopectin consists of more than 1000 glucose units linked by α-(1→4) linkages and α-(1→6) glycosidic bonds occurring at branch points, thus forming highly branched chains. The α-amylase present in pancreatic juice and saliva hydrolyzes the α-(1→4) glucosidic linkages randomly producing a mixture of glucose, maltose, and amylopectin. β-Amylase acts on the α-(1→4) linkages of the amylopectin from the nonreducing ends until a branch point is approached. β-Amylase lacks the capacity to hydrolyze α-(1→6) glycosidic bonds. The amylose and amylopectin are deposited in distinct granules of different shapes and sizes, depending on the plant, in the amyloplasts in storage organs. On the starch granular surface there may be polar lipids, mainly phospholipids. The fatty acid composition of wheat starch is composed of palmitic acid, linoleic acid, and oleic acid.[13–15]

2.1.3.2 Cellulose

Cellulose is one of the structural components of plant cell walls. Cotton fibers contain almost pure cellulose. Chemically cellulose is a linear polymer consisting of β-D-glucose monomers linked by β-glucoside (1→4) linkages[13] (Figure 2.8). The number of monomer units may vary between 200 to about 3000 or even as high as 10 000.

2.1.3.3 Hemicelluloses

Hemicelluloses are a heterogenous group of cell wall polysaccharides which can be grouped into xyloglucan, xylans, mannans, glucomannans, and β-(1→3,1→4)-glucans.[16]

Xyloglucans (Figure 2.9) are detected in all land plants. Xyloglucans are one of the major hemicelluloses in primary walls.[16] Xyloglucan is composed of β-(1→4)-linked glucosyl residues. Half or three quarters of the glucosyl residues are substituted by α-linked xylosyl residues on the 6-position. Depending on the plant species or tissue type, many of the xylose units are further replaced by sugars like galactose and fucose.[17]

Xylan polymer consists of repeating β-(1→4)-linked xylose residues. Xylans are present in the primary walls of grasses and the secondary walls of woody

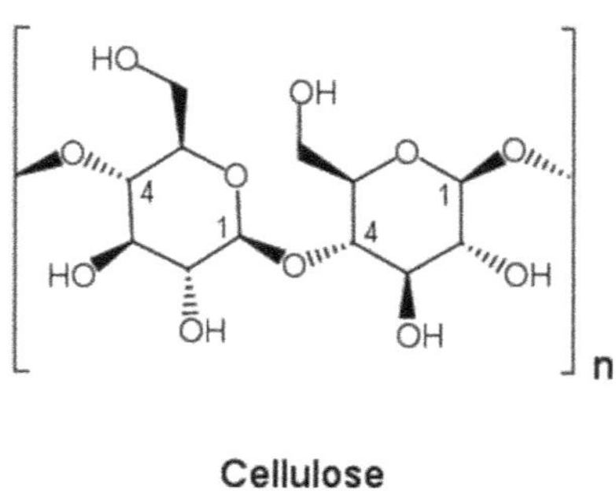

Figure 2.8 Cellulose structure.

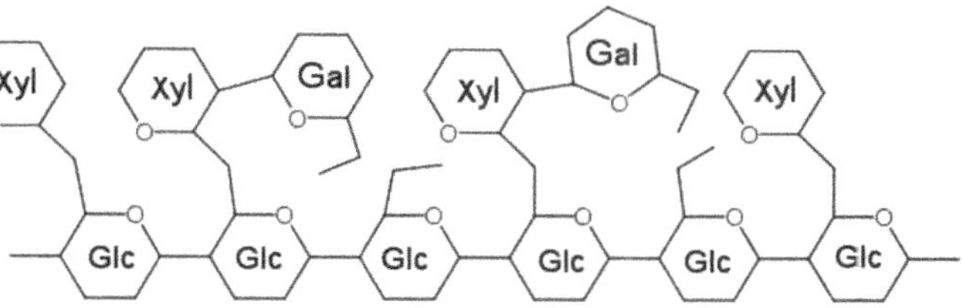

Figure 2.9 Schematic diagram of xyloglucan. Xyl: D-xylose; Glc: D-glucose; Gal: D-galactose.

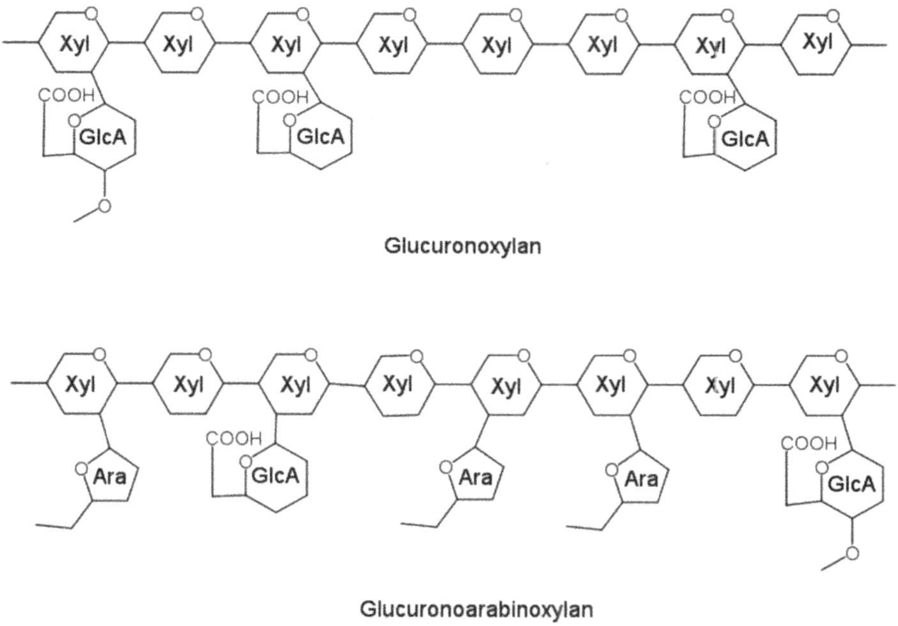

Figure 2.10 Schematic diagram of xylans. Xyl: D-xylose; GlcA: D-glucuronic acid; Ara: L-arabinose.

plants. Xylans are substituted by glucuronosyl and (4-*O*-methyl)-α-D-glucuronic acid side groups. Xylans with such substitutions are known as **glucuronoxylans** (Figure 2.10). Glucuronoxylans are the polysaccharides dominating the secondary walls of dicots. Xylans with attached arabinose residues are known as **arabinoxylans** and **glucuronoarabinoxylans** (Figure 2.10). Cell walls in gymnosperms contain high amounts of arabinoxylans. **Arabinofuranose** substituents in the cell walls of grasses are from *O*-3 of the backbone xylose residues. Grass xylans often have xylose residues substituted by **feruloyl-arabinofuranosyl** side chains.[16]

Mannans are polysaccharides containing β-(1→4)-linked mannose units. Mannans with 80–95% mannose sugar are found in seeds of Apiaceae.

The main hemicelluloses of Charophytes are mannans. Mannans with D-galactose branching at every C-6 position are known as **galactomannans.** Endosperms of the seeds in Fabaceae contain galactomannans. True mannose when interspersed by β-D-glucosyl residues is known as **glucomannan**. Glucomannans are found in corm of *Amorphophallus konjac* (Araceae), ferns such as *Equisetum giganteum* (Equisetaceae), *Selaginella kraussiana* (Selaginellaceae), *Adiantum raddianum* (Pteridaceae), and *Microgramma squamulose* (Polypodiaceae). **Galactoglucomannan** (Figure 2.11) consists of β-(1→4)-D-mannosyl, and glucosyl residues about 8% of which are substituted by α-(1→6)-D-galactosyl side groups. **Acetylated galactoglucomannans** are found in woods of gymnosperms, in some angiosperms (*Rubus fruticosus*, Rosaceae) and in ferns.[16,18]

β-(1→3, 1→4)-Glucans (Figure 2.12) are not found in dicots. β-(1→4)-Linked glucans with interspersed single β-(1→3) linkages are found in grasses (Poaceae), *Equisetum*, Charophytes, and red algae.[16]

2.1.3.4 *Pectin*

Pectins are polysaccharides found in the cell walls and intercellular layers in plants except in woody tissues. Chemically, the pectin structure is highly complex. It is a heterogeneous polysaccharide consisting of regions of homogalacturonans alternating with highly branched rhamnogalacturonan

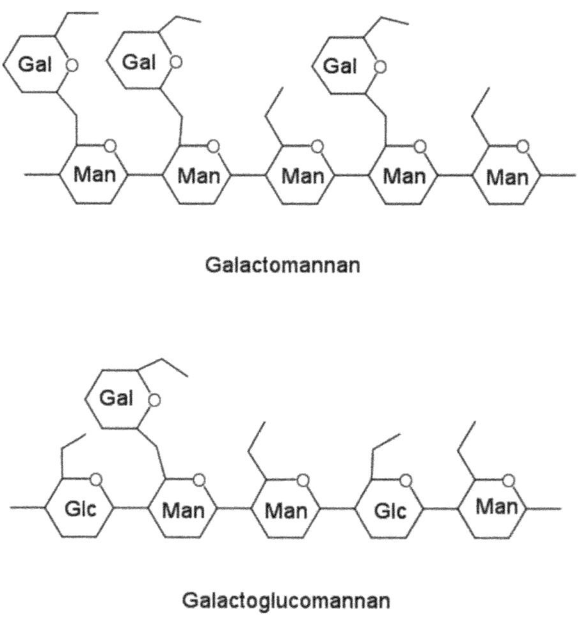

Figure 2.11 Schematic diagram of mannans. Man: D-mannose; Glc: D-glucose; Gal: D-galactose.

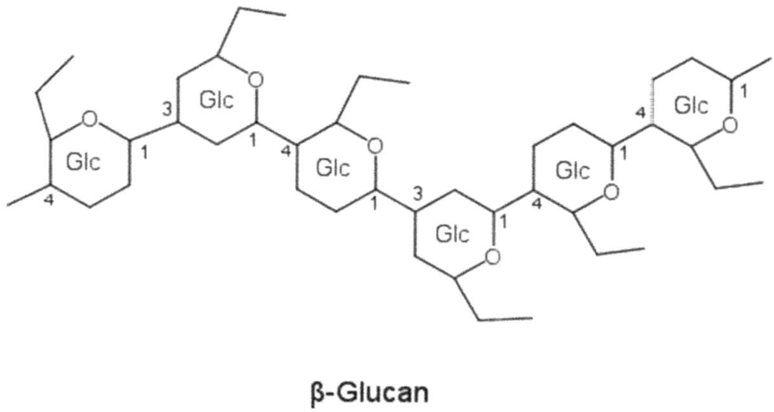

Figure 2.12 Schematic diagram of β-glucan: β-(1→3, 1→4)-glucan. Glc: D-glucose.

regions. Other polysaccharides such as xylogalacturonans, arabinogalactans, and arabinans are also included. **Homogalacturonans** are linear polymers of galacturonic acids joined by α-(1→4)-glycosidic linkages. The carboxylic groups at C-6 of galacturonic acid units are partially methyl esterified. The free acids are partly neutralized with sodium, potassium, or ammonium ions. Pectin may also be acetylated on *O*-2 or *O*-3 positions. Homogalacturonans are sometimes substituted with β-linked D-xylose units at *O*-3 and occasionally at *O*-4 when they are known as xylogalacturonans. The backbone of **rhamnogalacturonans** contains repeating [α-galacturonic acid-α-(1→2)-α-rhamnose-α (1→4)-]$_n$ units. Rhamnogalacturonan is also substituted by arabinan, galactan and arabinogalactans in the side chain attached to the *O*-4 position of rhamnose. Some of the rhamnogalacturonans are substituted by other unusual sugars such as apiose, 3-*O*-methyl-L-fucose, 2-*O*-methyl-D-xylose, 3-deoxy-D-manno-octulosonic acid or 3-deoxy-D-lyxo-heptulosaric acid. Ferulated pectins have also been reported in many plants.[19]

2.1.3.5 *Gums and Mucilages*

Both gums and mucilages are plant hydrocolloids yielding sugars and uronic acids on hydrolysis. Gums are pathological products, formed after injury to the plant and consist of calcium, potassium, magnesium salts of complex substances known as polyuronides. Mucilages are physiological products (products of metabolism) formed within the cells.[13,20] Gums and mucilages contain pentose, galactose or methyl pentose sugar joined to uronic acid residues by glycosidic linkages. The terminal carbon of the monosaccharide unit may occur in an oxidized (carboxylic acid) form.[21] Starch and cellulose and their derivatives are also considered as gums.[13]

Gums from plant and algal origin[22–30] may be classified as shown in Table 2.4. Gums are also obtained from plant extracts (*e.g.*, pectin), bacterial and fungal sources (xanthan, dextran, pullulan, zanflo, Baker's yeast gum, *etc.*) and animal sources (chitin, chitosan, hyaluronic acid, chondroitin sulfate).[13,20–22] Some mucilages of plant origin[22,29,31] are mentioned in Table 2.5.

2.1.4 Role of Carbohydrates in Plants

Carbohydrates are sources of energy, carbon skeletons of organic compounds, and storage components. Cellulose, hemicellulose, and pectin are structural components of cell walls. Sugars and starch are involved in transport, energy metabolism, and osmoregulation. Oligogalacturonides, released from plant cell walls upon partial degradation of homogalacturonan, elicit defense responses, accumulate Reactive Oxygen Species (ROS) and pathogenesis-related proteins, and protect plants against pathogen infections.[33] Carbohydrates act to communicate and coordinate metabolism with growth, development, and responses to environmental changes and stress. One of the adverse factors of various stressful conditions is a disturbance in osmoregulation. Osmotically active compounds accumulate in cells to resist water deficit. These compounds, known as compatible solutes, are polyols (glycerols, mannitol, sorbitol), trehalose, fructans, and methylated inositol in addition to amino acids and their derivatives, polyamines. Compatible solutes accumulate to avoid the toxic effect of stresses such as cold climate, saline soil, and water deficit.[34] Polyols are osmotically active metabolites produced in response to abiotic stress. They accumulate at high levels inside a cell to compensate reduced water potential of the stressed cells and protect enzyme activities. The osmoprotectant properties are due to the water-like hydroxyl groups in the polyols. Polyols also play a significant role in stress responses because of their antioxidant properties.[35] Plants also accumulate soluble sugars like hexoses (glucose and fructose), disaccharides (sucrose, trehalose), and oligosaccharides (raffinose, stachyose) as compatible osmolytes. Many such carbohydrates accumulate depending on stress type and type of resistance/sensitivity of plants under the stressed condition. Compatible osmolytes fulfill osmotic adjustment, stabilize the photosystem II complex, maintain membrane structure, and scavenge ROS. The raffinose family of oligosaccharides act as signaling molecules in response to different biotic and abiotic stresses in plants. Sugars also act as antioxidants as they possess properties to scavenge ROS.[35,36]

Starch is emerging as a key molecule in mediating plant responses to abiotic stresses such as salinity, drought and extreme temperature. The sugars and other metabolites released from starch support plant growth under stress and function as osmoprotectants and compatible solutes. Sugars also act as signaling molecules and crosstalk with ABA dependent signaling pathways to activate components in the stress response cascade. Different combinations of starch degrading enzymes operate under different stress conditions.[37]

Table 2.4 Gums of plant and algal origin.

Gums	Source	Structure	Uses[2,13,20,22]
Tree or shrub exudates			
Arabic gum	*Senegalia senegal* (synonym: *Acacia senegal*), (Fabaceae)	(1→3)-Linked β-D galactopyranosyl units with some side chains, both containing units of β-D-glucopyranosyl, 4-*O*-methyl-β-D-glucopyranosyl, α-L-arabinofuranosyl and α-L-rhamnopyranosyl.[23]	Used as a food emulsifier and stabilizer in the food industry; a dietary fibre ingredient in food; suspending and emulsifying agent, tablet binder, demulcent, and emollient in cosmetics; adhesive
Karaya gum (Sterculia gum)	*Sterculia urens*, and some other species of *Sterculia* (Malvaceae)	A heteropolymer composed of partially acetylated chains with high concentrations of uronic acid, galactose, rhamnose, and xylose.[24]	Suspending agent, emulsifying agent; dental adhesive; bulk laxative
Tragacanth gum	*Astragalus brachycalyx*, *Astragalus gummifer* (Fabaceae)	The gum is a mixture of two main fractions: tragacanthin and bassorin. Tragacanthin is composed of tragacanthic acid containing residues of D-galacturonic acid, D-xylose, L-fucose, D-galactose, and arabinogalactan. Bassorin is formed as an insoluble fraction of the gel.[25]	Suspending and emulsifying agent, sustained release agent in pharmacy; demulcent, emollient in cosmetics
Gum ghatti	*Anogeissus latifolia* (Combretaceae)	Complex structure of the gum consists of sugars like L-arabinose, D-galactose, D-mannose, D-xylose, D-glucuronic acid, and alternating 4-*O*-substituted and 2-*O*-substituted α-D-mannopyranose units and chains of 1→6 linked β-D-galactopyranose units with side chains of L-arabinofuranose residues.[26]	Emulsifier, suspending agent, coating material in tablets

(*continued*)

Table 2.4 (*Continued*)

Gums	Source	Structure	Uses[2,13,20,22]
Seed gums			
Guar gum	Seeds of *Cyamopsis tetragonoloba* (Fabaceae)	Mainly consists of high molecular weight polysaccharides of galactomannans that are a linear chain of (1→4)-linked β-D-mannopyranosyl units with (1→6)-linked α-D-galactopyranosyl residues.[27]	Tablet binder, disintegrant, thickening agent, suspending agent, sustained release agent, emulsifier, laxative, colon targeted drug delivery.
Locust bean gum	Endosperm of the seeds of *Ceratonia siliqua* (Fabaceae)	The gum contains D-galacto-D-mannoglycan, pentane, proteins, and cellulose.[22]	Used for coating tablets, disintegrant, suspending agent, acts as a suitable carrier for colonic drug delivery.
Algal gums			
Agar	Red algae (Rhodophyceae) including *Gelidium cartilagineum,* some species of *Gracilaria* (Gelidiaceae)	The linear polysaccharide consists of two fractions: neutral agarose and anionic agaropectin; **agarose** consists of alternating (1→4)-linked 3,6-anhydro-α-L-galactopyranose and (1→3)-linked β-D-galactose units; **agaropectin** has the same backbone but with many anionic groups such as sulfates, pyruvate, and glucuronate.[28]	Widely used in bakery products due to its thermo-reversible gelling property; disintegrating agent, gelling agent in suppositories, surgical lubricant, laxative.
Carrageenan	Red algae *Chondrus crispus* (Gigartinaceae)	Consists of three different types of polysaccharides: κ-carrageenan, ι-carrageenan and λ-carrageenan; κ-carrageenan and ι-carrageenan are sulfate esters of 1→3-linked β-D-galactose and 1→4-linked 3,6-anhydro-α-D-galactose (κ-carrageenan with one sulfate group per repeating disaccharide unit and ι-carrageenan with two sulfate groups); λ-carrageenan consists of β-D-galactopyranosyl residues sulfated at C-2 and 2,6-di-*O*-sulfato-α-D-galactopyranosyl units (carrying three sulfate groups).[29]	Gelling agent, stabilizer in emulsions and suspensions, used in tooth paste formulations, demulcent and laxative.

Alginic acid and alginates	All species of brown algae	Anionic polysaccharide consisting of (1→4)-linked β-D-mannuronic acid and α-L-guluronic acid residues.[28]	Food thickeners; suspending agent, gelation for dental films, stabilizer, sustained release agent, tablet coating.
Laminarin or laminaran	Species of *Phaeocystis* (Phaeocystaceae), *Laminaria* (Laminariaceae)	It is a highly water-soluble branched polysaccharide consisting of a linear β-(1→3)-linked glucose-based chains with some β-(1→6)-linked side chains consisting of one or more glucose moieties.[30]	Contributing to carbon export from surface waters in oceans

Table 2.5 Some mucilages of plant origin.

Mucilage source	Structure	Uses
Aloe mucilage from the leaves of *Aloe vera* (Asphodelaceae)	Arabinan, arabinorhamnogalactan, galactan, galactogalacturan, glucogalactomannan, galactoglucoarabinomannan[22]	Gelling agent, sustained release agent
Seed husks of Psyllium (*Plantago ovata*), Plantaginaceae (known as ispagol mucilage)	Highly branched arabinoxylan with high molecular weight; composed mainly of arabinose, xylose, uronic acids, and small amounts of galactose, rhamnose, glucose, and mannose.[29]	Used in beverages, confectionery and bakery products as a stabilizer and a gel filling. The mucilage is also used as a cathartic, lubricant, demulcent, laxative, binder, emulsifying and suspending agent.
Flax seed (*Linum usitatissimum*), Linaceae	A mixture of neutral and acidic polysaccharides: the neutral fraction containing a (1→4)-linked β-D-xylopyranosyl backbone chain together with terminal D-xylose; the acidic fraction containing a rhamnogalacturonan backbone composed of (1→2)-linked L-rhamnosyl residues and (1→4)-linked D-galacturonic acid.[29]	Used as a thickener, stabilizer, and water-holding agent.
Chan seed (*Hyptis suaveolens*, Lamiaceae)	Neutral and acidic polysaccharides in the ratio 1 : 1; neutral polysaccharides consisting of galactose, glucose, and mannose; acidic polysaccharide consisting of fucose, xylose, and 4-*O*-methylglucuronic acid residues.[31]	The mucilage has prebiotic potential.

Carbohydrates play an important role in plant immunity to defend plants against microorganisms (bacteria, oomycetes, and fungi). The first step is activation of defense reactions and perception of PAMPs (Pathogen Associated Molecular Patterns), MAMPs (Microbe Associated Molecular Patterns), and DAMPs (Damage-Associated Molecular Patterns). PAMPs and MAMPs are secreted by microorganisms during interaction with the plant. DAMPs are produced by plant cell walls by pathogen cell wall degrading enzymes. PAMPs, MAMPs, and DAMPs are recognized by pattern-recognition receptors (PRR) that are receptor-like kinases (RLKs) or receptor-like proteins (RLPs). The recognition initiates phosphorylation and subsequent complex cascade of signaling events. The signaling events lead to the generation of defense gene activation, followed by synthesis of pathogenesis related (PR) proteins (such as β-(1→3)-glucanase, chitinases, peroxidases, proteinase inhibitor *etc.*) and accumulation of phytoalexins and cell wall strengthening. Sugars are involved in plant immunity as signaling molecules. Di- and oligosaccharides like sucrose, trehalose, raffinose and galactinol are involved in activation of plant defense responses and increased resistance to pathogens.[38,39]

2.1.5 Role in Nutrition and Human Health

Dietary fibre is the plant cellular material resistant to salivary and small intestinal enzyme actions in humans. Dietary fibre includes structural polysaccharides (cellulose, hemicellulose, and pectin), nonstructural polysaccharides (non-digestible oligosaccharides, gums, and mucilages), and structural nonpolysaccharides (lignin). Dietary fibre intake improves glycemic index, reduces lipid levels, exhibits prebiotic effects, and is useful in weight management. Undigested carbohydrates including the NDOs act as substrates for fermentation by intestinal bacteria in the colon. As consequences, faecal output is increased, short chain fatty acids are produced as end products of the fermentation process and provide a prebiotic effect. The **prebiotics** (non-digestible carbohydrates and non-polysaccharides) are fermented and stimulate growth and/or activity of some beneficial bacteria, known as **probiotics**. The most important bacterial genera are *Bifidobacterium* and *Lactobacillus*. An increase in beneficial bacteria may reduce pathogenic bacteria. An increase in the proportion of such beneficial bacteria in the gastrointestinal tract preserves gut health and reduces disease risks. The growth and stimulation of these bacteria which have antimutagenic and anticarcinogenic properties reduces the risk of cancer.[29,40–44]

2.1.6 Other Pharmaceutical and Non-pharmaceutical Applications of Plant Carbohydrates

Pectins, obtained from fruits, are used widely in the food industry as emulsifiers, gelling agents, stabilizers, and/or thickeners. Commercial **cellulose** ethers, the semisynthetic polymers, are valuable in the pharmaceutical industry. **Sodium carboxymethyl cellulose** is used in pharmaceutical formulations as a thickener, stabilizer, disintegrating agent and bioadhesive material. Sodium carboxymethyl cellulose is also used in ophthalmic applications as a lubricant. At low concentration, it has been used as artificial tears and lens lubricant. **Methyl cellulose** has been used as a thickener, binder, stabilizer, suspending agent and film former in the pharmaceutical industry. **Ethyl cellulose** has been used in oral drug delivery systems as a carrier to improve drug bioavailability. It is a good film former. **Hydroxypropyl cellulose** is employed as a tablet binder. **Hydroxypropyl methyl cellulose** is used as a thickener, hydrophilic matrix material and tablet binder. Starch is a low-cost pharmaceutical excipient, diluent, disintegrant, binder and lubricant.[13]

Cellulose is an important renewable resource for textiles. **Nanocellulose** and **cellulose nanofibres** have applications in the paper and board industry, adsorbent products, paint and coatings, adhesives, packaging and cosmetic products. **Regenerated cellulose fibres** are obtained by dissolving cellulose, pure or derivatized, from wood or plant fibres. Examples of such regenerated cellulose fibres are viscose, and lyocell rayon.[45]

Different gums are used as pharmaceutic aids in the manufacture of tablets, syrups as tablet binders, disintegrants, stabilizers, lubricants, laxatives, cathartics *etc.* (see Tables 2.4 and 2.5). Plant based mucilages are also used in the food industry.[21]

Sugar acid *e.g.*, D-gluconic acid is produced commercially. Other sugar acids such as D-glucaric acid and *meso*-galactaric acid are produced at lower scale. The sugar acids are applied as sequestering agents, binders, corrosion inhibitors, biodegradable chelators for pharmaceuticals, and pH regulators. These molecules are of interest in the production of synthetic polymers, polyesters, and hydrogels.[4] Polyols are important in food industry being used as flavour enhancers, sweeteners in bakery goods, candies, *etc.*[5]

2.2 Glycosides

A **glycoside** is a glycoconjugate of one or more sugars with a non-carbohydrate metabolite. Glycosides are metabolized by glycosyltransferases and hydrolyzed by glycosidases. On glycosylation the hydrophobic material becomes more water soluble.[46] The non-carbohydrate part is known as genin or aglycone. The sugar part is known as glycone. The most common glycosides are *O*-glycosides. Here the sugar is linked through the oxygen atom of an alcohol or phenol. Sugars may be directly linked to the carbon atom when the glycoside is known as a *C*-glycoside (*e.g.*, mangiferin, barbaloin, vitexin). In the case of *S*-glycosides, the sugar is linked to a sulfur atom (*e.g.*, glucosinolates). Sugars are directly linked to the nitrogen atom in case of *N*-glycosides (*e.g.*, nucleosides).

The common sugars in glycosides are L-arabinose, D-glucose. D-galactose, D-glucuronic acid, D-galacturonic acid, D-fucose and D-xylose. Some unusual sugars found in cardiac glycosides are 6-deoxy-D-glucose and 2-deoxy-D-arabinohexose. In case of *O*-glycosides, sugars may remain linked to one oxygen atom or to two different hydroxyl groups of the genin (diglycoside), or to more hydroxyl groups (oligoglycosides). Complex glycosides consist of oligosaccharides with branched chains that are attached to two hydroxyl groups (*e.g.*, gypsoside in *Gypsophila pacifica*, Caryophyllaceae) (Figure 2.13).

Based on the chemical structure of the aglycones, glycosides may be of different types.

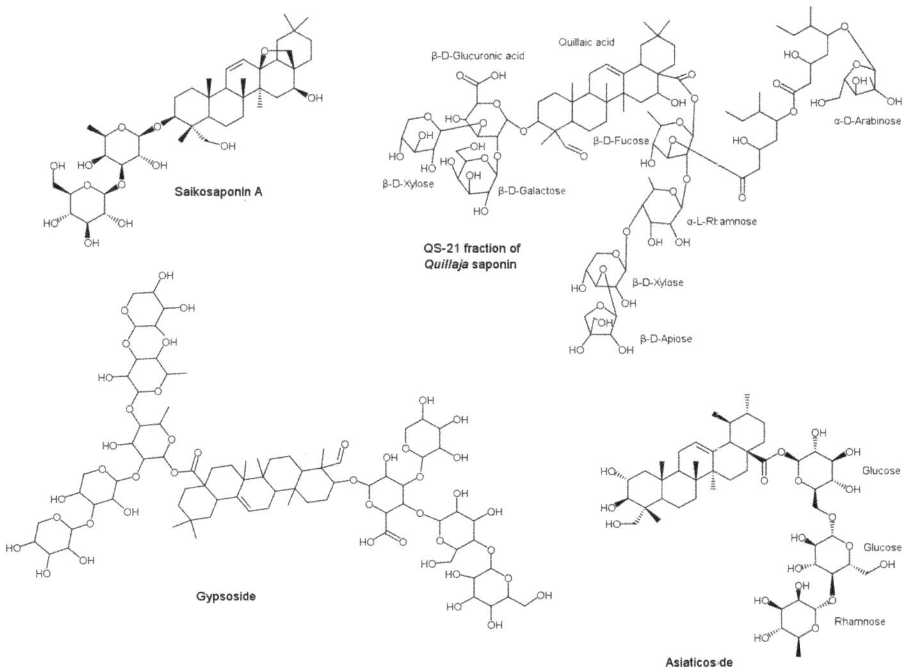

Figure 2.13 Examples of some saponins.

2.2.1 Iridoid Glycosides

Iridoids and secoiridoids are monoterpenes which are available in nature as glycosides, glucose being the sugar attached to the genin in most cases. Iridoid glucosides are reported from many plants including *Gentiana lutea, Swertia chirayita* belonging to Gentianaceae. Details are available in Chapter 4; Table 4.1; Figures 4.13 and 4.14.

2.2.2 Saponins

Saponins are glycosides containing triterpene or steroid backbones as the aglycone (also known as sapogenin) and one or more sugar chains attached at C-3 (when there is a single sugar chain) and at C-21 (triterpene saponins) or at C-26 (furostanol saponins). Saponins are surface active compounds with detergent, emulsifying and foaming properties because of the presence of both lipid soluble aglycones and water-soluble sugar chains in the structure.[47] Generally, saponins in food have been considered as antinutritional factors. However, recent research has provided evidence of the health beneficial properties of some saponins. Different subclasses of saponins are mentioned in Table 2.6. For details regarding the attached sugars please consult Chapter 4 [(Figure 4.49 for steroidal saponin dioscin), (Figure 4.39 for glycosides of

Table 2.6 Saponins.

Aglycone (sapogenin)	Saponin	Source
Pentacyclic triterpenoids		
Asiatic acid	Asiatocoside	*Centella asiatica* (Apiaceae)
Glycyrrhetinic acid	Glycyrrhizic acid	*Glycyrrhiza glabra* (Fabaceae)
Gymnemagenin	Gymnemic acid IV	*Gymnema sylvestre* (Apocynaceae)
Gypsogenin	Gypsoside	*Gypsophila pacifica* (Caryophyllaceae)
Oleanolic acid	Calendula saponin A	*Calendula officinalis* (Asteraceae)
Quillaic acid	Quillaja saponins	*Quillaja saponaria* (Quillajaceae)
Saikogenin F	Saikosaponin A	*Bupleurum* sp. (Apiaceae)
Soyasapogenol A	Soya saponins	*Glycine max* (Fabaceae)
Soyasapogenol B	Soya saponins	*Glycine max* (Fabaceae)
Soyasapogenol E	Soya saponins	*Glycine max* (Fabaceae)
Tetracyclic triterpenoids (Dammarane type)		
Protopanaxatriol	Ginsenosides	*Panax ginseng* (Araliaceae)
Jujubogenin	Bacoside A, bacoside X	*Bacopa monnieri* (Plantaginaceae)
Tetracyclic triterpenoids (Steroidal type)		
Diosgenin	Dioscin	*Dioscorea* sp. (Dioscoreaceae)

pentacyclic triterpenoids glycyrrhizic acid and gymnemic acid IV), (Figures 4.36 and 4.37 for ginsenosides and bacosides respectively)], and Figure 2.13. (Quillaja saponin, saikosaponin, gypsoside, and asiaticoside). The widely studied saponin fraction of *Quillaja Saponaria* (Quillajaceae) is QS-21.It is composed of one branched trisaccharide, and a fatty acid group attached to it.

2.2.3 Cardiac Glycosides

These glycosides have effect on heart muscles. Details are available in Chapter 4, Figures 4.55–4.59, Tables 4.5–4.7.

2.2.4 Steryl Glycosides

Phytosterols are often found attached to a sugar moiety at the C-3 hydroxyl group of the sterol. These glycosides are known as **steryl glycosides** (also known as **sterolins**). One example is β-sitosteryl-D-glucoside (Figure 2.14). The acylated form of steryl glycoside is known as **acyl steryl glycoside**. Steryl glycosides and acyl steryl glycosides play important roles for the maintenance of plasma membrane organization and in adapting to environmental stresses.[46]

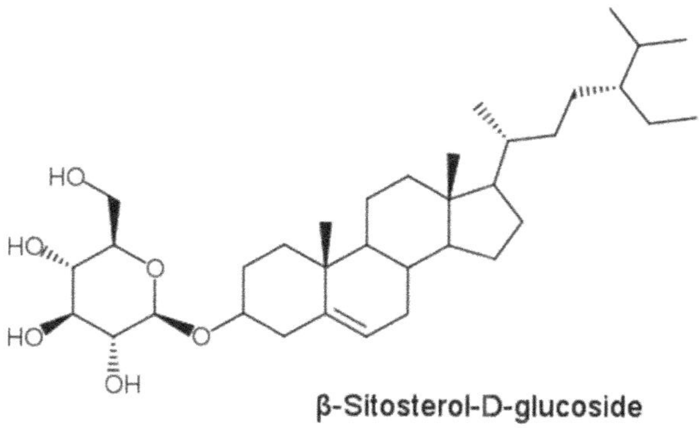

Figure 2.14 Steryl glycoside.

Table 2.7 Pregnane glycosides[a].

Plant	Glycoside	Aglycone	Sugar moiety
Asclepias incarnata (Apocynaceae)	Metaplexigenin 3-*O*-trioside	Metaplexigenin	3-*O*-β-D-Ole$_p$-(1→4)-β-D-Dgx$_p$-(1→4)-β-D-Dgx$_p$
Baseonema acuminatum (Apocynaceae)	Baseonemoside A	Pregn-5-ene-3β,16α,20(*S*)-triol	3-*O*-β-D-Cym$_p$-(1→4)-β-D-Dgx$_p$ 20-*O*-β-D-Glu$_p$-(1→2)-β-D-Dgt$_p$
Caralluma lasiantha (Apocynaceae)	Lasianthoside A	Caralasigenin	3-*O*-β-D-Glu$_p$-(1→4)-β-D-Dgt$_p$ 20-*O*-α-L-Rha$_p$-(1→6)-β-D-Glu$_p$
Dioscorea collettii (Dioscoreaceae)	Hypoglaucin G	Pregn-5,16-dien-3β-ol-20-one	16β-(4′-Me-5′-*O*-β-D-glu$_p$-pentanoxyl)-3-*O*-α-L-Rha$_p$-(1→2)-α-L-Rha$_p$-(1→4)-β-D-Glu$_p$

[a]Cym: cymarose; Dgt: digitalose; Dgx: digitoxose; Glu: glucose; Ole: oleandrose; Rha: rhamnose; p: pyranose.

2.2.5 Pregnane Glycosides

Many pregnane glycosides were isolated and identified from different plants. Many of them have useful biological activities including anticarcinogenic or cancer inhibitory properties. A few examples are given in Table 2.7. In general, sugar moieties of pregnane glycosides are oligosaccharides consisting of different kinds of sugar units *e.g.*, D-glucose, L-rhamnose, D-cymarose, oleandrose, D-allose, and D-digitoxose.[48]

2.2.6 Phenolic Glycosides

Different classes of phenolic compounds commonly exist in the form of glycosides. Sugars are linked to the phenolic compound *via* α- or β-linkages. Some examples are shown in Table 2.8 and Figure 2.15. More detailed description is available in Chapter 5.

Table 2.8 Different classes of phenolic glycosides.

Aglycones	Glycosides	Sources
Simple phenols	Hydroquinone-*O*-β-D-glucopyranoside (arbutin)	*Arctostaphylos* spp. (Ericaceae)
	2-(Hydroxymethyl)phenyl hexopyranoside (salicin)	*Salix* spp. (Willow bark), (Salicaceae)
Phenylethanoid	Bacomoside A	*Bacopa monnieri* (Plantaginaceae)
	Bacomoside B1	*Bacopa monnieri*
	Purpureaside D	*Digitalis purpurea* (Plantaginaceae)
	Magnoloside D	*Magnolia officinalis* (Magnoliaceae)
Anthraquinones	Sennosides A, B, the dianthrone-*O*-glycosides of sennidin A and sennidin B	*Senna alexandrina* (Fabaceae)
	Rhein-8-diglucoside	
	Aloe-emodin-8-glucoside	
Flavonoids		
Flavonols	Quercetin-4′-*O*-β-D-glucoside	Onion
	Quercetin-3,4′-*O*-β-D-glucoside	
	Quercetin-3-*O*-rutinoside (Rutin)	Citrus fruits
Flavones	Apigenin-6-*C*-glucoside (isovitexin)	*Vitex agnus-castus*
	Apigenin-8-*C*-glucoside (vitexin)	(Lamiaceae)
	Luteolin-7-*O*-rutinoside	Citrus fruits
	Luteolin-6,8-di-*C*-glucoside	
Isoflavones	Genistein-7-glucoside (genistin)	Soy products
Flavanones	Hesperetin-7-rutinoside (Hesperidin)	Citrus fruits
Anthocyanidins	Anthocyanins such as Cyanidin-3-*O*-rutinoside, Malvidin-3-*O*-β-D-glucoside, Malvidin-3-*O*-β-D-galactoside	Blueberry
Betacyanins	Betanidin-5-*O*-β-glucoside (betanin)	Beet root
Xanthone	Mangiferin	*Mangifera indica* (Anacardiaceae)
Stilbene	Piceatannol-3′-*O*-glucoside	Several plants, rhubarb rhizome
	Rhaponticin (a glucoside of rhapontigenin)	
Tannins	Complex glycosides of gallic acid and/or ellagic acid and/or catechin	Tea leaves

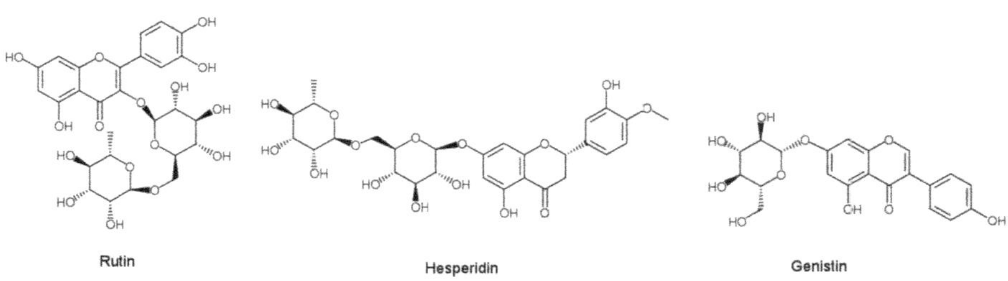

Figure 2.15 Flavonoid glycosides.

Phenylethanoid glycosides are products of different phenolic acids (*e.g.*, caffeic acid, coumaric acid, cinnamic acid, ferulic acid, and isoferulic acid) and various saccharides[49] (*e.g.*, rhamnose, xylose, apiose, glucose, lyxose, allose, and arabinose). Structures of a few phenylethanoid glycosides are presented in Figure 2.16.

2.2.7 Glycosylated Lipids

Glycosylated lipids such as monogalactosyldiacylglycerol (MGDG) and digalactosyldiacylglycerol (DGDG) (Figure 2.17) are common in plants particularly in the thylakoid membranes of the chloroplasts.[46]

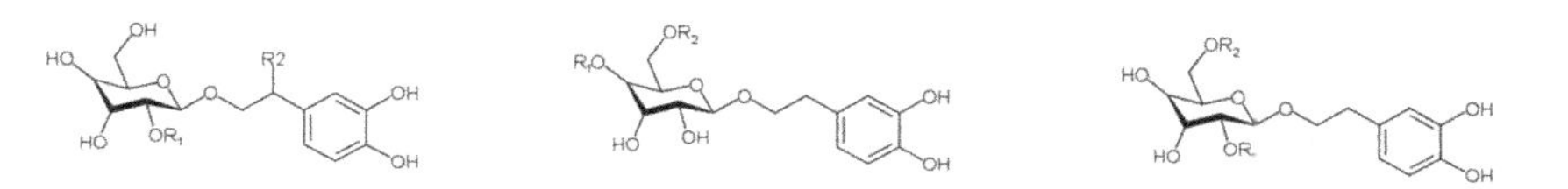

Bacomoside A R_1: *p*-Hydroxybenzoyl; R_2: =O
Bacomoside B1 R_1: Caffeoyl; R_2: OCH_3

Purpureaside D R_1: Feruloyl; R_2: Rhamnose

Magnoloside D R_1: Rhamnose; R_2: Caffeoyl

Figure 2.16 Phenylethanoid glycosides.

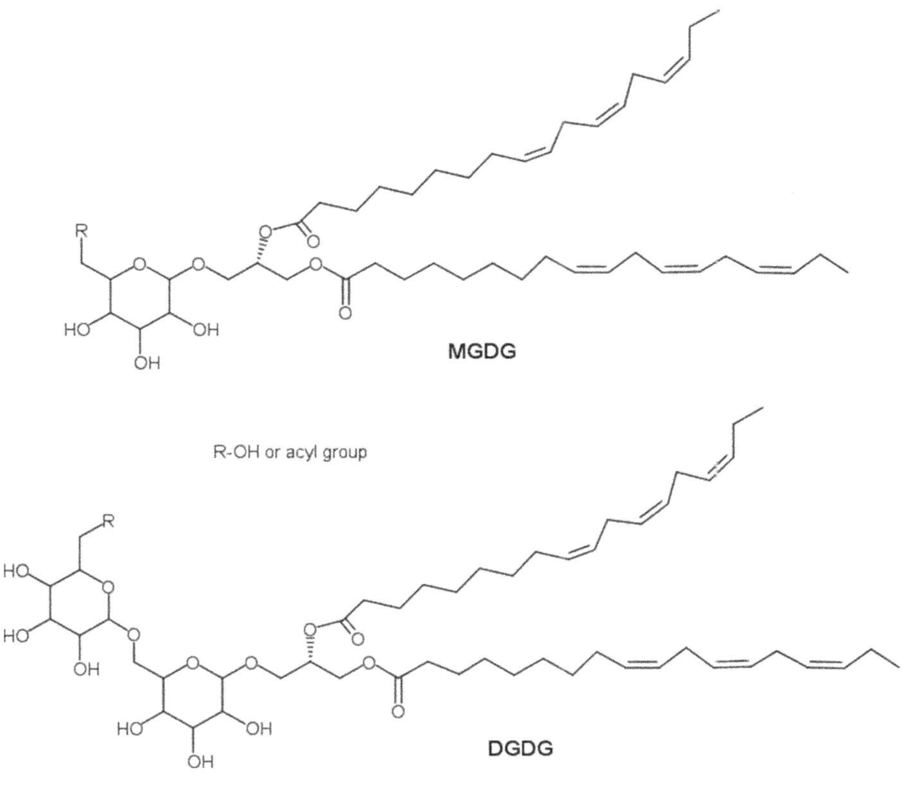

Figure 2.17 Glycosylated lipids.

2.2.8 Cyanogenic Glycosides

Cyanogenic glycosides[50] are specialized plant metabolites derived from oximes and cyanohydrins (α-hydroxynitriles) as intermediates. The labile cyanohydrin is stabilized after glucosylation. α-Amino acids are hydroxylated to form *N*-hydroxyamino acids from which oximes are produced. Oximes are then converted to a nitrile. Nitrile is again hydroxylated to form α-hydroxynitrile. Glycosylation of α-hydroxynitrile forms the corresponding cyanogenic glycosides (Figure 2.18). In the presence of β-glycosidase, the unstable cyanohydrin dissociates to release hydrogen cyanide (HCN). The process is known as cyanogenesis. Part of HCN may be reincorporated into primary metabolism. The aglycones (α-hydroxynitriles) may be aliphatic or aromatic compounds. Diversity in chemical structure of cyanogenic glycosides based on oximes and the amino acid origin is shown in Figure 2.19 and Table 2.9 respectively.

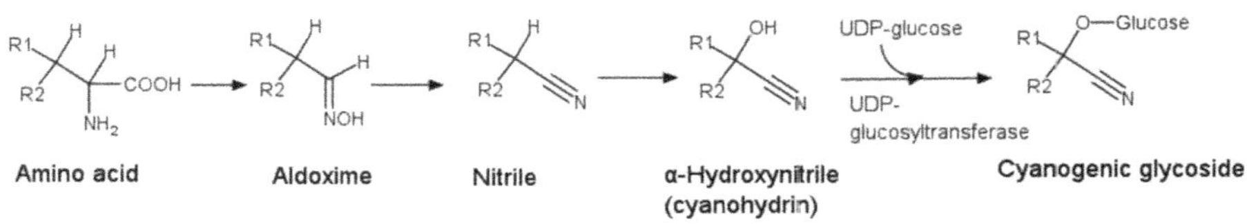

Figure 2.18 Biosynthesis of cyanogenic glycosides. (linamarin: $R_1 = R_2 = CH_3$; prunasin: $R_1 = H$, $R_2 =$ Phenyl).

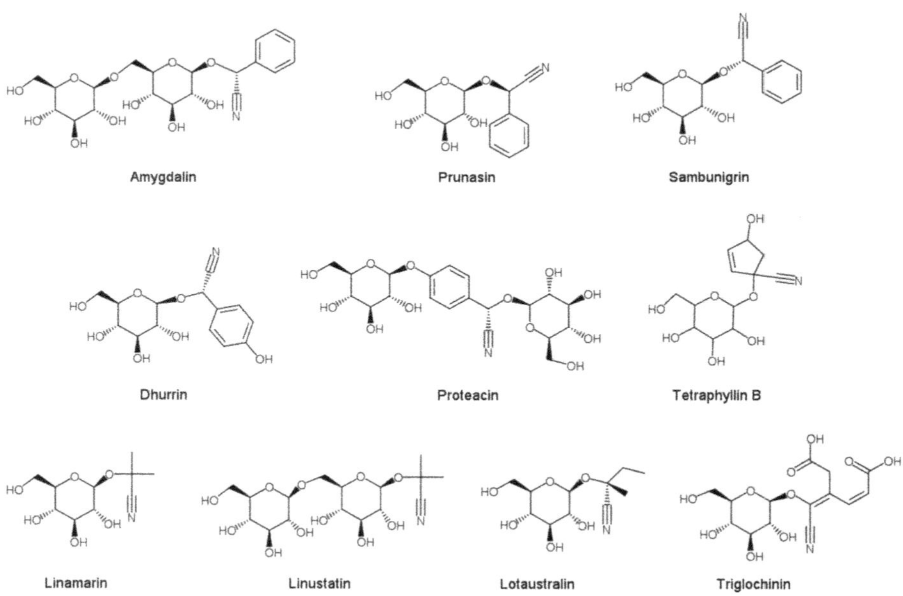

Figure 2.19 Structural diversity of cyanogenic glycosides based on the amino acid precursors.

Table 2.9 Some cyanogenic glycosides and their sources.

Amino acid	Glycoside	Source
Phenylalanine	Amygdalin	*Prunus amygdalus* (Rosaceae)
	Prunasin	*Prunus serotina* (Rosaceae)
	Sambunigrin	*Sambucus nigra* (Viburnaceae)
Valine	Linamarin	*Linum usitatissimum* (Linaceae)
	Linustatin	*Hevea brasiliensis* (Euphorbiaceae)
Isoleucine	Lotaustralin	*Manihot esculenta* (Euphorbiaceae)
		Phaseolus lunatus (Fabaceae)
Tyrosine	Proteacin	*Polyscias australiana* (Araliaceae)
	Triglochinin	*Triglochin maritima* (Juncaginaceae)
	Dhurrin	*Sorghum vulgare* (Poaceae)

Cyanogenic glycosides are widespread in plants being found in the ferns, gymnosperms, and angiosperms. Cyanogenic glycosides are deterrents of herbivores. In some plants, they control germination. Cyanogenic glycosides modulate oxidative stress.[50]

2.2.9 Glucosinolates

Glucosinolates are sulfur containing specialized metabolites found mainly in plants belonging to Brassicaceae such as broccoli (*Brassica oleracea* var. *italica*), cabbage (*Brassica oleracea* var. *capitata*), cauliflower (*Brassica oleracea* var. *botrytis*), brussels sprouts (*Brassica oleracea* var. *gemmifera*), radish (*Raphanus sativus*), turnip (*Brassica rapa* var. *rapa*), mustard (*Brassica nigra*) and rapeseed (*Brassica napus*). Glucosinolates are anions formed by a thiohydroximate-*O*-sulfonate group linked to glucose and an alkyl, aralkyl, or indolyl side chain (R). The R chain originates from any one of the eight amino acids alanine, leucine, isoleucine, methionine, valine, phenylalanine, tyrosine and tryptophan. Myrosinase (β-thioglucosidase) is confined in scattered myrosin cells or in different intracellular compartments depending on plant species. Upon tissue damage due to chewing or insect attack, myrosinase is discharged. Myrosinase breaks down glucosinolates producing β-D-glucose and the unstable aglycone thiohydroximate-*O*-sulfonate. The unstable aglycone reorganizes with the release of a sulfate ion and the metabolites depending on the R chain of the glucosinolates. The different breakdown products are isothiocyanates, thiocyanates, nitriles, epithionitriles, oxazolidine-2-thiones, and indole-3-carbinol (Figure 2.20). Isothiocyanate has a pungent taste. Myrosinase is inactivated when plants are consumed after cooking. Glucosinolates can be absorbed partly by the stomach. The rest of the glucosinolates are hydrolyzed by intestinal microbiota or bacterial myrosinase. Isothiocyanates are absorbed in the small intestine, caecum, and colon.[51–54]

More than ninety glucosinolates are known from the genus *Brassica*. These are grouped into three subgroups as olefinic (sinigrin, gluconapin), methylsulfinyl alkyl (glucoraphanin), and aryl (gluconasturtiin). The condiment

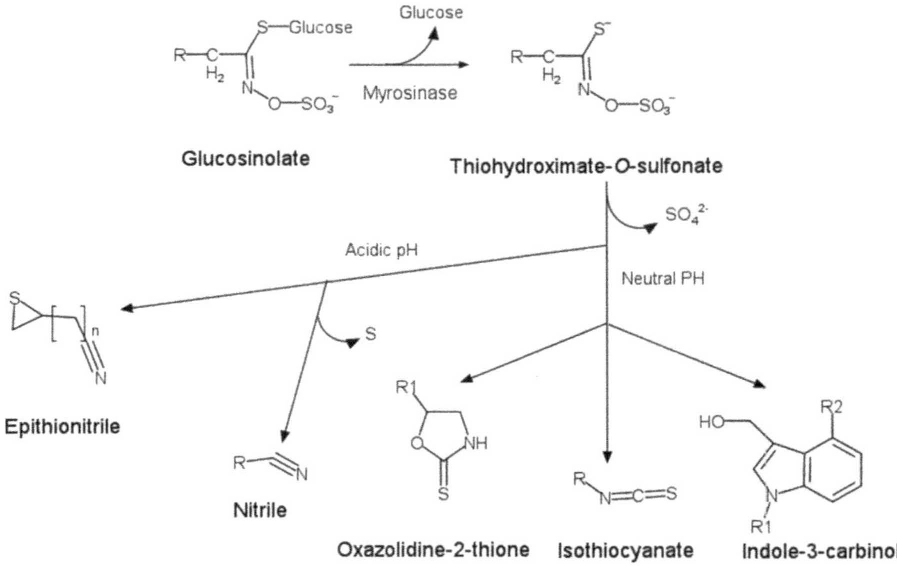

Figure 2.20 Hydrolysis of glucosinolates. (R, R_1, R_2 are variable side chains).

mustard has different forms *e.g.*, black mustard (*Brassica nigra*), white mustard (*Sinapis alba*) and oriental mustard (*Brassica juncea*) which is a hybrid of *Brassica rapa* and *Brassica nigra*. The main glucosinolate of *Brassica juncea*, *Brassica nigra*, broccoli (*Brassica oleracea* var. *italica*), and brussels sprouts is sinigrin (allyl-glucosinolate or 2-propenyl-glucosinolate) which on hydrolysis yields allyl-isothiocyanate. The main glucosinolate of *Sinapis alba* is sinalbin, the hydrolyzed product being *p*-hydroxybenzyl isothiocyanate.[52]

Isothiocyanates inhibited mitosis and stimulated apoptosis in human tumour cells. Isothiocyanates also exhibited fungicidal, nematicidal, and bactericidal activities.[54]

References

1. R. T. Ahnen, R. Mottet, M. Omolo and J. Slavin, Carbohydrates, in *Carbohydrates: Present Knowledge in Nutrition*, ed. B. Marriott, D. Bert, V. Stalling and A. Yates, Elsevier Inc., 2020, **vol. 1**, pp. 37–50.
2. P. M. Dewick, in *Medicinal Natural Products: A Biosynthetic Approach*, John Wiley & Sons Ltd, 2009, pp. 485–508.
3. S. C. Moldoveanu, Analytical pyrolysis of polymeric carbohydrates, in *Analytical Pyrolysis of Natural Organic Polymers*, Elsevier BV., 2021, pp. 111–245.
4. T. Mehtio, M. Toivari, M. G. Wiebe, A. Harlin, M. Penttila and A. Koivula. Production and applications of carbohydrate derived sugar acids as generic biobased chemicals, *Crit. Rev. Biotechnol.*, 2015, **36**, 904–916.

5. T. Rice, E. Zannini, E. K. Arendt and A. Coffey, A review of polyols – biotechnological production, food applications, regulation labelling and health effects, *Crit. Rev. Food Sci. Nutr.*, 2020, **60**, 2034–2051.
6. C. Martinez-Villaluenga and J. Frias, Production and bioactivity of oligosaccharides in plant foods, in *Food OligoSaccharides: Production, Analysis and Bioactivity*, ed. E. Moreno and M. L. Sanz, John Wiley and Sons, Ltd, 2014, pp. 35–54.
7. J. Courtois, Oligosaccharides from land plants and algae: production and applications in therapeutics and biotechnology, *Curr. Opin. Microbiol.*, 2009, **12**, 261–273.
8. G. Eggleston and G. L. Cote, Oligosaccharides in food and agriculture, *ACS Symposium Series*, American Chemical Society, Washington DC, 2003, pp. 1–14.
9. P. D. Meyer, Nondigestible oligosaccharides as dietary fibre, *J. AOAC Int.*, 2004, **87**, 718–726.
10. S. I. Mussatto and I. M. Mancilha, Non-digestible oligosaccharides: A review, *Carbohydr. Polym.*, 2007, **68**, 587–597.
11. A. S. Althubiani, S. B. Al-Ghamdin, F. A. Qais, M. S. Khan, I. Ahmad and H. A. Malak, Plant-derived prebiotics and its health benefits, in *New Look to Phytomedicine*, ed. M. S. A. Khan, I. Ahmed and D. Chattopadhyay, Elsevier Inc., 2019, ch. 4, pp. 63–88.
12. M. Sabater-Molina, E. Larque, F. Torrela and S. Zamora, Dietary fructooligosaccharides and potential benefits on health, *J. Physiol. Biochem.*, 2009, **65**, 315–328.
13. V. E. Tyler, L. R. Brady and J. E. Roberts, *Pharmacognosy*, Lee & Berger, 1988.
14. Y. I. Cornezo-Ramirez, O. Martinez-Cruz, C. L. D. Tolo-Sanchez, F. J. Wong-Corral, J. Borboa-Flores and F. J. Cinco-Moroyoqui, The structural characteristics of starches and their functional properties, *CyTA--J. Food*, 2018, **16**, 1003–1017.
15. Q. Chen, H. Yu, L. Wang, Z. ul Adin, Y. Chen, J. Wang, W. Zhou, X. Yang, R. U. Khan, H. Zhang and X. Chen, Recent Progress in chemical modification of starch and its applications, *RSC Adv.*, 2015, **5**, 67459–67474.
16. H. V. Sceller and P. Ulvskov, Hemicelluloses, *Annu. Rev. Plant Biol.*, 2010, **61**, 263–289.
17. S. Kim, B. Chandrasekar, A. C. Rea, L. Danhof, S. Zemelis-Durfee, N. Thrower, Z. S. Shepard, M. Pauly and F. Brandizzi, The synthesis of xyloglucan, an abundant plant cell wall polysaccharide, requires CSLC function, *Proc. Natl. Acad. Sci. U. S. A.*, 2020, **117**, 20316–20324.
18. S. Singh, G. Singh and S. K. Arya, Mannans: An overview of properties and application in food products, *Int. J. Biol. Macromol.*, 2018, **119**, 79–95.
19. C. Lara-Espinoza, E. Carvajal-Millan, R. Balandran-Quintana, Y. Lopez-Franco and A. Rascon-Chu, Pectin and pectin based composite materials: beyond food texture, *Molecules*, 2018, **23**, 942.
20. M. S. Amiri, V. Mohammadzadeh, M. E. T. Yazdi, M. Barani, A. Rahdar and G. Kyzas, Plant-based gums and mucilages applications in pharmacology and nanomedicine: A review, *Molecules*, 2021, **26**, 1770.

21. M. M. Tosif, A. Najda, A. Bains, R. Kaushik, S. B. Dhull, P. Chawla and A. Walasek-Janusz, Comprehensive review on plant derived mucilage: characterization, functional properties, applications, and its utilization for nanocarrier fabrication, *Polymers*, 2021, **13**, 1066.
22. P. D. Choudhury and H. A. Pawar, Recently investigated natural gums and mucilages as pharmaceutical excipients. An overview, *J. Pharm.*, 2014, Article ID 204849.
23. H. H. Musa, A. A. Ahmed and T. H. Musa, Chemistry, biological, and pharmacological properties of gum Arabic, Bioactive Molecules in Food, in *Reference Series in Phytochemistry*, ed. J.-M. Merillon and K. G. Ramawat, Springer International Publishing AG, 2018, pp. 1–18.
24. A. A. R. Freitas, A. J. Ribeiro, A. C. Santos, F. Veiga, L. C. C. Nunes, S. Soares-Sobrinho and E. C. Silva-Filho, *Sterculia striata* gum as a potential oral delivery system for protein drugs, *Int. J. Biol. Macromol.*, 2020, **164**, 1683–1692.
25. M. E. T. Yazdi, S. Nazarnezhad, S. H. Mousavi, M. S. Amiri, M. Darroudi, F. Baino and S. Kargozar, Gum tragacanth (GT): A versatile biocompatible material beyond borders, *Molecules*, 2021, **26**, 1510.
26. S. Al-Assaf, V. Amar and G. O. Phillips, Characterisation of gum ghatti and comparison with gum Arabic, in *Gums and Stabilisers for the Food Industry*, ed. P. A. Williams and G. O. Phillips, Royal Society of Chemistry, Special Publication, 2008, **vol. 316**, pp. 280–290.
27. D. Mudgil, S. Barak and B. S. Khatkar, Guar gum: processing, properties and food applications, *J. Food Sci. Technol.*, 2014, **51**, 409–418.
28. E.-H. Song, J. Shang and D. M. Ratner, Polysaccharides, in *Polymer Science: A Comprehensive Reference*, ed. M. Moeller and K. Matyjaszewski, Elsevier BV, 2012, **vol. 9**, pp. 137–155.
29. S. W. Cui, Y. Wu and H. Ding, The range of dietary fibre ingredients and comparison of their technical functionality, in *Fibre Rich and Wholegrain Foods: Improving Quality*, ed. J. A. Delcour and K. Poutanen, Woodhead Publ Ltd, 2013, **vol. 237**, pp. 96–119.
30. S. Becker, J. Tebben, S. Coffinet, K. Wiltshire, M. H. Iversen, T. Harder, K. Hinrichs and J. Hehemann, Laminarin is a major molecule in the marine carbon cycle, *Proc. Natl. Acad. Sci. U. S. A.*, 2020, **117**, 6599–6607.
31. M. Mueller, A. Cavarkapa, F. M. Unger, H. Viernstein and W. Praznik, Prebiotic potential of neutral oligo- and polysaccharides from seed mucilage of *Hyptis suaveolens*, *Food Chem.*, 2017, **221**, 508–514.
32. H. Hartmann and S. Trumbore, Understanding the roles of nonstructural carbohydrates in forest trees – from what we can measure to we want to know, *New Phytol.*, 2016, **211**, 386–403.
33. S. Ferrari, D. V. Savatin, F. Sicilia, G. Gramegna, F. Cervone and G. D. Lorenzo, Oligogalacturonides: plant damage-associated molecular patterns and regulators of growth and development, *Front. Plant Sci.*, 2013, **4**, Article 49.
34. M. S. Krasavina, N. A. Burmistrova and G. N. Raldugina, The role of carbohydrates in plant resistance to abiotic stresses, in *Emerging Technologies*

and Management of Crop Stress Tolerance, ed. P. Ahmad and S. Rasool, Elsevier Inc., 2014, **vol. 1**, pp. 229–270.
35. I. Slama, C. Abdelly, A. Bouchereau, T. Flowers and A. Savoure, Diversity, distribution and roles of osmoprotective compounds accumulated in halophytes under abiotic stress, *Ann. Bot.*, 2015, **115**, 433–447.
36. S. Sengupta, S. Mukherjee, P. Basak and A. L. Majumder, Significance of galactinol and raffinose family oligosaccharide synthesis in plants. *Front. Plant Sci.*, 2015, **6**, Article 656.
37. M. Thalmann and D. Santelia, Starch as determinant of plant fitness under abiotic stress, *New Phytol.*, 2017, **214**, 943–951.
38. S. Trouvelot, M. Heloir, B. Poinssot, A. Gauthier, F. Paris, C. Guillier, M. Combier, L. Trda, X. Daire and M. Adrian, Carbohydrates in plant Immunity and Plant protection: roles and potential application as foliar sprays, *Front. Plant Sci.*, 2014, **5**, Article 592.
39. N. Guarnizo, D. Oliveros, W. Murillo-Arango and M. B. Bermudez-Cardona. Oligosaccharides: Defense inducers, their recognition in plants, commercial uses and perspectives. *Molecules*, 2020, **25**, 5972.
40. M. Cruz-Requena, S. Escobedo-Garcia, J. A. Salas-Tovar, Y. Mora-Cura, M. L. Chavez-Gonzalez, F. Castillo-reyes, A. C. Flores-Gallegos and R. Rodriguez-Herrera, Definitions and regulatory perspectives of dietary fibres, in *Dietary Fibre: Properties, Recovery, and Applications*, ed. C. M. Galanakis, Academic Press. Elsevier Inc., 2019, pp. 1–25.
41. M. B. Roberfroid, Health benefits of non-digestible oligosaccharides, in *Dietary Fibre in Health and Disease*, ed. D. Kritchevsky and C. Bonfield, Plenum Press, New York. 1997, pp. 211–219.
42. J. L. Slavin, Health benefits of oligosaccharides. *J. Nutraceuticals Funct. Med. Foods*, 1999, **1**, 43–55.
43. J. L. Slavin, Nutritional value of soybean fiber, *J. Am. Diet Assoc.* 1991, **91**, 816–819.
44. I. A. A. Kassem, T. J. Ashaolu, R. Kamel, N. A. Elkasabgy, S. M. Afifi and M. A. Farag, Mucilage as functional food hydrocolloid: ongoing and potential applications in prebiotics and nutraceuticals. *Food Funct.*, 2021, **12**, 4738–4748.
45. C. Felgueiras, N. G. Azonia, C. Gonclaves, M. Gama and F. Dourado, Trends on the cellulose-based textiles: Raw materials and technologies. *Front. Bioeng. Biotechnol.* 2021, **9**, Article 608826.
46. K. Kytidou, M. Artola, H. S. Overkleeft and J. M. F. G. Aerts, Plant glycosides and glycosidases: A treasure-trove for therapeutics. *Front. Plant Sci.*, 2020, **11**, Article 357.
47. O. Guclu-Ustundag and G. Mazza, Saponins: properties, applications, and processing. *Crit. Rev. Food Sci. Nutr.*, 2007, **47**, 231–258.
48. N. Panda, S. Banerjee, N. B. Mandal and N. P. Sahu, Pregnane Glycosides. *Nat. Prod. Commun.* 2006, **1**, 665–695.
49. Z. Xue and G. Yang, Phenylethanoid glycosides: Research advances in their phytochemistry, pharmacological activity and pharmacokinetics. *Molecules* 2016, **21**, 991.

50. R. M. Gleadow and B. L. Moller, Cyanogenic glycosides: synthesis, physiology, and phenotypic plasticity. *Annu. Rev. Plant Biol.*, 2014, **65**, 155–185.
51. F. J. Barba, N. Nikmaram, S. Roohinejad, A. Khelfa, Z. Zhu and M. Koubaa, Bioavailability of glucosinolates and their breakdown products: impact of processing. *Front. Nutr.*, 2016, **3**, article 24.
52. K. Cools and L. A. Terry, The effect of processing on the glucosinolate profile of mustard seed. *Food Chem.*, 2018, **252**, 343–348.
53. A. Mazumder, A. Dwivedi and J. Plessis, Sinigrin and its therapeutic benefits. *Molecules*, 2016, **21**, 416.
54. M. A. Prieto, C. J. Lopez and J. Simal-Gandara, Glucosinolates: Molecular structure, breakdown, genetic, bioavailability, properties and healthy and adverse effects, in *Advances in Food and Nutrition Research*, ed. I. C. F. R. Ferreira and L. Barros, Elsevier Inc., 2019, **vol. 90**, pp. 305–355.

CHAPTER 3

Acetate–Malonate Pathway: Fatty Acids and Polyketides

The major products produced by the acetate–malonate pathway are (1) fatty acids and different types of lipids and (2) aliphatic and aromatic compounds produced through the formation of polyketides. Both these two groups of metabolites, although produced by the acetate–malonate pathway, are differentiated by their methods of biogenesis. Basically, fatty acids consist of only methylene groups whereas a polyketide consists of a chain with alternate keto and methylene groups at the early stages of biosynthesis. Acetyl-CoA and malonyl-CoA initiate the acetate–malonate pathway.

3.1 Fatty Acids

A **fatty acid** consists of an aliphatic carbon chain with a carboxylic acid group at one end and a methyl group at the other end. Fatty acids are accumulated in plants as triacylglycerols. These fatty acids and triacylglycerols are major sources of renewable carbon used as food, feedstocks, and fuel.[1] Free fatty acids are sometimes found in nature but they are found mostly esterified to phospholipid, glycerolipid, or sterol backbones. Fatty acids are important constituents of membrane lipids[2] (to provide the correct molecular environment to allow channeling of metabolism), signaling molecules and energy-storage reserves. There are more than 450 different fatty acids reported from the plant kingdom. However, only five fatty acids occur widely as membrane lipids including phospholipids, galactolipids and triacylglycerols available as storage oils. These are palmitic acid, stearic acid,

Chemical Diversity of Plant Specialized Metabolites: A Biosynthetic Approach
By Bratati De

Published by the Royal Society of Chemistry, www.rsc.org

oleic acid, linoleic acid, and α-linolenic acid which are often called common fatty acids.[3]

Fatty acids vary in chain length as well as degree of unsaturation. Based on the degree of unsaturation, the fatty acids may be of two types, saturated (Figure 3.1) and unsaturated (Figure 3.2). Very long chain fatty acids (VLCFAs) are fatty acids with acyl chain length longer than 18 carbons.

A shorthand notation of fatty acids indicates chain length with number, position, and stereochemistry of double bonds. The carbon atoms are usually numbered from the carboxyl group which is C1.For example, palmitic acid and oleic acid are represented by 16:0 and 18:1 (9*c*) respectively. The first number denotes the total number of carbon atoms in the chain. The second number denotes the degree of unsaturation (0 indicates no unsaturation *i.e.*, a saturated fatty acid). The number(s) in brackets indicates the position of unsaturation and *c* or *t* represents *cis* or *trans*. The notation 'Δ' (delta) is also used to indicate double bonds. Thus, 18:2 Δ9,12 for linolenic acid indicates the fatty acid has double bonds between carbons 9 and 10, 12 and 13 counting from the carboxylic acid end. Occasionally, numbering starts from the methyl end (the ω end). The major groups of fatty acids are designated as ω-3 (omega-3), ω-6, ω-9 *etc.*, depending on the first double bond from the methyl end group of the fatty acid. Alternatively, the double bonds of fatty acids may also be designated as *n*-3, *n*-6, *n*-9 *etc.*, by this method.[4]

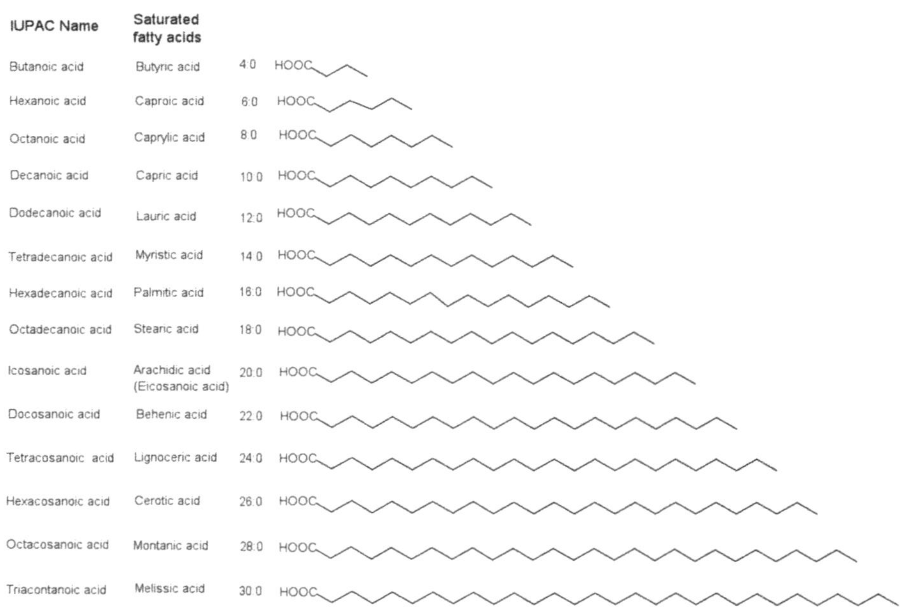

Figure 3.1 Saturated fatty acids. IUPAC: International Union of Pure and Applied Chemistry.

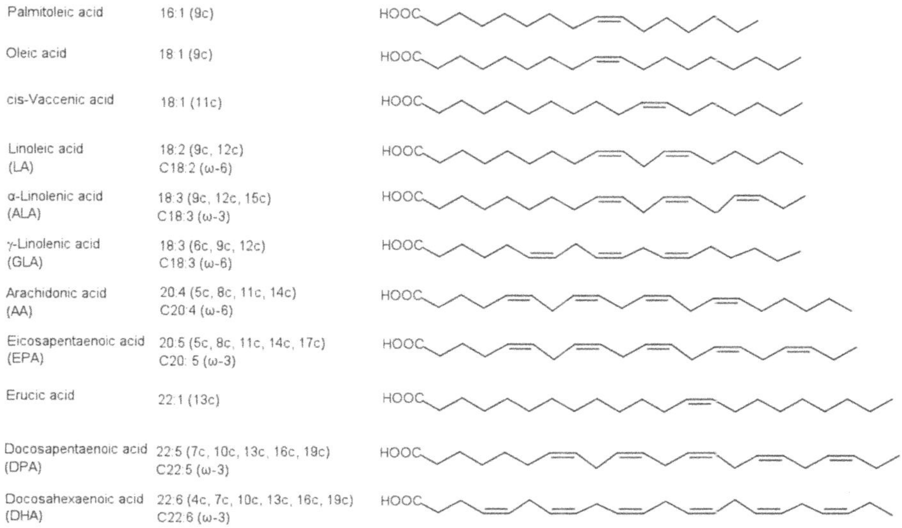

Figure 3.2 Unsaturated fatty acids.

3.1.1 Fatty Acid Biosynthesis

Fatty acid biosynthesis in plants occurs in the chloroplasts in green tissues and in the plastids of non-photosynthetic tissues. Fatty acid biosynthesis also occurs in mitochondria. In the plastidial compartment, fatty acid biosynthesis starts from acetyl-CoA, a direct product of photosynthesis. Acetyl CoA may also be synthesized from free acetate by the activity of the enzyme acetyl-CoA synthase. Two enzyme systems are required for biosynthesis of fatty acids up to a chain length of C18.[5,6] These are:

a. **Acetyl-CoA carboxylase (ACCase)**: This enzyme complex catalyzes ATP dependent carboxylation of acetyl-CoA to malonyl-CoA. This is the first committed step in the biosynthesis of fatty acids. In non-photosynthetic cells or organs, plastids depend on the import of metabolites from the cytosols to metabolize acetyl-CoA. In chloroplasts, photosynthesis supplies an endogenous source of carbon.
b. **Fatty acid synthase (FAS)**: This enzyme complex is involved in the biosynthesis of fatty acids.

Acetyl-CoA carboxylase (ACCase) may be of two molecular forms **type I and type II**. The heteromeric **type II** multiprotein or **multisubunit complex** (MS complex), analogous to that found in prokaryotes, is present in the plastids of all plants (except Poaceae). This heteromeric MS complex consists of **four** different polypeptides *i.e.*, biotin carboxyl carrier protein (BCCP), biotin carboxylase (BC), α-carboxyltransferase (α-CT), and β-carboxyltransferase (β-CT). These enzymes are involved in the conversion of acetyl-CoA to malonyl-CoA required for formation of fatty acids. The homomeric **type I multifunctional** (MF) ACCase consists of a single large polypeptide with BCCP, BC, and CT domains. In the cytosol of plants (extra plastidial) there is type I MF ACCase that is involved in the biosynthesis of very long chain fatty acids and flavonoid biosynthesis in the cytosol (Figure 3.3). In the grasses the homomeric form is present in both plastids and cytosol.[2,5,6]

Fatty acid synthase (FAS) is again divided into two groups. **Fatty acid synthase type I** consists of a large, multifunctional protein. FAS type I is found in eukaryotic cytoplasm of yeast and mammals (also in some bacteria).[7] **Fatty acid synthase type II** is composed of seven discrete monofunctional proteins. Six proteins are **catalytic proteins**. The seventh one is a low molecular weight **acyl-carrier protein (ACP)**. Fatty acid biosynthesis by an FAS type II system is found in most bacteria and plastids of eukaryotic cells such as plant cells. The ACP contains a thiol group essential for fatty acid biosynthesis as it forms thioesters. The thiol is present on a 4′ phosphopantetheine prosthetic group that is transferred from CoA to the serine residue. A series of reactions take place on the ACP linked substrate.[8] The plastidial FAS in plants (ptFAS) produces fatty acids for most lipids, including membrane lipids, signaling lipids, and storage lipids. The mitochondrial FAS (mtFAS) synthesizes an acyl precursor for the biosynthesis of lipoic acids.[9]

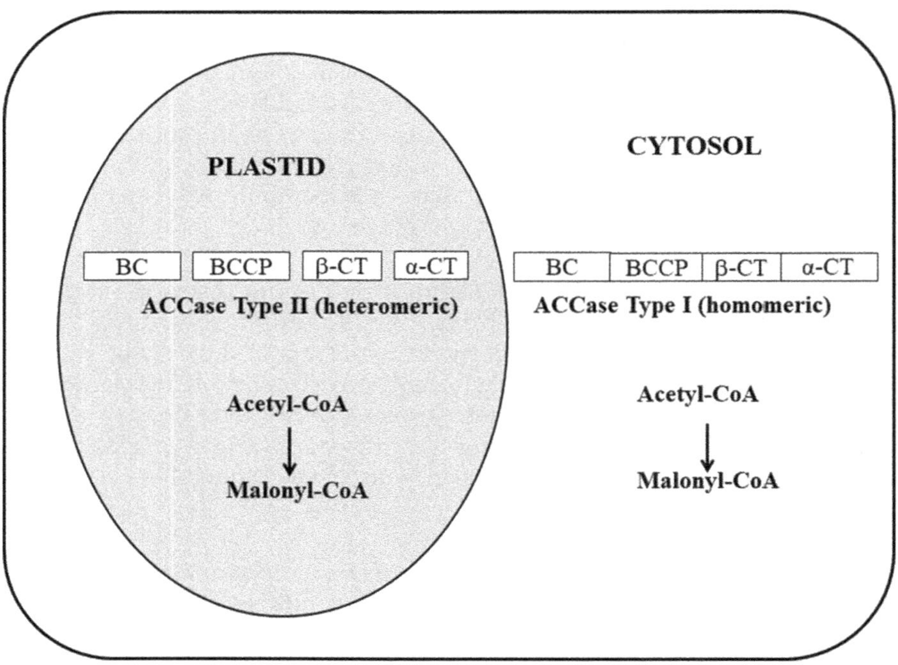

Figure 3.3 Compartmentation of two types of ACCase. BC: biotin carboxylase; BCCP: biotin carboxyl carrier protein; α-CT: α-carboxyltransferase; β-CT: β-carboxyltransferase.

During fatty acid synthesis, the malonyl group from malonyl-CoA is transferred on to the thiol of ACP to form malonyl-ACP. The reaction is catalyzed by **malonyl-CoA: ACP transacylase (MCAT or MAT)**. Then a condensation reaction of acetyl-CoA (or acetyl-ACP produced from acetyl-CoA) and malonyl-ACP leads to the formation of a four-carbon **3-ketoacyl-ACP**, releasing carbon dioxide. The reaction is catalyzed by the enzyme **β-ketoacyl-ACP synthase III (KAS III)** which links acetyl-CoA to a cysteine-thiol moiety then uses malonyl-ACP for condensation. The 3-ketoacyl-ACP subsequently undergoes a **reduction** (by **β-ketoacyl-ACP reductase**) to generate **3-hydroxyacyl-ACP**. 3-Hydroxyacyl-ACP generates **enoyl-ACP** after **dehydration** (by **β-hydroxyacyl-ACP dehydratase**). Further reduction of enoyl-ACP (by **enoyl-ACP reductase**) produces **butyryl ACP.** Butyryl-ACP is the product of the first cycle and acts as the substrate for another round of chain elongation. Fatty acid chains will be extended by two further C2 units after condensation with another malonyl-ACP. Each malonyl ACP provides the carbon atoms (C2 units) for fatty acid elongation. Butyryl-ACP is transferred on to the cysteine. For conversion of C4 to C16, the condensation reactions with malonyl-ACP are catalyzed by **β-ketoacyl-ACP synthase I (KAS I)**. After seven rounds of such cycles, C16 palmitoyl-ACP (hexadecanoyl-ACP) is formed. Palmitoyl-ACP may be extended by two new units to form C18 stearoyl-ACP (octadecanoyl-ACP) by **KAS II, β-ketoacyl-ACP reductase, hydroxyacyl-ACP dehydratase, and enoyl-ACP reductase** (Figure 3.4). The KAS I, II, and

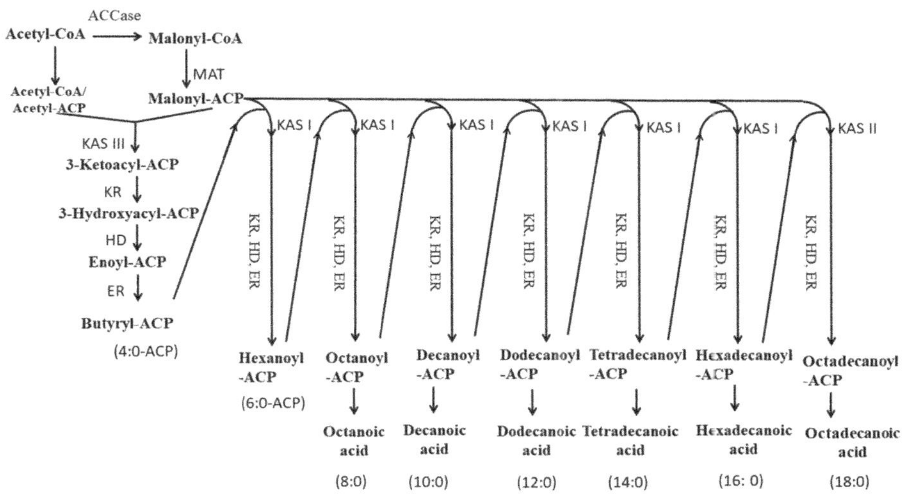

Figure 3.4 Biosynthesis of fatty acids. ACCase: Acetyl-CoA carboxylase; MAT: malonyl-CoA: ACP transacylase; KR: β-ketoacyl-ACP reductase; HD: β-hydroxyacyl-ACP dehydratase; ER: enoyl-ACP reductase; KAS III: β-ketoacyl-ACP synthase III; KAS II: β-ketoacyl-ACP synthase II; KAS I: β-ketoacyl-ACP synthase I; [fatty acid (IUPAC name): caprylic acid (octanoic acid); capric acid (decanoic acid); lauric acid (dodecanoic acid); myristic acid (tetradecanoic acid); palmitic acid (hexadecenoic acid); stearic acid (octadecanoic acid)].

III enzymes are also known as FabB, FabF, and FabH, respectively, after their gene names. Similarly, β-ketoacyl-ACP reductase, β-hydroxyacyl-ACP dehydratase, and enoyl-ACP reductase are also known as FabG, FabZ, and FabI, respectively. Plastidial soluble stearoyl desaturase converts stearoyl-ACP to oleoyl-ACP (18:1-ACP, C9).[2,10–13]

After the formation of C16 or C18 fatty acids, the following alternative reactions may occur:

1. Introduction of a double bond, usually at Δ9 position, may lead to the production of mono-unsaturated 18: 1-ACP. The reaction is catalyzed by stearoyl-ACP desaturase.
2. The acyl group may be cleaved to form free fatty acids and ACP. This reaction is catalyzed by FAT (fatty acyl-ACP thioesterase).
3. Very long chain fatty acids may be produced.

3.1.1.1 Elongation of Fatty Acids

Very long chain fatty acids (VLCFAs) are elongated in the endoplasmic reticulum. The elongase complex adds two carbons to long chain fatty acids with 16- and 18-carbons sequentially. C16 and C18 fatty acids produced in the plastid are exported to the cytosol where they are activated as acyl-CoAs by long-chain fatty acyl-CoA synthetase. The endoplasmic reticulum contains the elongase complex consisting of 3-keto-acyl-CoA synthase (KCS), β-ketoacyl-CoA reductase (KCR), 3-hydroxyacyl-CoA dehydratase (HCD), and *trans*-2,3-enoyl reductase (ECR). KCS is involved in the condensation of fatty acyl-CoA with malonyl-CoA to form a 3-ketoacyl-CoA intermediate. 3-Ketoacyl-CoA intermediate is then reduced by KCR to 3-hydroxyacyl-CoA. 3-Hydroxyacyl-CoA is then dehydrated by HCD to *trans*-2,3 enoyl-CoA. Finally, *trans*-2,3 enoyl-CoA is reduced by ECR to yield elongated fatty acyl-CoA (Figure 3.5). The fatty acyl-CoA, thus elongated, may further enter the elongation cycle.[14] The reaction cycle thus can be repeated to produce very long-chain fatty acids (VLCFA) of carbon chain length 20–38.

Very long-chain fatty acids (VLCFAs) have roles in plant development and cellular processes such as cell division, cell differentiation, membrane trafficking *etc.* They are also components of membrane lipids, sphingolipids, storage lipids, cuticular waxes and suberin. VLCFAs are non-cell autonomous regulators of plant development by repressing cytokinin synthesis. Some plants store VLCFAs as triacylglycerols in the seeds where they later play a role in seed germination. The VLCFAs also display nutritive and energetic properties.

3.1.1.2 Unsaturated Fatty Acids

The long chain saturated fatty acids such as palmitic acid (16:0) and stearic acid (18:0) are converted into monounsaturated fatty acids (MUFAs) at Δ9 by desaturases to produce 16:1 (palmitoleic acid) and 18:1 (oleic acid),

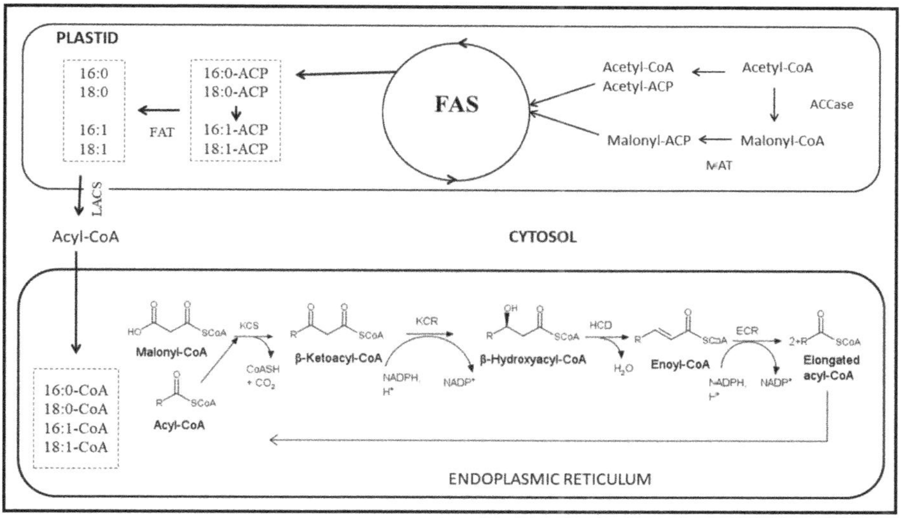

Figure 3.5 Biosynthesis of very long chain fatty acids. FAT: fatty acyl-ACP thioesterase; LACS: long-chain acyl-CoA synthetases; KCS: β-keto-acyl-CoA synthase; KCR: β-keto-acyl-CoA reductase; HCD: 3-hydroxyacyl-CoA dehydratase; ECR: enoyl-CoA reductase.

respectively, in plastids. Other plant desaturases use complex lipid substrates. In such cases, both the substrates and desaturases are found in the membranes in the endoplasmic reticulum. Enzymes for Δ12 and Δ15 desaturases convert oleic acid [18:1 (9*c*)] to linoleic acid [18:2 (9*c*, 12*c*)] and α-linolenic acid [18:3 (9*c*, 12*c*, 15*c*)], the polyunsaturated fatty acids. In case of unsaturated fatty acids, the stereochemistry of double bonds is *Z* (*cis*) creating a bend in the alkyl chain.

Important polyunsaturated fatty acids (PUFAs) can be categorized to two main families, ω6 and ω3, based on the position of first double bond from the methyl end group of the fatty acid. Main ω3 PUFAs are ALA (18:3 α-linolenic acid), DHA (docosahexaenoic acid), EPA (eicosatetraenoic acid), and DPA (docosapentaenoic acid). Main ω6 PUFAs are LA (linoleic acid), and AA (arachidonic acid) (Figures 3.2 and 3.6).[15]

Mammals cannot synthesize LA and ALA from oleic acid. Conversions of LA and ALA to AA, DHA, and EPA are low so direct intake of these PUFAs from food sources is important. PUFAs play an important role in immune system regulation, regulation of blood clots, neurotransmitters, cholesterol metabolism (including inhibition of synthesis of low-density lipoprotein or LDL, acceleration of LDL elimination), structure of membrane phospholipids in the brain and retina, prevention of cardiovascular diseases, diabetes, vasodilation, platelet aggregation, blood pressure *etc.* ω3 PUFAs also have beneficial effects on skin diseases, asthma, arthritis *etc.* DHA is concentrated in the retina and brain and is essential for visual and brain function. DHA regulates

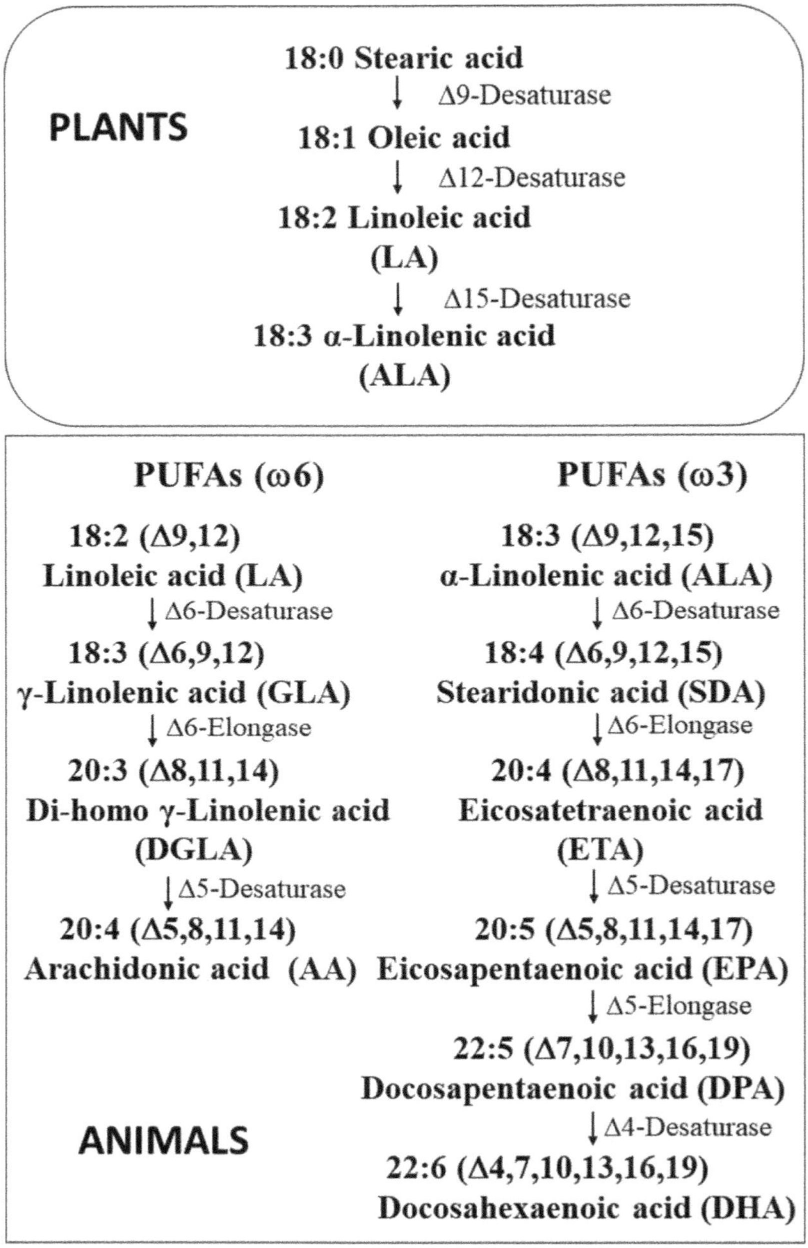

Figure 3.6 Pathway for the synthesis of PUFAs in plants and animals.

neurotransmitter biosynthesis, signal transduction and uptake of serotonin. PUFAs regulate neurotransmission, cell survival and neuroinflammation thereby mood and cognition. PUFAs have attracted attention for prevention of food allergies. Marine fish contains longer chain ω3 fatty acids, specially EPA and DHA, which are not found in seed oils of higher plants. Regular consumption of oily fish is claimed to reduce the risk of heart attacks and atherosclerosis.[16–18]

GLA (γ-linolenic acid) and SDA (stearidonic acid, an intermediate produced during PUFA biosynthesis) are produced in certain plants such as primrose seed, hemp and blackcurrant seed oil. GLA is present in evening primrose (*Oenothera biennis*, Onagraceae), borage seeds (*Borago officinalis*, Boraginaceae), blackcurrant seeds (*Ribes nigrum*, Grossulariaceae) and other *Ribes* species. LA and ALA are available in the seed oils of black raspberry (*Rubus occidentalis*, Rosaceae), red raspberry (*Rubus ideaus*, Rosaceae), blueberry (*Vaccinium corymbosum*, Ericaceae), cranberry (*Vaccinium macrocarpon*, Ericaceae), sea buckthorn (*Hippophae rhamnoides*, Elaeagnaceae), basil (*Ocimum basilicum*, Lamiaceae), soybean (*Glycine max*, Fabaceae), walnut (*Juglans regia*, Juglandaceae), *etc.* Hemp (*Cannabis sativa*, Cannabaceae) seed oil contains LA, ALA, and GLA.[18]

3.1.1.3 Mitochondrial Biosynthesis of Lipoic Acid

Lipoic acid (6,8-dithiooctanoic acid), a sulfur containing coenzyme, is essential for mitochondrial metabolism. It is an essential cofactor required for catalysis by multiple mitochondrial 2-ketoacid dehydrogenase complexes. It is metabolized *de novo* using mitochondrial FAS II utilizing multiple acyl carrier protein isoforms. The biosynthesis is initiated by octanoyl-ACP (8:0-ACP) (Figure 3.7). The E2 subunit is a dihydrolipoamide acyltransferase.[19,20]

Lipoic acid is used as an antioxidant as dietary supplementation. Lipoic acid has shown ability to inhibit the proliferation of breast tumour cells, thus having potential application as an anticancer drug. It is also a potential anti-diabetic compound. It has now been possible to produce *R*-lipoic acid from lipoate-modified proteins in *Saccharomyces cerevisiae* by metabolic engineering.[21]

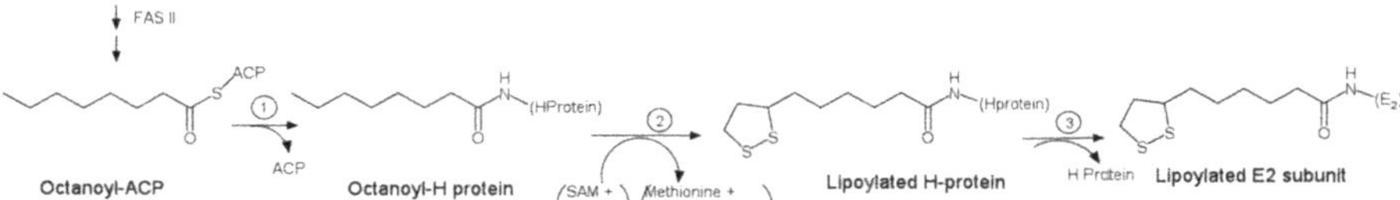

Figure 3.7 Lipoic acid biosynthesis in mitochondria of eukaryotes. 1: octanoyl transferase; 2: lipoate synthase; 3: lipoyl transferase; E2 subunit: dihydrolipoamide acyltransferase.

3.1.2 Unusual Fatty Acids

Many structurally diverse fatty acids, often mentioned as unusual fatty acids, are found in the plant kingdom. Changes in chemical structure[3] of fatty acids may be due to *cis/trans* orientation, acetylenic or triple bonds, carbon side chain modifications *e.g.*, epoxidation, hydroxylation and incorporation of cyclopropane and furan rings. A few examples are given below.

Ricinoleic acid (12-hydroxy-9-*cis*-octadecenoic acid), an unusual **mono-hydroxy** fatty acid (Figure 3.8), is the major fatty acid of castor oil obtained from the seeds of *Ricinus communis* (Euphorbiaceae). Ricinoleic acid is formed after 12-hydroxylation of oleic acid. Castor oil is a mixture of tri-glycerides about 75% of which is triricinolein and the rest consists of diri-cinoleoglycerides with the third acyl group being either oleic, linoleic, dihydroxystearic or a saturated (palmitic or stearic) acid. Castor oil is used as a domestic purgative. Ricinoleic acid is released after hydrolysis of triri-cinolein by lipase. Ricinoleic acid exerts a cathartic effect. Ricinoleic acid is also used as engine oil because of its exceptional lubricity properties. It has recently been found that *Hiptage benghalensis* (Malpighiaceae) seed oil con-tains a high amount of ricinoleic acid.[4,22,23]

Vernolic acid is an **epoxy** fatty acid obtained from the seeds of *Vernonia galamensis* and *Stokesia laevis* (Asteraceae). It is formed by the epoxidation of linoleic acid. Another epoxy fatty acid is **coronaric acid** (an isomer of vernolic acid) reported from the seed oil of *Chrysanthemum coronarium* (Asteraceae) (Figure 3.8).[4,24]

Crepenynic acid is an unusual fatty acid reported from the seed oil of *Crepis foetida* (Asteraceae). It is a polyunsaturated fatty acid containing **acetylenic**

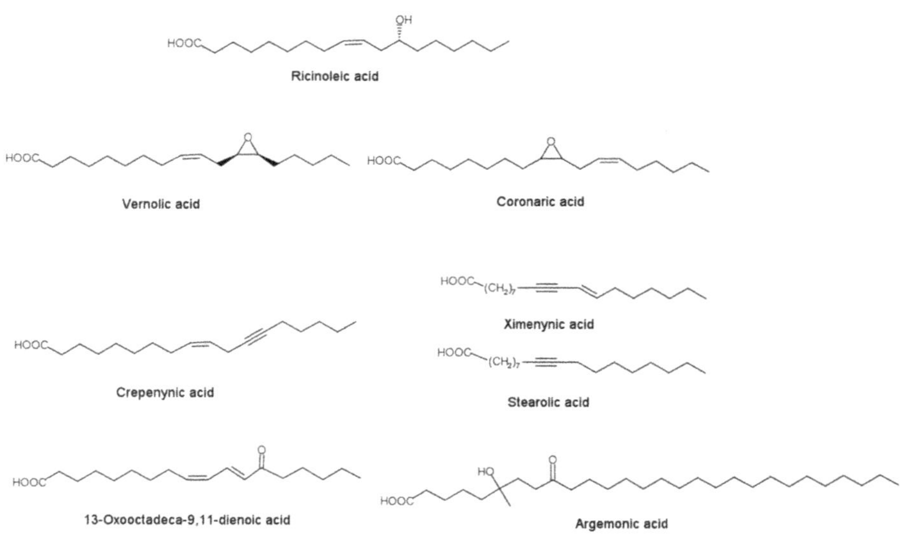

Figure 3.8 Unusual fatty acids.

unsaturation. Two other acetylenic fatty acids are reported from the seed oils of *Exocarpus cupressiformis* and *Santalum acuminatum*, both belonging to the family Santalaceae. These are **ximenynic acid** and **stearolic acid** (Figure 3.8).[25]

There are some fatty acids containing **ketone** group (Figure 3.8) *e.g.*, **13-oxooctadeca-9,11-dienoic acid**, reported from *Monnina emerginata* (Polygalaceae) seed oil[24] and **argemonic acid** from the seed oil of *Argemone mexicana* (Papavaraceae).[26,27]

Chaulmoogric acid and **hydnocarpic acid** (Figure 3.9) are **cyclopentenyl** fatty acids of chaulmoogra oil obtained from the seeds of *Hydnocarpus wightiana* (Achariaceae). These fatty acids are derived from malonate chains as extender units and coenzyme A ester of 2-cyclopentenyl carboxylic acid as the starter unit. Chaulmoogra oil has long been used for the treatment of leprosy caused by *Mycobacterium leprae*. Chaulmoogric acid and hydnocarpic acid have strong bactericidal effects. This oil has now been replaced by more effective synthetic agents.[4]

Sterculic acid (Figure 3.10), a highly strained **cyclopropene** fatty acid with a double bond between C9 and C10, present in the seeds of *Sterculia foetida* (Sterculiaceae) is the inhibitor of stearoyl-CoA desaturase 1 (also known as Δ9-desaturase). Δ9-Desaturase is active in the conversion of saturated fatty acids palmitic acid and stearic acid into monounsaturated fatty acids (MUFAs) like palmitoleic acid and oleic acid. This action is harmful to humans. However, because of its stearoyl-CoA desaturase inhibitory properties, sterculic acid has potential in preventing pathologies where this enzyme is involved, such as playing a protective role in age-related macular degeneration of the retina. It has a predominant role in lipid metabolism and body weight control. Another cyclopropane fatty acid **malvalic acid** (Figure 3.10), produced by chain shortening of sterculic acid, is also present in *Sterculia* oil. Malvalic acid is also found in the cotton seed oil from *Gossypium* species (Malvaceae). Malvalic acid, like that of sterculic acid, has stearoyl-CoA desaturase inhibitory property.[4,28]

Furan fatty acids (F-acids) are heterocyclic fatty acids with a furan ring in the middle of the fatty acid structure and are produced in algae, plants and microorganisms. F-acids have been detected in many plants of economic

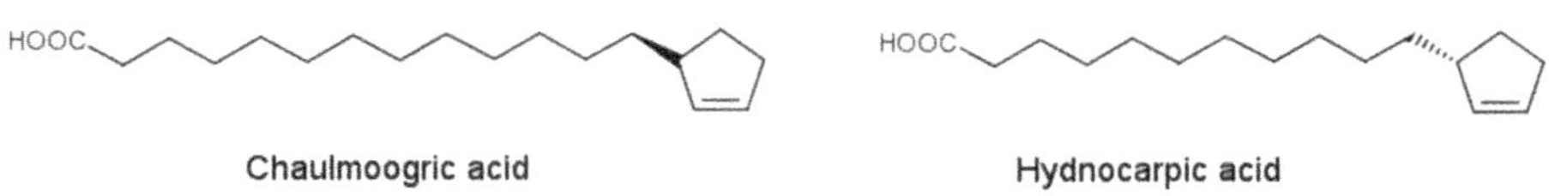

Figure 3.9 Unusual cyclopentenyl fatty acids.

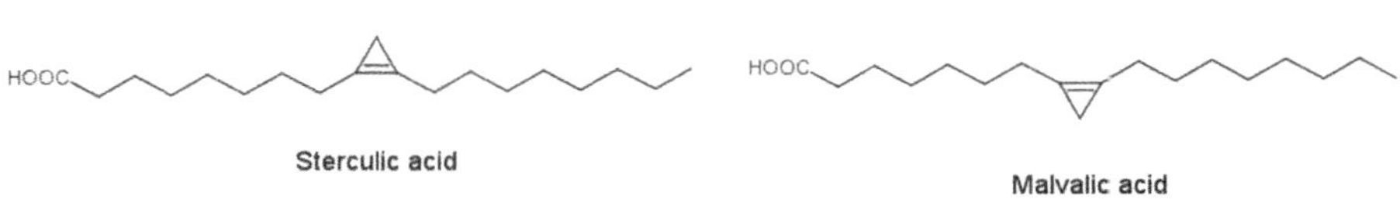

Figure 3.10 Unusual cyclopropene fatty acids.

importance such as fruits, vegetables, and oilseed crops. In plants, the F-acids are bound to phospholipids by substituting PUFAs and function as free radical scavengers. F-acids are unstable because of their antioxidative properties. Their photooxidative products impart flavour to the edible oils. One typical example of F-acid is **9-(3-methyl-5-pentylfuran-2-yl)-nonanoic acid (9M5)** (Figure 3.11) reported in the latex of *Hevea brasiliensis* (Euphorbiaceae). Several studies have detected F-acids in vegetables [such as, garlic (*Allium sativum*, Amaryllidaceae), onion (*Allium cepa*, Amaryllidaceae), potato (*Solanum tuberosum*, Solanaceae), cabbages (*Brassica oleracea*, Brassicaceae)], food grains like rice and wheat, fruits [such as, lemon (*Citrus limon*, Rutaceae), strawberries (*Fragaria* spp., Rosaceae), oranges (*Citrus sinensis*, Rutaceae)], soya oil, corn oil, tea leaves *etc.*[29–31]

Cyanolipids are common in the seed oils of Sapindaceae. In addition to the acylglycerols, cyanolipids consist of different hydroxynitrile fatty esters derived from amino acid metabolism. There are four types of structures reported. These are type I (1-cyano-2-hydroxymethyl-prop-2-en-1-ol-diester), type II (1-cyano-2-methyl-prop-1-en-3-ol-ester), type III (1-cyano-2-hydroxymethyl-prop-1-en-3-ol-diester) and type IV (1-cyano-2-methyl-prop-2-en-1-ol-ester) with fatty acids esterified to mono- or dihydroxynitrile moieties (Figure 3.12).[32]

Tuberculostearic acid (Figure 3.13) is a *C*-methyl derivative of stearic acid. It is found in the cell wall of *Mycobacterium tuberculosis* and is used as a marker for the diagnosis of tuberculosis disease. **Lipstatin** (Figure 3.13) is a β-lactone derivative isolated from *Streptomyces toxytricini* cultures. Lipstatin is formed by two fatty acids octanoic acid and tetradeca-5,8-dienoic acid. Lipstatin is an inhibitor of pancreatic lipase. Lipase inhibitors are often used to reduce body mass in obese patients. Tetrahydrolipstatin or orlistat has been developed as an anti-obesity drug.[4]

3.1.3 Triacylglycerols

Lipids (fats, oils, and waxes) are obtained from plant and animal sources. Fats and oils are composed of **triacylglycerols** (Figure 3.14). Fats are usually semisolid at room temperature while oils are liquid. Usually, glycerides of

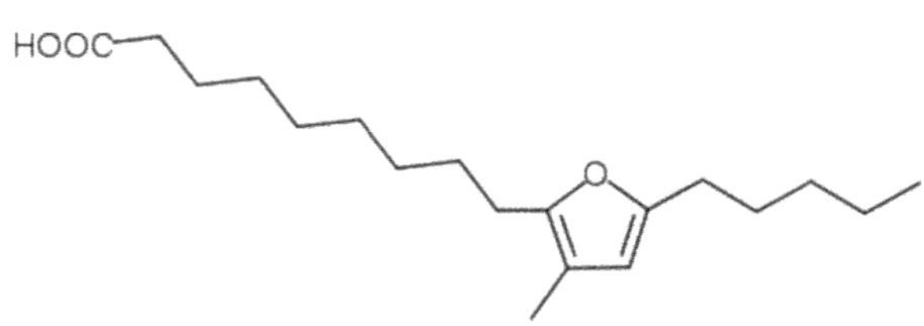

9-(3-Methyl-5-pentylfuran-2-yl) nonanoic acid
(9M5)

Figure 3.11 Unusual furan fatty acid.

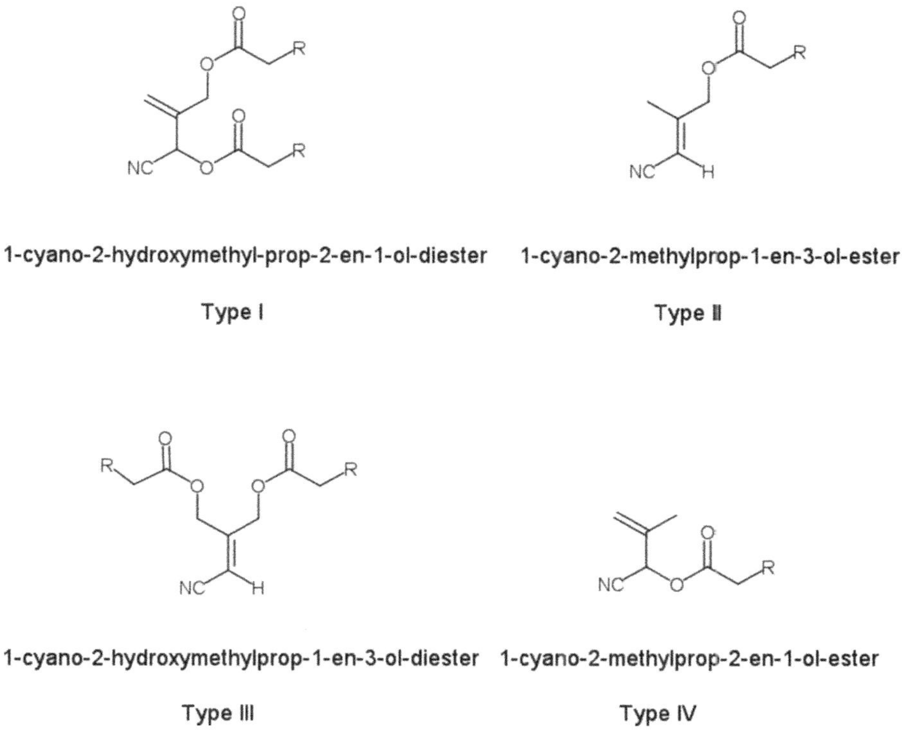

Figure 3.12 Unusual cyanolipids, R = fatty acyl.

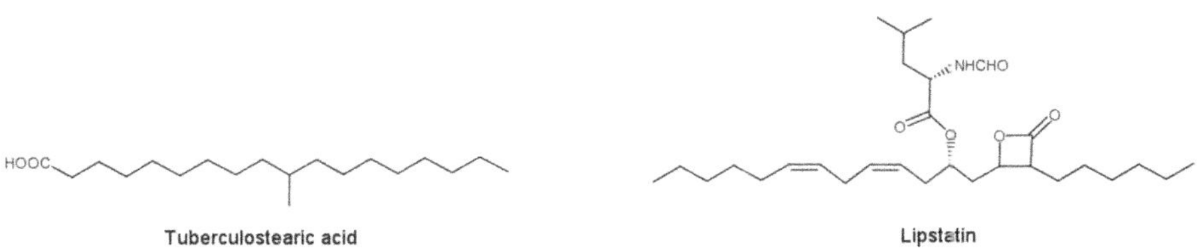

Figure 3.13 Unusual fatty acids from microbial sources.

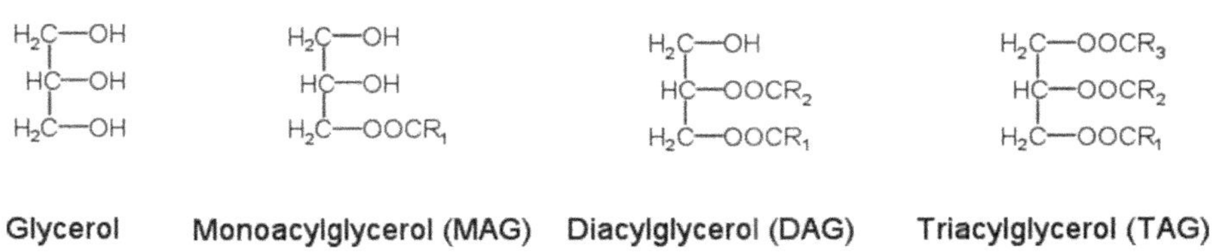

Figure 3.14 Acylglycerols. R_1, R_2, R_3 = fatty acids.

unsaturated fatty acids are liquid whereas glycerides of saturated fatty acids of sufficient length are solid. Thus, cocoa butter (*Theobroma* oil, from the beans of *Theobroma cacao*, Malvaceae), is solid as it contains relatively high concentration of saturated fatty acids. Similarly, coconut oil is semisolid as it contains a high proportion of lauric acid, a saturated fatty acid. Olive oil has a high content of oleic acid (18:1) and rapeseed oil has high concentrations of the long chain unsaturated fatty acid erucic acid (22:1).[23] Fatty acids of fish oil have high unsaturation and long chain length such as eicosapentaenoic acid or EPA (20:5). Docosahexaenoic acid or DHA (22:6) is present in cod liver oil. Vegetable oils, fish oils and animal fats are the dietary sources of lipids. Edible vegetable oils are triacylglycerols of plant origin and include sunflower oil, soybean oil, safflower oil, peanut oil, corn oil, olive oil, palm oil, and canola oil *etc.* Oils and fats are important nutritional components. They regulate body temperature, provide metabolic energy and membrane structure. Vegetable oils provide α-linolenic acid (ALA) that is the precursor of EPA and DHA synthesis in the human body. Soybean oil is one of the primary sources of the PUFAs linoleic acid (LA) and ALA. Fish oil provides EPA and DHA readily available for the human body. Saturated fats are considered to have adverse effects like promoting insulin resistance, increasing low-density lipoprotein (LDL) cholesterol and consequently increasing the risk of diabetes and cardiac problems.[16,33]

Triacylglycerol (TAG) synthesis occurs predominantly in the endoplasmic reticulum (ER) in most plant species. The enzymes responsible for TAG biosynthesis are available in the membrane of ER. Fatty acids, mostly with 16- and 18- carbon acyl groups, are exported from the plastid to the endoplasmic reticulum. There are two proposed pathways for triacylglycerol biosynthesis: acyl-CoA dependent pathway (also known as Kennedy pathway) and acyl-CoA independent pathway phospholipid:diacylglycerol acyltransferase (PDAT).[34,35]

Following the Kennedy pathway, fatty acids synthesized in the plastids are exported to the cytosol where they are activated to form acyl-CoA before entering endoplasmic reticulum. In the endoplasmic reticulum, glycerol 3-phosphate (G3-P) is acylated into lysophosphatidic acid (LPA) using acyl-CoA as an acyl donor catalyzed by glycerol-3-phosphate acyltransferase (GPAT). LPA is then acylated again to yield phosphatidic acid (PA) by acyl-CoA:lysophosphatidic acid acyltransferase (LPAAT). Phosphatidic acid phosphatase (PAP) dephosphorylases PA into diacylglycerol (DAG). DAG can be directly acylated to TAG by diacylglycerol acyltransferase (DGAT). Acyl-CoA can directly be incorporated in the membrane lipid phosphatidylcholine (PC) by the PC:acyl-CoA acyl editing cycle with ultimate formation of DAG and TAG (Figure 3.15).[35]

The second pathway of TAG formation utilizes *de novo* DAG to first synthesize PC. DAG is regenerated from PC by the removal of the phosphocholine head-group to generate the substrate for TAG synthesis (Figure 3.15).[34,35]

Synthesized TAGs are accumulated within the membrane bilayer. Then lipid droplets (an organelle) are produced and are deposited in the cytosol.

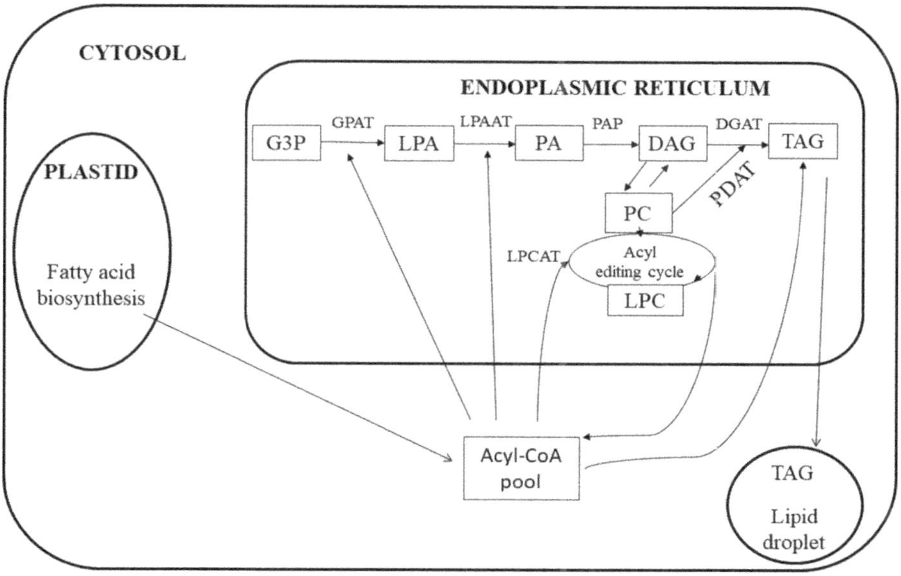

Figure 3.15 Triacylglycerol biosynthesis. G3P: glycerol 3-phosphate; LPA: lysophosphatidic acid; PA: phosphatidic acid; DAG: diacylglycerol; TAG: triacylglycerol; PC: phosphatidylcholine; LPC: lysophosphatidylcholine; GPAT: glycerol-3-phosphate acyltransferase; LPAAT: acyl-CoA:lysophosphatidic acid acyltransferase; PAP: phosphatidic acid phosphatase; DGAT: diacylglycerol acyltransferase; PDAT: phospholipid:diacylglycerol acyltransferase; LPCAT: lysophosphatidylcholine acyltransferase.

Eventually fatty acids are removed from the TAGs by the actions of lipases. They are then either reused for membrane lipid synthesis or transported into peroxisomes for β-oxidation.[35]

3.1.3.1 Waxes

Plant waxes are lipid components present in the epicuticular and intra-cuticular regions of the epidermal cells. Plant waxes are complex mixtures of very long chain aliphatic lipids and small amounts of terpenoids, sterols and flavonoids. Major aliphatic wax components are n-alkanes (chain length C21–C35), secondary alcohols (chain length C21–C35), ketones (chain length C21–C35), fatty alcohols (chain length C22–C34), fatty acids (chain length C16–C34), aldehydes (chain length C21–C35) and wax esters (chain length C32–C64). Wax composition varies from plant to plant, organ to-organ, cell to cell, and by developmental stages and environmental conditions.[36,37]

Wax biosynthesis starts with fatty acid synthesis in the plastid to produce very long chain fatty acids which undergo subsequent processing to produce other components like alkanes, secondary alcohols, ketones, primary alcohols, and wax esters (Figure 3.16).[37–39]

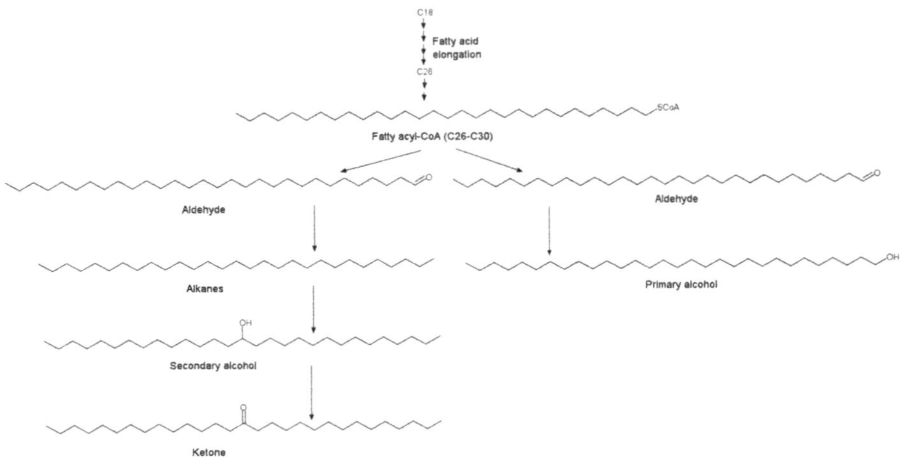

Figure 3.16 Wax components produced from very long chain fatty acids.

Cuticular wax is important to plants. It restricts non-stomatal water loss, protects plants against ultra-violet radiation and reduces water retention on the surface thus minimizing deposition of dust, pollen, and pollutants. The wax also protects the plant against bacterial and fungal pathogens and participates in plant-insect interactions.[38,39]

There are some important plant waxes such as carnauba wax, candelilla wax, jojoba oil, rice bran wax and sunflower wax, in addition to the animal waxes like beeswax, shellac wax and wool wax. **Carnauba** wax is obtained from the frond cuticle of the Brazilian palm *Copernicia cerifera* (Arecaceae). This wax is used in cosmetics, automobile and furniture wax, molds for semiconductor devices, coating for dental floss, and as food. **Candelilla wax** is extracted from the Mexican plant *Euphorbia antisyphilitica* (Euphorbiaceae). Microemulsions of this wax are used as coatings for fruits and as food additive. It is also used in cosmetics, and as shoe and furniture polishes. **Jojoba oil** is the wax produced from the seeds of *Simmondsia chinensis* (Simmondsiaceae). Melting point of the oil range from 15 °C to 70 °C. The wax is used in the cosmetic industry. **Rice bran wax** is produced from rice oil. This wax is used in cosmetics, fruit and vegetable coatings, paper coatings, candles, waterproofing, crayons, chewing gum, pharmaceuticals *etc.* **Sunflower wax** is obtained from different parts of the sunflower plant *Helianthus annuus* (Asteraceae). It is used in cosmetics (lipsticks, lip balms *etc.*).[36]

3.1.4 Lipid Polyesters

Cutin and suberin, the two hydrophobic glycerolipid polyester components found on the cell walls of specialized plant tissues or cell types, are mostly composed of oxygenated fatty acids. Cutin and suberin help plants to adapt to the environment of the terrestrial habitat limiting transpiration and movement of gases, water absorption, and protecting the plants against different biotic and abiotic stress factors. Both cutin and suberin consist of saturated and non-saturated fatty acids, fatty alcohols, hydroxycinnamic acids and glycerol that are linked by ester bonds into complex matrices. However, the proportions and compositions of these monomers differ substantially between species, and among organs of the same species. Cutin monomeric units traverse the cell wall to the cell surface before polymerization whereas suberin is polymerized immediately after cell export.[40]

Cutin is the lipid, insoluble in both water and organic solvents, covering the outside of epidermal cells. Cutin monomers of the polyester compound are composed of the hydroxy/epoxy/oxo derivatives of C16 and C18 fatty acids. Hydroxyl groups are present in the mid-chain and at the ω-end. ω-Hydroxylated 10,16-dihydroxy-palmitic acid (10,16-dihydroxyhexadecanoic acid), 9,10,18-trihydroxystearic acid (9,10,18-trihydroxyoctadecanoic acid) and 9,10-epoxy-18-hydroxystearic acid (9,10-epoxy-18-hydroxyoctadecanoic acid) are characteristic of the C16 and C18 cutins. Cutin also contains glycerol, dicarboxylic acids, and small amounts of phenylpropanoids such as ferulic acid and coumaric acid (Table 3.1; Figure 3.17). Monomers are held

Table 3.1 Monomer components of cutin and suberin.

Monomers (common in cutin)	Monomers (common in suberin)
Unsubstituted fatty acids (minor components)	**Unsubstituted fatty acids** (minor components)
C16–C18	C16–C26
Hydroxy/epoxy/oxo substituted fatty acids (major components) 16-Hydroxyhexadecanoic acid 18-Hydroxyoctadec-9-enoic acid 10,16-Dihydroxyhexadecanoic acid 9,10,18-Trihydroxyoctadecanoic acid 16-Hydroxy-10-oxo-hexadecanoic acid 18-Hydroxy-9,10-epoxyoctadecanoic acid	**Hydroxy/epoxy/oxo substituted fatty acids** (minor components) 18-Hydroxyoctadec-9-enoic acid 9,10,18-Trihydroxyoctadecanoic acid 18-Hydroxy-9,10-epoxyoctadecanoic acid ω-Hydroxyalkanoic acids (n = 1 to 6)
Dicarboxylic acid (minor components)	**Dicarboxylic acid** (substantial components)
10-Hydroxyhexadecanedioic acid	Octadec-9-enedioic acid 9,10-Epoxy-octadecanedioic acid 9,10-Dihydroxyoctadecanedioic acid α,ω-Aalkanodioic acids (n = 1 to 6)

(*continued*)

Table 3.1 (*Continued*)

Monomers (common in cutin)	Monomers (common in suberin)
Fatty alcohols (rare/minor components)	**Fatty alcohols** (substantial components)
	n-Alkanols (n = 1 to 8)
Glycerol (substantial component)	**Glycerol** (substantial component)
Aromatics (low)	**Aromatics** (high)
Ferulic acid Coumaric acid	Ferulic acid Coumaric acid Cinnamic acid

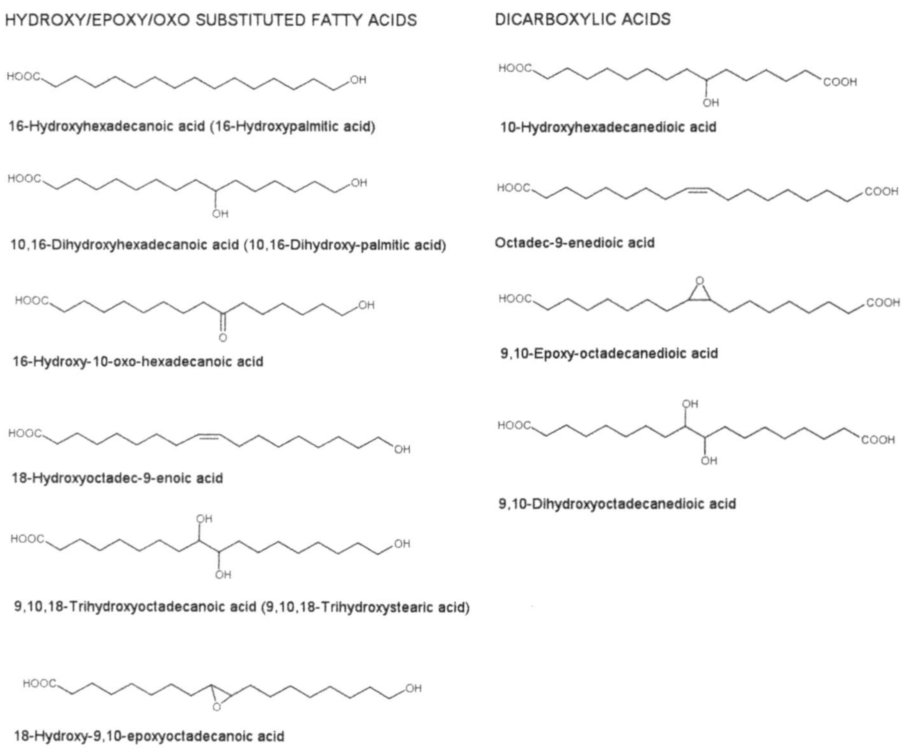

Figure 3.17 Structural composition of some cutin and suberin monomers.

by primary alcohol ester linkages in the polymer with branching or cross linking. Secondary hydroxyls may also be esterified to fatty acid monomers forming branched structures.[41–45]

Suberin is hydrophobic in nature and is associated with cork cells in plants. The aliphatic monomer components of suberin are ω-hydroxy (C16–C28) and α,ω-dicarboxylic acids (C16–C26). High levels of α,ω-dicarboxylic acids are characteristics for suberin. Minor quantities of very long chain fatty acids, and long chain alcohols (C18–C30) are also present. Significant quantities of hydroxycinnamic acids such as ferulic acid, cinnamic acid and *p*-coumaric acid are also detected. Glycerol is a constituent of suberin (Table 3.1; Figure 3.17). Monomers are cross linked to form polymers.[41–45]

3.1.5 Sphingolipids

Sphingolipids contain fatty acid linked to a long chain amino alcohol.[46] Sphingolipids are one of the major membrane lipid components in eukaryotic cells and a few bacteria. Some sphingolipids are conserved between animals, plants, and fungi. Some sphingolipid structures are found only in plants and fungi.[47] Over 300 sphingolipids have been reported from plants. In plant cells, a small portion (0.5% to 10%) of the total lipid content is sphingolipid. However, in some cases, sphingolipids are major components of cell membranes. For example, sphingolipids are available in relatively large quantities (up to 20%) in the tonoplast, ER and Golgi in plants. Over 40% of membrane lipids in plasmodesmata and extracellular vesicles are sphingolipids. But sphingolipids are either undetectable or present in trace amounts in chloroplasts and mitochondria. The proportion of different types of sphingolipids also varies.[46] Sphingolipids, identified from plants, are divided into four groups (Figure 3.18) as free long chain bases (LCBs) (the simplest form), ceramides (the backbones of sphingolipids), glycosylceramides (GlcCers), and glycosyl inositol phosphoceramides (GIPCs). Long chain bases include sphingosine (Sph), dihydrosphingosine (dh-Sph) and phyto-sphingosine (phyto-Sph). Long chain base phosphates (LCBPs) are the phosphate compounds of the corresponding LCBs. Examples are sphingosine-1-phosphate (S1P), dihydro-sphingosine-1-phosphate (dh-S1P) and phyto-sphingosine-1-phosphate (Phyto-S1P). About 40% of plant plasma membrane lipids are GlcCers and GIPCs.[46,47]

Biosynthesis of sphingolipids begins with condensation of serine and palmitoyl-CoA. The reaction is catalyzed by serine palmitoyl transferases in the ER. Ceramides are synthesized in the ER. GIPCs, the major plant sphingolipids, are synthesized in the Golgi. After synthesis, sphingolipids are translocated to the membranes.[46]

Sphingolipids are required for plant membrane integrity and help plant membranes in intracellular signaling to cope with biotic and abiotic stresses. Ceramides, LCBs or their phosphorylated forms may act as signaling molecules for the regulation of plant development and stress responses. Ceramides and free LCBs were found to induce hypersensitive response or

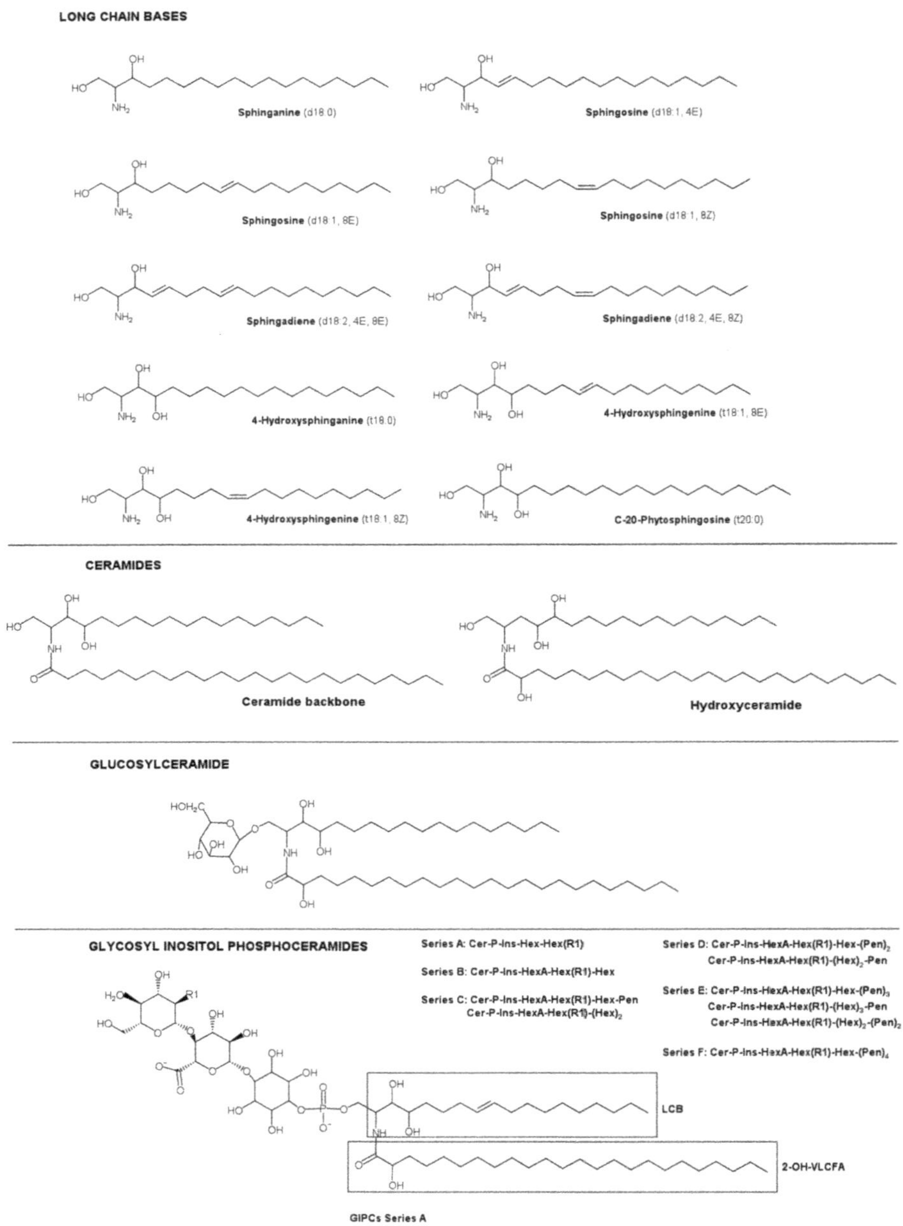

Figure 3.18 Plant Sphingolipids. R1: OH/NH2/N(Ac); Cer: ceramide; Hex: hexose; Pen: pentose; HexA: hexuronic acid; LCB: long chain base; 2-OH-VLCFA: 2-hydroxylated very long chain fatty acid; Ins: inositol.

programmed cell death in plants. Phyto-sphingosine phosphate (phyto-S1P) can initiate the salicylic acid signaling pathway. S1P signaling can trigger a Ca^{2+} spike and then stomatal closure. Cold treatment elevated the level of phyto-S1P and ceramide-1-phosphate suggesting their role in a signaling response to low temperature.[46]

3.1.6 Phyto-oxylipins

Oxylipins are oxygenated unsaturated fatty acids and metabolites derived from them. Plant oxylipins are important signaling molecules for defense and development. More than 500 such molecules have been reported from eukaryotic plants ranging from algae, mosses, to angiosperms. Jasmonic acid (JA) and derivatives collectively called jasmonates are the most studied lipid derived signals among oxylipins in plants. Oxylipins may exist in plants either as free form or as esterified form. In plants, the lipoxygenase (LOX) pathway seems to be the most prevalent one for biosynthesis of oxylipins. The LOX pathway may be further divided into a 9-LOX-derived branch and a 13-LOX-derived branch depending on product specific LOX enzymes. Either esterified or free fatty acids may serve as substrates in 13-LOX branch. All 13-LOXs are involved in wound induced JA formation. In the 9-LOX branch only free fatty acid derivatives serve as substrates. LOXs are nonheme iron containing dioxygenases that form fatty acid hydroperoxides from polyunsaturated substrates. Release of α-linolenic acid (ALA) from galactolipids of chloroplast membranes by phospholipase A1 (PLA1) initiates biosynthesis of JA. The biosynthesis of JA involves two compartments, plastid and peroxisome (Figure 3.19).[48,49] In the chloroplast *cis*-(+)-12-oxo-phytodienoic acid (OPDA) is produced. OPDA is exported out of the chloroplast to the peroxisome *via* cytosol at pH ~7.2.[50]

Jasmonic acid is a core hormone signaling molecule that interacts with multiple hormone signaling pathways (such as auxin, ethylene, abscisic acid, salicylic acid, brassinosteroids and gibberellin) to regulate plant growth, biotic and abiotic stress tolerance.[51] Conversion of JA into active and inactive compounds or compounds with partial activity takes place by twelve metabolic pathways. JA-Ile is formed by jasmonoyl amino acid conjugate synthase (JAR1). JA-Ile is the most biologically active JA compound. Other endogenous bioactive JA derivatives identified are (+)-7-*iso*-JA-Leu, (+)-7-*iso*-JA-Ala, (+)-7-*iso*-JA-VAL, and (+)-7-*iso*-JA-Met.[49] The methyl ester of JA (MeJA) has been known as a component of floral scent active in pollination and as a signaling molecule. However, recent studies show that MeJA is not active itself. MeJA is to be cleaved by esterases and then converted to JA-Ile. Tuberonic acid (12-OH-JA) and its glucosyl derivative were described as tuber-inducing factors of potato. 12-*O*-glucosyl-JA is accumulated abundantly in wounded leaves of tomato and *Arabidopsis*.[49] Oxylipins esterified in complex lipids such as arabidopsides and linolipins are thought to be involved in plant defense and developmental processes. These molecules are produced under stress conditions. Very few such oxylipins are reported, mainly from *Arabidopsis* and a few other plants. The esterified oxylipins might act as a pool of free oxylipins

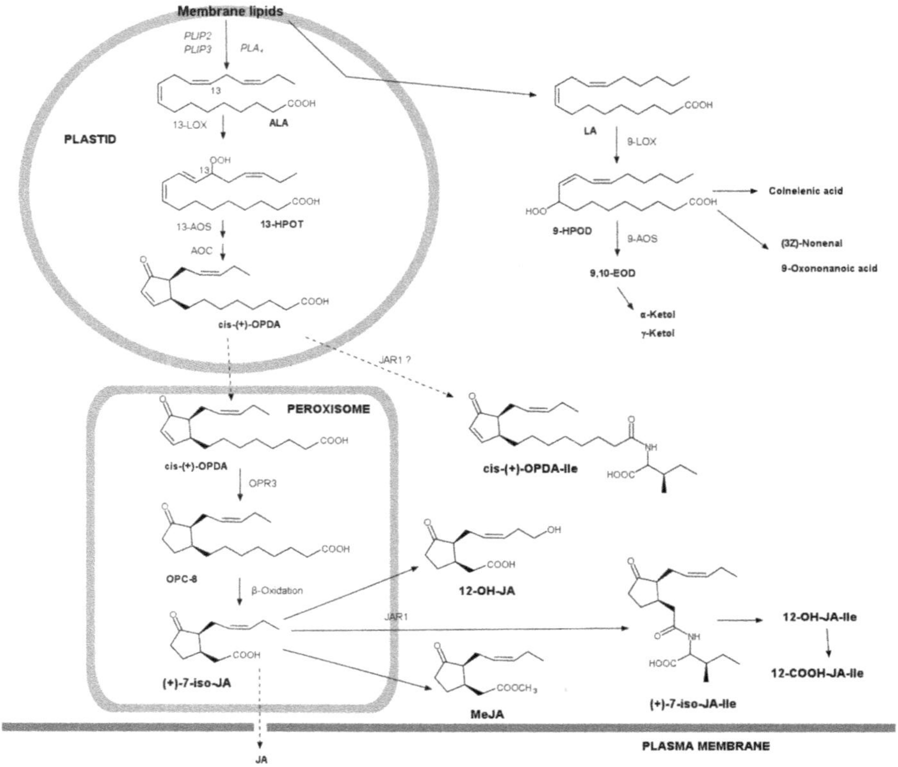

Figure 3.19 Biosynthesis of oxylipins. ALA: α-linolenic acid; LA: linoleic acid; 13-HPOT: (13S)-hydroperoxy octadecatrienoic acid; *cis*-(+)-OPDA: *cis*-(+)-12-oxo-phytodienoic acid; *cis*-(+)-OPDA-Ile: *cis*-(+)-12-oxo-phytodienoic acid isoleucine conjugate; OPC-8: 3-oxo-2-(2-pentenyl)-cyclopentane-1-octanoic acid; JA: jasmonic acid; 12-OH-JA: 12-hydroxyjasmonic acid; MeJA: methyl jasmonate; JA-Ile: jasmonic acid-isoleucine conjugate; HPOD: hydroperoxy octadecatrienoic acid; EOD: epoxy octadecaenoic acid; PLA_1: phospholipase A1; PLIP2: plastid lipase 2; PLIP3: plastid lipase 3; LOX: lipoxygenase; AOS: allene oxide synthase; AOC: allene oxide cyclase; OPR3: 12-oxophytodienoic acid reductase 3; JAR1: jasmonoyl amino acid conjugate synthase.

that could be released when necessary. In addition, esterified oxylipins could also have direct function. Arabidopsides E and G could decrease pathogen growth *in vitro*.[52]

3.2 Polyketides

Polyketides are a large class of structurally complex natural products. Most of the polyketides are aromatic derivatives although acyclic compounds also occur. Initially a chain of alternating keto and methylene groups is formed and later by different reactions a structurally complex compound is created.

3.2.1 Polyketide Biosynthesis

The polyketide pathway for metabolite biosynthesis starts with a starter unit and an extender unit. The starter unit is usually acetyl-CoA or propionyl-CoA. The extender unit is either malonyl-CoA or methylmalonyl-CoA. The enzymes that make the polyketide backbone are **polyketide synthases (PKSs)**. The starter unit and the extender units are bound through Claisen reactions involving thioesters to produce a basic chain consisting of C2 units. The condensation is induced by decarboxylation of the extender unit. The diketide, thus formed, may undergo further condensation employing another extender unit to form a triketide. The polyketide chain is thus extended by extender units. Further condensation of extender unit may result in formation of tetraketides, pentaketides, hexaketides, octaketides, decaketides, dodecaketides *etc.* If other functional enzymes are present, ketoreduction (by ketoreductase), dehydration (by dehydratase) or enoylreduction (by enoylreductase) may take place to form a compound with a hydroxyl group, double bond *etc.* The acyltransferases deliver the extender unit to the PKS complex for the condensation cycle.

Triacetic lactone and 6-methylsalicylic acid are the smallest polyketides consisting of 6–8 carbons. Maitotoxin is the biggest compound containing 164 carbons. The PKS determines the number of cycles that take place.[53]

Based on domain architecture three PKS enzymes have been distinguished. These are type **I PKSs, type II PKSs**, and **type III PKSs**. However, all the types share the similar synthetic process.

Type I PKSs are large multimodular complexes. Each module contains multifunctional proteins with individual functional domains, each domain being used only once in the biosynthetic process. They have multiple active sites. The type I PKS systems can be subdivided into 'iterative' and 'non-iterative' categories. In case of 'iterative' category (iPKSs), only a single multienzyme acts repeatedly. These types of iPKSs are found in many fungi. In case of 'non-iterative' modular category (mPKSs), the protein has multiple active sites. mPKSs are found in bacteria. This type of PKS uses acyl carrier protein (ACP) to activate an acyl CoA substrate followed by growing the chain of polyketides. The PKSs of type I consist of different units such as ketosynthase (KS), acyltransferase (AT), ketoreductase (KR), dehydratase (DH), enoylreductase (ER), and acyl carrier protein (ACP) (involved in condensation, selection, and transfer of units as well as retaining the sequentially growing polyketide chain). Different enzyme units contribute to the structural diversity of compounds. Chain elongation depends on ketosynthase (KS), acyltransferase (AT), and acyl carrier protein (ACP). Non-iterative multimodular type PKSs (mPKSs) biosynthesize macrolides like erythromycin, rifampicin *etc.*[4,53,54]

Erythromycin is produced through intermediate formation of 6-deoxyerythronolide B by PKS type I (modular type) (Figure 3.20). The

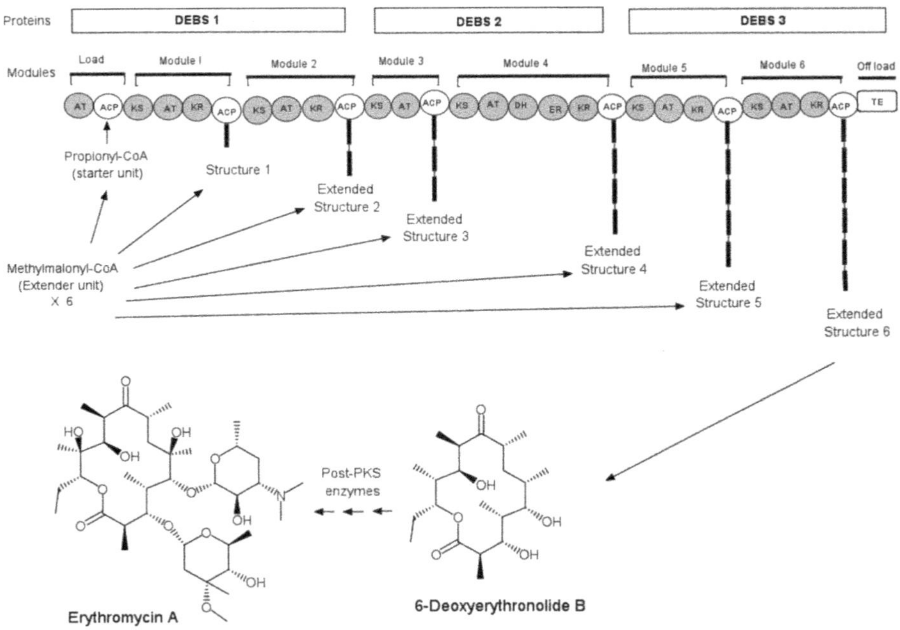

Figure 3.20 Schematic diagram of biosynthetic pathway to 6-deoxyerythronolide B and erythromycin using non-iterative modular category PKS of type I. ACP: acyl carrier protein; AT: acyltransferase; DH: dehydratase; ER: enoyl reductase; KR: β-ketoacyl reductase; KS: β-ketoacyl synthase; TE: thioesterase.

enzyme 6-deoxyerythronolide B synthase (DEBS) consists of three subunits (DEBS-I, DEBS-2, and DEBS-3). It has a linear sequence of 6 modules. Each module contains the enzymes required for one cycle of chain extension (two carbon chain extension). The load domain consists of AT and ACP. AT determines the extender unit and the starter unit. Here the starter unit is propionyl-CoA, and the extender unit is methylmalonyl-CoA. The starter unit (propionyl-CoA) is loaded and extended by the extender unit (methylmalonyl-CoA). After the Claisen reaction, the other units of the module (KS/AT/KR) modify the oxidation status. The modified chain is then transferred to the next module for the further Claisen reaction allowing chain length extension by the extender unit and other reactions to be repeated to attain an appropriate chain length. Finally, the full-length polyketide is released from the PKS by off load module by the thioesterase domain. 6-Deoxyerythronolide B is formed. 6-Deoxyerythronolide B is processed further in the post-PKS steps to produce erythromycin A. Hydroxylations, glycosylations and methylations are post-PKS changes.[4,6,55]

iPKS in *Aspergillus nidulans*, a filamentous fungus, contains KS, AT, and ACP. The KS catalyzes C–C bond formation through Claisen condensations between thioesters. The acyl groups are transferred from CoA onto KS and ACP by AT. Additionally, the iPKSs contain an N-terminal starter unit – ACP transacylase (SAT) domain which is responsible for selecting the starter unit to be extended by the enzyme. The product template (PT) domain is responsible for folding and cyclization of the non-reduced polyketide backbone. A proposed pathway of orsellinic acid biosynthesis in *Aspergillus nidulans* is presented in Figure 3.21.[57]

Type II PKSs are found in bacteria only. The activity of type II PKSs is less understood. Type II PKSs (Figure 3.22) are composed of a complex of individual multifunctional proteins that conduct iterative chain elongation followed by reduction or cyclization during the biosynthetic process to produce aromatic polyketides. This type II PKS consists of ketosynthase (KS), chain length factor (CLF) that catalyzes chain initiation and elongation, and acyl carrier protein (ACP) that delivers malonyl building blocks to KS-CLF. Products of this type of PKS reaction are aromatic polyketides of chain length C10 to C30 *e.g.*, tetracyclins, anthracyclins like daunorubicin, doxorubicin *etc.*[58]

Type III PKSs are found in plants, fungi, and bacteria. Type III PKSs differ from the other types being homodimeric proteins. They use coenzyme A, instead of ACPs, and employ a single active site to perform a series of decarboxylation, repeated condensation of acetate units, cyclization, and aromatization reactions. Despite the structural simplicity, type III PKS systems may form a diverse group of compounds in plants. The enzyme system can use a noncanonical starter substrate and an extender substrate without canonical CoA activator. Flavonoids, stilbenes, xanthones, curcuminoids and quinoids are produced by these PKSs of type III (Figure 3.23).[56]

In the PKSs, each ketoacyl synthase (KS) domain conserves the three "Cys-His-Asn" catalytic sites. The CoA linked starter unit binds to the active site Cys. Then decarboxylation and C–C Claisen type condensation with the extender substrate (usually malonyl CoA) takes place. Further linear carbon chain extension occurs on the resulting polyketide. The polyketide is then cyclized by intramolecular Claisen and/or aldol condensation within the

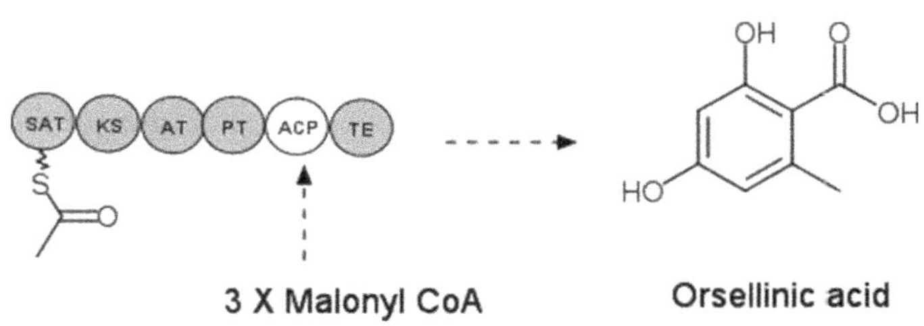

Figure 3.21 Pathway to orsellinic acid using iPKS of type I. SAT: N-terminal starter unit – ACP transacylase; KS: β-ketoacyl synthase; AT: acyltransferase; PT: product template; TE: thioesterase.

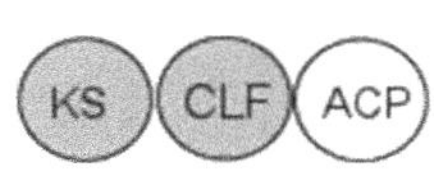

Figure 3.22 Type II PKS (iterative). KS: ketosynthase; CLF: chain length factor; ACP: acyl carrier protein.

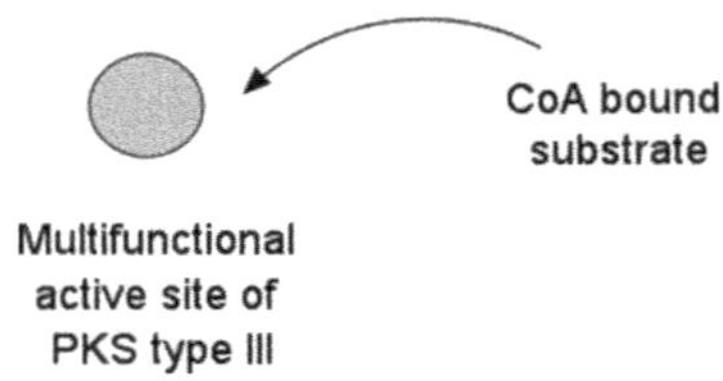

Figure 3.23 Type III PKS (iterative and ACP-independent).

same single active site. There are different PKSs which are 30–90% identical in amino acid sequences but differ in the selection of substrate, the number of condensations, and the cyclization reactions.[56]

3.2.2 Polyketide Types

Generally, polyketides are divided into two groups. These are (1) **complex polyketides** that are subdivided into macrolides, polyenes, polyethers, and ansamycins and (2) **aromatic polyketides** (containing one to six aromatic rings)

3.2.2.1 Complex Polyketides

These are produced by type I PKSs.

Macrolides are macrocyclic lactones (sometimes lactam) of different ring sizes to which one or more deoxy-sugar or amino sugar residues are attached.[59] Macrolides are hydrophobic compounds and are important because of their antibiotic, antiparasitic, antifungal, and immunosuppressant activities. Examples of different macrolides are provided in Table 3.2.

Macrolide antibiotics contain 12–16 membered rings and are active against the Gram-positive bacteria and Gram-negative bacteria as well as Gram indeterminate bacteria. Because of their lowest toxicity, macrolide antibiotics are often used for antibacterial treatment. These macrolides bind to the bacterial 50S ribosomal subunit (specific binding to 23S ribosomal RNA molecule within the 50S subunit) interfering with the protein synthesis.[59] **Erythromycin A, oleandomycin, spiramycin I, and tylosin** are some examples (Figure 3.24).

Avermictins (Figure 3.25), produced by *Streptomyces avermitilis*, are a class of macrocyclic lactones which have anthelmintic, antiparasitic, and antiviral but generally no antibacterial and antifungal activities. Avermectins are a mixture of closely related structures. **Ivermectin,** a semisynthetic mixture of two avermectin derivatives such as 80% 22,23-dihydroavermectin B1a and 20% 22,23-dihydroavermectin B1b, is used against parasites.[59,60]

Polyene macrolides (Figure 3.26) have antifungal properties. The important examples are **amphotericin B, nystatin, natamycin, candicidin.** Polyene macrolides are ionophores that bind with ergosterol, a sterol component of fungal cell membrane. The complex formed due to such binding between a polyene antibiotic and ergosterol results in the formation of a transmembrane pore with polar portion facing the center of the channel. These

Table 3.2 Macrolides of different groups.

Macrolides	Source	Activity
Macrolide antibiotics		
Erythromycin A (14 membered lactone ring)	*Streptomyces erythraeus*	Antibiotics
Derivatives of erythromycin		
Clarithromycin		
Azithromycin		
Oleandomycin (14 membered lactone ring)	*Streptomyces antibioticus*	Antibiotics
Spiramycin I (16 membered lactone ring)	*Streptomyces ambofaciens*	Antibiotics
Tylosin (16 membered lactone ring)	*Streptomyces fradiae*	Veterinary antibiotics
Avermectins	*Streptomyces avermitilis*	Antiparasitic
Ivermectin	Semisynthetic derivative of avermectins	Used against parasites
Rapamycin (sirolimus) (31 membered macrocyclic ring)	*Streptomyces hygroscopicus*	Antifungal, immunosuppressant
Tacrolimus	*Streptomyces tsukabaensis*	Immunosuppressant
Polyene (26–38 membered rings)		
Amphotericin B	*Streptomyces nodosus*	Antifungal candida infections in the mouth or on the skin
Nystatin	*Streptomyces noursei*	Oral treatment of intestinal candidiasis
Natamycin	*Streptomyces natalensis*	
Candicidin	*Streptomyces griseus*	
Polyethers		
Lasalocid	*Streptomyces lasaliensis*	Anticoccidial
Monensin	*Streptomyces cinnamonensis*	Anticoccidial
Salinomycin	*Streptomyces albus*	Antibacterial, active against filamentous fungi
Ansamycin		
Rifamycin (A–E)	*Streptomyces mediterranei (Syn: Amycolatopsis mediterranei)*	
Semisynthetic rifampicin		Antibacterial

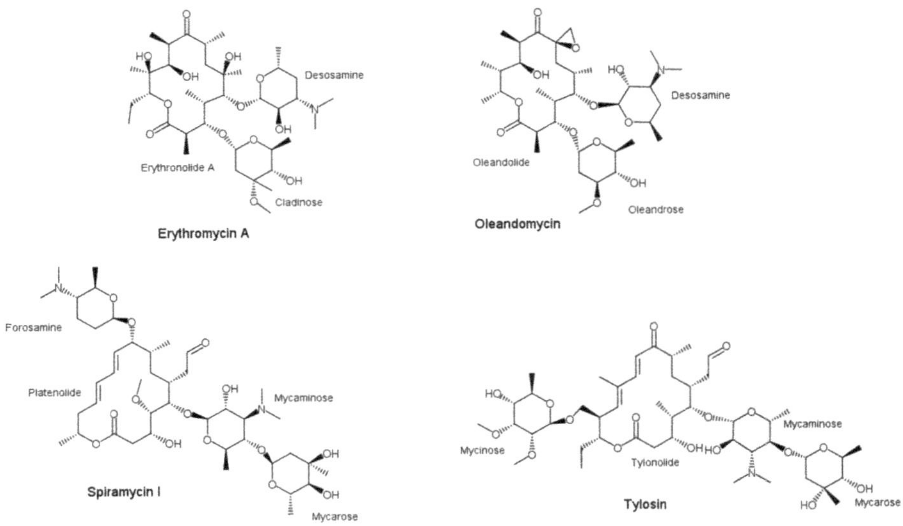

Figure 3.24 Macrolide antibiotics.

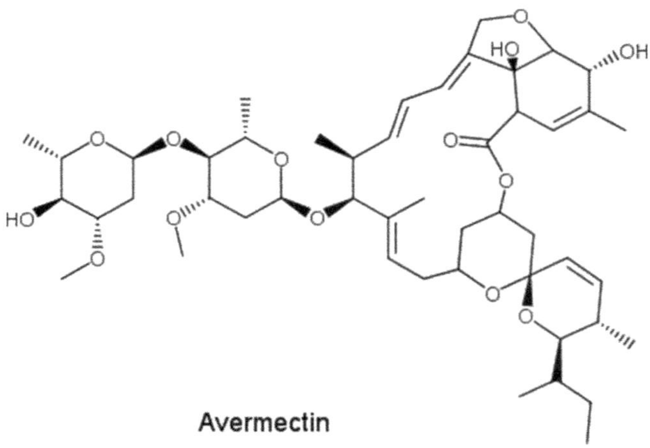

Figure 3.25 Avermectin.

polyene-ergosterol complexes cause the loss of ions and other small molecules from the fungal cell and subsequent cell death.[59,60]

Tacrolimus produced from *Streptomyces tsukubaensis* is an immunosuppressant used to prevent kidney, liver, and bone marrow rejection in transplant patients. Rapamycin, isolated from *Streptomyces hygroscopicus*, displayed antifungal properties. Rapamycin was later found to be immunosuppressant developed for prevention of renal, pancreatic islet cell, liver, and heart transplant rejection.[59,60]

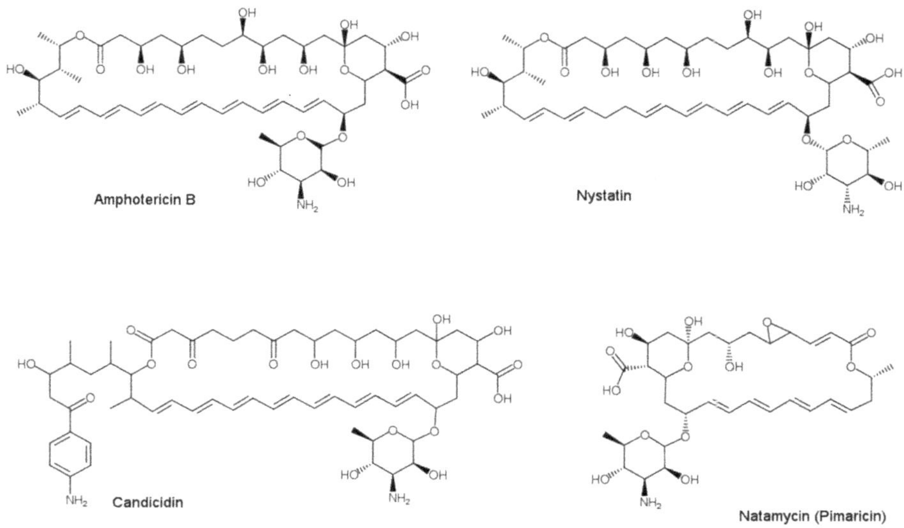

Figure 3.26 Polyenes.

3.2.2.1.1 Polyether Antibiotics. Polyether ionophores (Figure 3.27) are very large group of naturally occurring biologically active compounds produced by *Streptomyces* spp. These lipid soluble antibiotics transport metal cations across cell membranes. The antibiotics have a wide range of activities *e.g.*, antibacterial, antifungal, antiparasitic, antiviral, and tumour cell cytotoxicity. The main features of the polyether ionophores is their antibiotic activity, including activity against drug-resistant strains. Some of these antibiotics have anticoccidial activities targeting ruminal bacterial population. Several ionophores show anticancer activities (monensin, salinomycin and laidlomycin).[61,62]

3.2.2.1.2 Ansamycin. Rifamycins are ansamycin antibiotics. The crude mixture consists of five closely related molecules rifamycins A-E produced by *Amycolatopsis mediterranei*. Rifamycin B (with no antibiotic activity) is used to produce rifampicin, a semi-synthetic derivative that has antibiotic activity.[4]

3.2.2.2 Aromatic Polyketides

Aromatic polyketides are produced by type II and type III PKSs, except for a few examples produced by type I PKSs.

6-Methylsalicyclic acid (Figure 3.28) is produced by many fungi including *Penicillium patulum*. The enzyme 6-methylsalicyclic acid synthase enzyme is a type I PKS (iPKS) of iterative type. One acetyl CoA and three molecules of malonyl CoA generates 6-methylsalicyclic acid. 6-Methylsalicyclic acid is a key structural moiety of the promising antibiotic and anticancer agents chlorothricin, maduropeptin, and neocarzinostatin.[63] 6-Methylsalicyclic acid is the precursor of the toxin **patulin** (Figure 3.28). Patulin is the mycotoxin

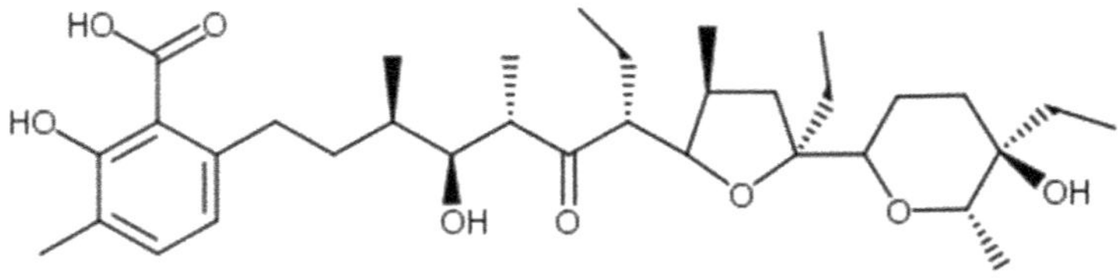

Lasalocid

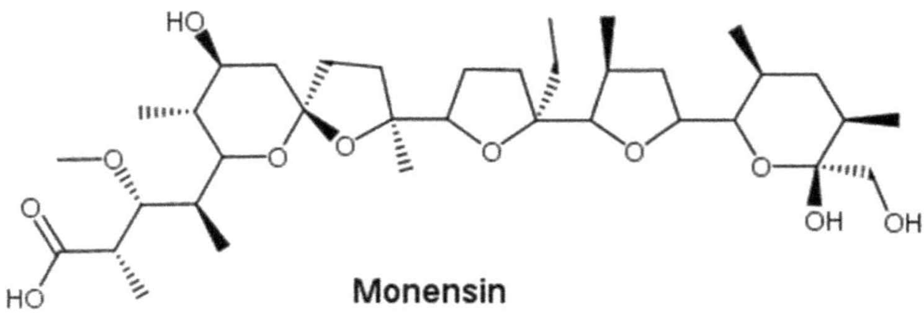

Monensin

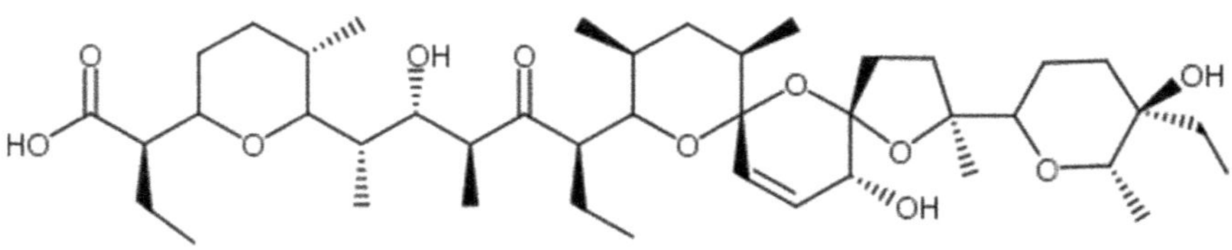

Salinomycin

Figure 3.27 Polyether antibiotics.

produced by moulds belonging to several species of *Aspergillus*, *Penicillium* and *Byssochlamys*. This mycotoxin is found in apple and apple-derived products (juice, cider, and other food for young children). Consumption of this mycotoxin contaminated food affects the immunological, neurological, and gastrointestinal systems. Patulin is toxic. Several countries proposed regulation of patulin level in food.[64]

Orsellinic acid (Figure 3.21) is a phenolic acid of polyketide type and is a constituent of a rare lichen (*Omphalodiscus spodochrous*), *Aspergillus nidulans*. However, derivatives of orsellinic acid are common lichen metabolites.

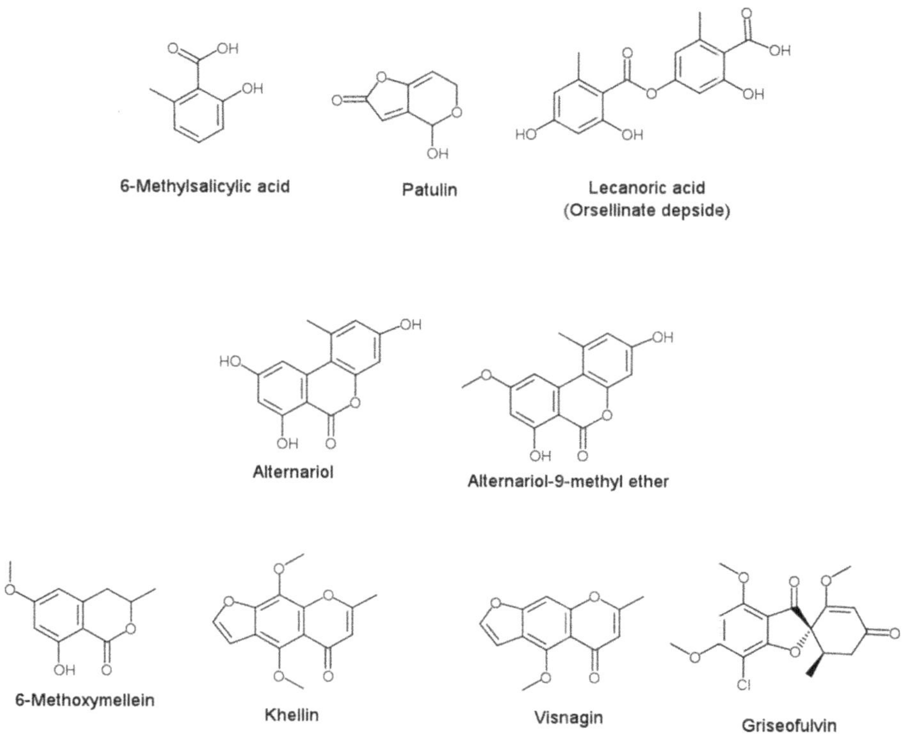

Figure 3.28 Some aromatic polyketides.

One acetate starter unit and three malonate extender units produce the simplest tetraketide orsellinic acid. Poly-β-ketoester arising after condensation is folded after enolization and hydrolysis of the thioester bond to produce this compound. **Lecanoric acid** is a **depside** (an ester of two phenolic acids) (Figure 3.28) formed by the combination of two orsellinic acid molecules, found in the cultured mycobionts of the lichens *Umbilicaria arctica* and *Umbilicaria nylanderiana*.[65]

Usnic acid is a toxic molecule produced by several lichens such as *Usnea barbata, Usnea deffractas, Usnea laevis, Usnea longissima, Lecanora pseudogangalevides, Roccella montagnei, Sticta weigelia*, and *Cladonia substellata*. Usnic acid shows promising antimicrobial, antitumour, antiviral, and antiparasitic properties. But it is a toxic molecule. Usnic acid is biosynthesized *via* the polymalonate pathway through intermediate formation of methyl-fluoro-acetophenone[66] (Figure 3.29). Coupling of two units of methyl-fluoroacetophenone leads to the formation of usnic acid.

Alternariol and **alternariol-9-methyl ether** are two major polyketide mycotoxins (Figure 3.28) produced by *Alternaria alternata* and are common contaminants of food (fruits, fruit juices, cereals) and feed. Alternariol has cytotoxic, foetotoxic and teratogenic effects.[67]

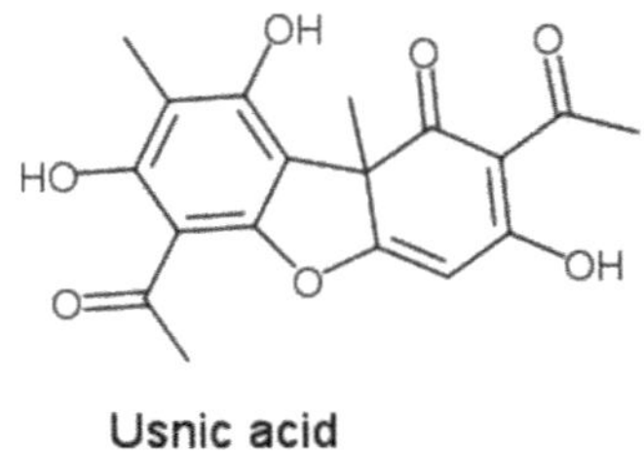

Figure 3.29 Usnic acid.

Aflatoxins are polyketide derived furanocoumarins produced by several fungal species mostly belonging to *Aspergillus* section *Flavi. Aspergillus flavus* and *Aspergillus parasiticus* are the most important producers of this group of toxins in crops like peanut and maize. The fungi of this section synthesize mainly four aflatoxins B1, B2, G1, and G2 which are carcinogenic and toxic compounds. Other derivatives are produced when metabolized by animals after consumption. Thus, totally 26 types of different aflatoxins have been identified in plants and animals. Aflatoxin B1, having carcinogenic properties, is one of the most important compounds detected in many food stuffs. Aflatoxin contamination in the food chain is difficult to control. Chronic exposure to aflatoxin B1 may lead to issues such as malabsorption of nutrients, immune suppression, infertility, endocrine problems, teratogenic effect, and hepatocellular carcinoma. The biosynthetic intermediates of aflatoxins such as versicolorin A and sterigmatocystin also have cytotoxic effects. Sterigmatocystin is also produced by *A. nidulans*. Aflatoxins are polyketide derivatives which require hexanoate units formed from acetyl-CoA and malonyl-CoA. The first stable metabolite formed is norsolorinic acid.[68,69] The biosynthetic pathway of aflatoxins is shown in Figure 3.30.

Griseofulvin (Figure 3.28) is also a malonyl-CoA derived polyketide produced by *Penicillium griseofulvum* and other species of *Penicillium* such as *Penicillium aethiopicum, Penicillium janczewskii, Penicillium nigrum* and *Penicillium patulum*. Griseofulvin is the antibiotic used to treat fungal dermatophytes. It can disrupt mitotic spindles. It is a potential inhibitor of centrosomal clustering in tumour cells.[70]

6-Methoxymellein (Figure 3.28) is a stress compound produced by *Daucus carota*.[71] The enzyme 6-methoxymellein synthase required for the compound's metabolism is a type III plant PKS. 6-Methoxymellein is an antifungal metabolite.[4] **Khellin** and **visnagin** (Figure 3.28) are malonyl-CoA derived furochromones found in the fruits of *Ammi visnaga*. These two compounds can be detected in formulations traditionally used in Asia and Middle East for the treatment of coronary diseases, bronchial asthma, renal colic, and muscle spasms. Both these metabolites possess vasodilatory activities. Khellin is used as the raw material in the pharmaceutical industry for the development of the anti-asthmatic agent sodium cromoglycate or the anti-arrhythmic drug amiodarone.[72]

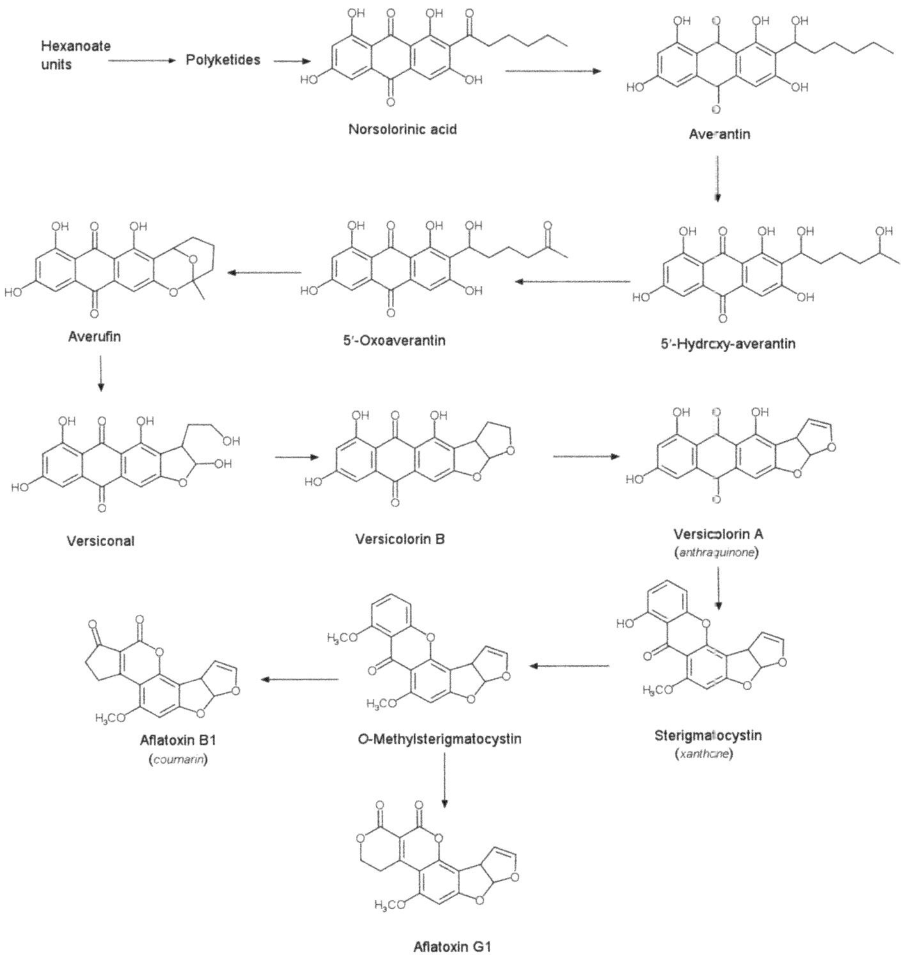

Figure 3.30 Biosynthetic pathway of aflatoxins.

Anthraquinones could be detected in the Gram-negative bacterium *Photorhabdus luminescens*. Thorough study using this bacterium revealed that type II PKSs are responsible for anthraquinone biosynthesis in this organism. Biosynthesis of anthraquinone **emodin** in plants is proposed to be catalyzed by octaketide synthase, a type III PKS enzyme (Figure 3.31). *O*-Methylation of emodin would lead to **physcion. Chrysophanol**, a polyketide anthraquinone, has been reported from fungi, lichens, plants, and some insects. It has also been detected in small amounts in a strain of *Streptomyces*. Chrysophanol is produced in fungi *via* the polymalonate pathway and in plants *via* the shikimic acid and polymalonate pathways. Emodin can form chrysophanol. Oxidation of chrysophanol anthrone yields chrysophanol. *Aloe* species produce the anthraquinones **aloenin** (hexaketide), **aloesin** (heptaketide), and

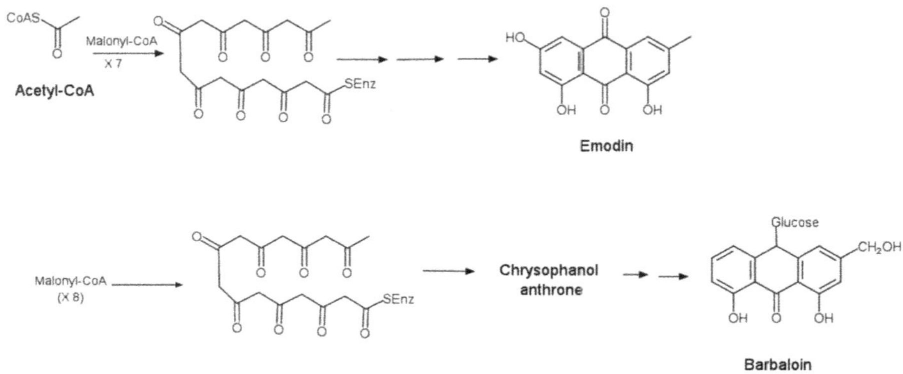

Figure 3.31 Biosynthesis of emodin and barbaloin.

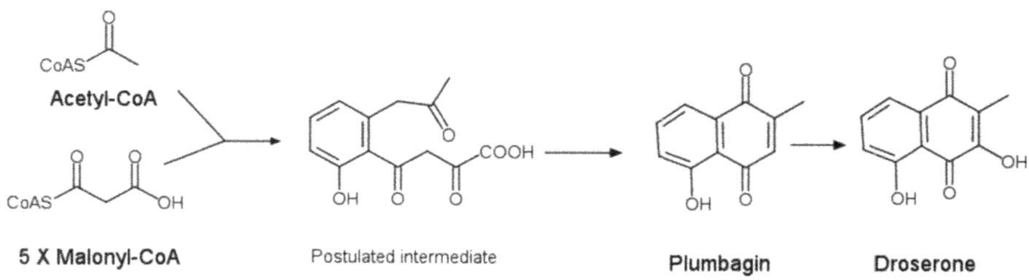

Figure 3.32 Biosynthesis of some naphthoquinones.

the anthraquinone **barbaloin** (octaketide). Plant specific type III PKSs pentaketide chromone synthase (PCS) and octaketide synthase (OKS), cloned and sequenced from *Aloe arborescens* (Asphodelaceae), catalyzed iterative condensation of 5–8 molecules of malonyl-CoA.[73–75]

Some of the **naphthoquinones** are of mixed biogenesis produced either *via* chorismic acid or *via* 4-hydroxybenzoic acid with incorporation of geranyl PP during the biosynthetic process (Figure 5.42). However, the naphthoquinones **plumbagin** and **droserone** (Figure 3.32) are produced using type III PKSs. The starter unit acetyl-CoA and 5 malonyl-CoA extender units are involved. Plumbagin was identified in species of *Plumbago*. Plumbagin is a toxin that inhibits insect ecdysis. Plumbagin and droserone along with 5-*O*-methyl droserone were also identified from the insectivorous pitcher plant *Nepenthes khasiana* (Nepenthaceae). These three naphthoquinones act as molecular triggers in prey capture and digestion of the insects in this pitcher plant.[76,77]

Flavonoids, stilbenes, curcuminoids: The reactions catalyzed by plant type III PKSs are shown in chapter 5 for biosynthesis of naringenin chalcone (Figure 5.31), resveratrol (Figure 5.45) and curcuminoids (Figure 5.51). In these cases, cinnamyl-CoA (produced by a shikimic acid pathway) or derivatives act as the starter unit that lead to the formation of flavonoids, stilbenes, and curcuminoids after addition of malonyl CoA.

Cannabinoids, present in *Cannabis sativa* (Cannabaceae), also known as marijuana, are responsible for the psychoactive property of the plant. Cannabis products are consumed as an illicit drug worldwide. Cannabinoids consists of more than 70 related metabolites including Δ^9-tetrahydrocannabinol (Δ^9-THC), Δ^9-tetrahydrocannabinolic acid, cannabigerolic acid, cannabidiolic acid, cannabinol, and cannabidiol. Δ^9-THC has pharmacological properties like antiemetic, analgesic and appetite stimulant. So medical marijuana is increasingly being used to treat diseases like multiple sclerosis, glaucoma, and chronic pain. Δ^9-THC is used to alleviate nausea and vomiting in cancer patients undergoing chemotherapy. In the cannabinoid pathway, type III PKSs catalyze the condensation of hexanoyl-CoA with three molecules of malonyl-CoA to produce olivetolic acid which in turn produces other cannabinoids[78] (Figure 3.33).

3.2.2.2.1 Acetate Derived Alkaloids. **Coniine** and other related alkaloids reported from poison hemlock (*Conium maculatum*, Apiaceae) have a piperidine nucleus. Instead of having lysine as the precursor, these alkaloids are polyketide derived alkaloids. Examples of these group of alkaloids are coniine, γ-coniceine, pseudoconhydrine, *N*-methylconiine, *N*,*N*-dimethylconiine, 1′-oxo-γ-coniceine, conhydrine, *N*-methylpseudoconhydrine, conmaculatin *etc.* Apart from hemlock, some of these alkaloids have also been reported from *Sarracenia* spp. (Sarraceniaceae), and different species of *Aloe* (*Aloe ballyi, Aloe deltoideodonta, Aloe descoingsii, Aloe globuligemma, Aloe gracilicaulis, Aloe ibitiensis, Aloe krapholiana, Aloe ortholopha, Aloe ruspoliana, Aloe sabaea, Aloe viguieri*) of family Asphodelaceae. Coniine is poisonous to humans and animals. It is nicotinic acetylcholine receptor antagonist leading to inhibition of the nervous system and eventual death. Socrates was the famous

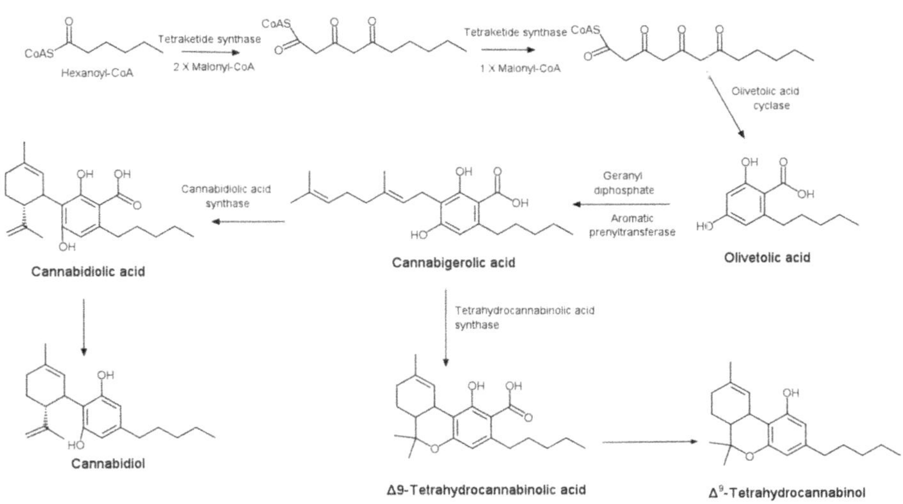

Figure 3.33 Biosynthetic pathway for cannabinoids.

victim of hemlock poisoning. He was sentenced to death by offering a chalice containing hemlock.[79]

Instead of lysine, the piperidine structure of these alkaloids originates from one butyryl-CoA and two malonyl-CoAs with formation of intermediate polyketides. The reaction is catalyzed by polyketide synthase. Nitrogen is incorporated by transamination reaction from L-alanine. Non-enzymatic cyclization leads to formation of alkaloids[79] (Figure 3.34).

3.2.2.2.2 Aromatic Polyketide Antibiotics. Several aromatic polyketide antibiotics (Table 3.3; Figure 3.35) are produced in microorganisms using type II PKSs.[80] These are anthracyclines (**doxorubicin, daunorubicin**), tetracyclines *etc.* Doxorubicin, a decaketide, is used for the treatment of cancers. It acts through intercalation of DNA, interference of topoisomerase-II mediated DNA repair, and formation of free radicals to deteriorate cell components such as cell membrane, DNA, and proteins. **Tetracyclines** are broad spectrum antibiotics clinically used against Gram-positive and Gram-negative bacteria. However, their use has been restricted due to emergence of microbial resistance to tetracyclines.

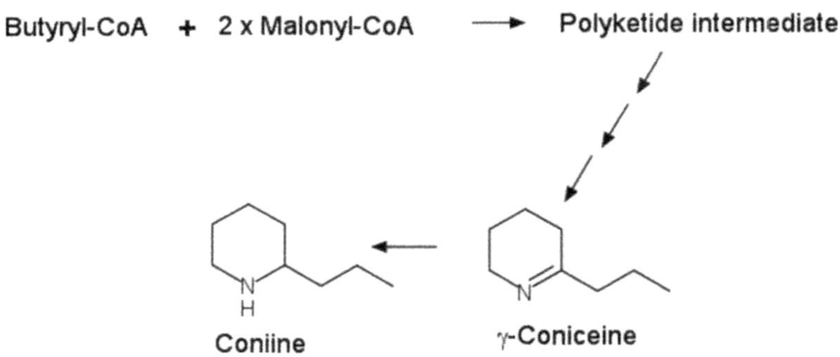

Figure 3.34 Biosynthesis of hemlock alkaloids.

Table 3.3 Aromatic polyketide antibiotics.

Antibiotics	Source	Activity
Anthracyclins		
Doxorubicin	*Streptomyces peucetius*	Doxorubicin is used for the treatment of cancers (breast cancer, childhood solid tumours, soft tissue sarcomas and aggressive lymphomas).
Daunorubicin	*Streptomyces peucetius*	
Tetracyclines	Soil borne actinomycetes, *Streptomyces aureofaciens*	Active against both Gram-positive and Gram-negative bacteria
Oxotetracycline	*Streptomyces rimosus*	
Chlortetracycline	*Streptomyces aureofaciens*	

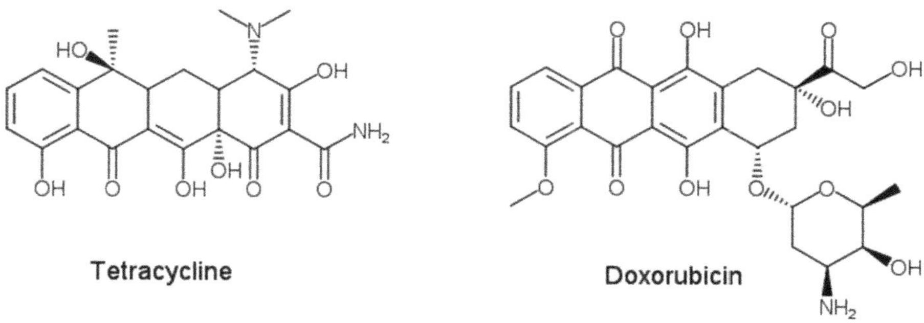

Figure 3.35 Aromatic polyketide antibiotics.

References

1. P. D. Bates, S. R. Johnson, X. Cao, J. Li, J. Nam, J. G. Jaworski, J. B. Ohlrogge and J. Browse, Fatty acid synthesis is inhibited by inefficient utilization of unusual fatty acids for glycerolipid assembly, *Proc. Natl. Acad. Sci. U. S. A.*, 2014, **111**, 1204–1209.
2. A. P. Brown and A. R. Slabas, Fatty acid biosynthesis in plants- metabolic pathways, structure and organization, in *Lipids in Photosynthesis: Essential and Regulatory Functions*, ed. H. Wada, and N. Murata, Springer Science + Business Media B.V, 2009, pp. 11–34.
3. E. B. Cahoon and Y. Li-Beisson, Plant unusual fatty acids: learning from the less common, *Curr. Opin. Plant Biol.*, 2020, **55**, 66–73.
4. P. M. Dewick, The acetate pathway: fatty acids and polyketides, in *Medicinal Natural Products: A Biosynthetic Approach*, John Wiley & Sons, 3rd edn, 2009, pp. 39–135.
5. S. Rawsthorne, Carbon flux and fatty acid synthesis in plants, *Prog. Lipid Res.*, 2002, **41**, 182–196.
6. Y. Sasaki and Y. Nagano, Plant acetyl-CoA carboxylase: structure, biosynthesis, regulation, and gene manipulation for plant breeding, *Biosci., Biotechnol., Biochem.*, 2004, **68**, 1175–1184.
7. I. M. Lopez-Lara and O. Geiger, Formation of fatty acids, in *Handbook of Hydrocarbon and Lipid Microbiology*, ed. K. N. Timmis, Springer-Verlag Berlin Heidelberg, 2010, pp. 386–393.
8. J. L. Blatti, J. Beld, C. A. Behnke, M. Mendez and S. P. Mayfield, Manipulating fatty acid biosynthesis in microalgae for biofuel through protein-protein interactions, *PLoS One*, 2012, **7**, e42949.
9. X. Guan, Y. Okazakki, R. Zhang, K. Saito and B. J. Nikolau, Dual-localized enzymatic components constitute the fatty acid synthase systems in mitochondria and plastids, *Plant Physiol.*, 2020, **183**, 517–529.
10. J. E. Conan and J. Thomas, Bacterial fatty acid synthesis and its relationships with polyketide synthetic pathways, *Methods Enzymol.*, 2009, **459**, 395–433.
11. N. Yu, W. F. Xiao, J. Zhu, X. Y. Chen and C. C. Peng, The Jatropha curcas KASIII gene alters fatty acid composition of seeds in *Arabidopsis thaliana*, *Biol. Plant*, 2015, **59**, 773–782.

12. I. Gonzalez-Thuillier, M. Venegas-Caleron, A. J. Moreno-Perez, J. J. Salas, R. Garces, P. von Wettstein-Knowles and E. Martinez-Force, Sunflower (*Helianthus annuus*) fatty acid synthase complex: β-ketoacyl-[acyl carrier protein] reductase genes, *Plant Physiol. Biochem.*, 2021, **166**, 689–699.
13. J. Beld, D. J. Lee and M. D. Burkart, Fatty acid biosynthesis revisited: Structure elucidation and metabolic engineering, *Mol. BioSyst.*, 2015, **11**, 38–59.
14. M. Batsale, D. Bahammou, L. Fouillen, S. Mongrand, J. Joubes and F. Domergue, Biosynthesis and functions of very-ling-chain fatty acids in the responses of plants to abiotic and biotic stresses, *Cells*, 2021, **10**, 1284.
15. J. A. Napier, Plumbing the depths of PUFA biosynthesis: a novel polyketide synthase-like pathway from marine organisms, *Trends Plant Sci.*, 2002, **7**, 51–54.
16. N. Ruiz-Lopez, S. Usher, O. V. Sayanova, J. A. Napier and R. P. Haslam, Modifying the lipid content and composition of plant seeds: engineering the production of LC-PUFA, *Appl. Microbiol. Biotechnol.*, 2015, **99**, 143–154.
17. T. Hoppenbrouwers, J. H. C. Hogervorst, J. Garssen, H. J. Wichers and L. E. M. Willemsen, Long chain polyunsaturated fatty acids (LCPUFAs) in the prevention of food allergy, *Front. Immunol.*, 2019, **10**, 1118.
18. E. Abedi and M. A. Sahari, Long-chain polyunsaturated fatty acid sources and evaluation of their nutritional and functional properties, *Food Sci. Nutr.*, 2014, **2**, 443–463.
19. A. Solmonson and R. J. De Berardinis, Lipoic acid metabolism and mitochondrial redox regulation, *J. Biol. Chem.*, 2018, **293**, 7522–7530.
20. X. Fu, X. Guan, R. Garlock and B. J. Nikolau, Mitochondrial fatty acid synthase utilizes multiple acyl carrier protein isoforms, *Plant Physiol.*, 2020, **183**, 547–557.
21. B. Chen, J. L. Foo, H. Ling and M. W. Chang, Mechanism driven metabolic engineering for biobased production of free R-lipoic acid in Saccharomyces cerevisiae mitochondria, *Front. Bioeng. Biotechnol.*, 2020, **8**, 965.
22. B. Tian, T. Lu, Y. Xu, R. Wang and G. Chen, Identification of genes associated with ricinoleic acid accumulation in *Hiptage benghalensis via* transcriptome analysis, *Biotechnol. Biofuels*, 2019, **12**, 16.
23. V. E. Tyler, L. R. Brady and J. E. Roberts, Lipids, in *Pharmacognosy*, Lee & Berger, 1988, ch. 4, pp. 82–1002.
24. R. C. Badami and K. B. Patil, Structure and occurrence of unusual fatty acids in minor seed oils, *Prog. Lipid Res.*, 1981, **19**, 119–153.
25. S. Okada, X. Zhou, K. Dancevski, N. Gibb, C. Wood and M. Hamberg, Diversity of Δ12 fatty acid desaturases in Santalaceae and their role in production of seed oil acetylenic fatty acids, *J. Biol. Chem.*, 2013, **288**, 32405–32413.
26. F. D. Gunstone, J. A. Holliday and C. M. Scrimgeour, Fatty acids, Part 51. The long-chain oxo acids (argemonic acid) in Argemone Mexicana seed oil, *Chem. Phys. Lipids*, 1977, **20**, 331–335.

27. C. Rukmini, New, unusual long chain fatty acid (argemonic acid) from Argemone Mexicana, *J. Am. Oil Chem. Soc.*, 1975, **52**, 171–173.
28. R. Pelaez, A. Pariente, A. Perez-sala and I. M. Larrayoz, Sterculic acid: the mechanisms of action beyond stearoyl-CoA desaturase inhibition and therapeutic opportunities in human diseases, *Cells*, 2020, **9**, 140.
29. M. Muller, M. Hogg, K. Ulms and W. Vetter, Concentrations, stability and isolation of the furan fatty acid 9-(3-methyl-5-pentylfuran-2-yl)-nonanoic acid (9M5) from disposable latex gloves, *J. Agric. Food Chem.*, 2017, **65**, 7919–7925.
30. X. Yu and J. Shanklin, Solving a furan fatty acid biosynthesis puzzle, *J. Biol. Chem.*, 2020, **295**, 9802–9803.
31. I. Mawlong, M. S. S. Kumar and D. Singh, Furan fatty acids: their role in plant systems, *Phytochem. Rev.*, 2016, **15**, 121–127.
32. A. Tava and P. Avato, Analysis of cyanolipids from Sapindaceae seed oils by gas-chromatography-EI-mass spectrometry, *Lipids*, 2014, **49**, 335–345.
33. C. Vaisali, S. Charanyaa, P. D. Belur and I. Regupathi, Refining edible oils: a critical appraisal of current and potential technologies, *Int. J. Food Sci. Technol.*, 2014, **50**, 13–23.
34. P. D. Bates and J. Browse, The pathway of triacylglycerol synthesis through phosphatidylcholine in Arabidopsis produces a bottleneck for the accumulation of unusual fatty acids in transgenic seeds, *Plant J.*, 2011, **68**, 387–399.
35. J. Lu, Y. Xu, J. Wang, S. D. Singer and G. Chen, The role of triacylglycerol in plant stress responses, *Plants*, 2020, **9**, 472.
36. W. E. Tinto, T. O. Elufioye and J. Roach, Waxes, in *Pharmacognosy*, Elsevier Inc, 2017, pp. 443–455.
37. P. Von Weettstein-Knowles, Plant waxes, in *eLS*, John Wiley & Sons, Ltd, Chichester, 2012, pp. 1–11.
38. L. Kunst and A. L. Samuels, Biosynthesis and secretion of plant cuticular waxes, *Prog. Lipid Res.*, 2003, **42**, 51–80.
39. C. Buschhaus and R. Jetter, Composition differences between epiculicular and intracuticular wax substructures: How do plants seal their epidermal surfaces? *J. Exp. Bot.*, 2011, **62**, 841–853.
40. G. Philippe, I. Sorensen, C. Jiao, X. Sun, Z. Fei, D. S. Domozych and J. K. C. Rose, Cutin and suberin: assembly and origins of specialized lipidic cell wall scaffolds, *Curr. Opin. Plant Biol.*, 2020, **55**, 11–20.
41. E. A. Fich, N. A. Segerson and J. K. C. Rose, The plant polyester cutin: biosynthesis, structure, and biological roles, *Annu. Rev. Plant Biol.*, 2016, **67**, 207–233.
42. Y. Li-Beisson, Cutin and suberin, in *eLS*, John Wiley & Sons, Ltd, Chichester, 2011, pp. 1–9.
43. M. Pollard, F. Beisson, Y. Li and J. B. Ohlrogge, Building lipid barriers: biosynthesis of cutin and suberin, *Trends Plant Sci.*, 2008, **13**, 236–245.
44. C. Nawrath, *The Biopolymers Cutin and Suberin. The Arabidopsis Book*, The American Society of Plant Biologists, 2002, p. e0021.

45. J. Graca, Suberin: the biopolyester at the frontier of plants, *Front. Chem.*, 2015, **3**, 62.
46. N. Liu, L. Hou, J. Bao, L. Wang and X. Chen, Sphingolipid metabolism, transport, and functions in plants: Recent progress and future perspectives, *Plant Commun.*, 2021, **2**, 100214.
47. A. M. Cassim, M. Grison, Y. Ito, F. Simon-Plas, S. Mongrand and Y. Boutte, Sphingolopids in plants: a guidebook on their function in membrane architecture, cellular processes, and environmental or developmental responses, *FEBS Lett.*, 2020, **594**, 3719–3738.
48. C. Wasternack and I. Feussner, The oxylipin pathways: Biochemistry and function, *Annu. Rev. Plant Biol.*, 2018, **69**, 4.1–4.24.
49. C. Wasternack and S. Song, Jasmonates: biosynthesis, metabolism, and signaling by proteins activating and repressing transcription, *J. Exp. Bot.*, 2017, **68**, 1303–1321.
50. C. Wasternack and B. Hause, The missing link in jasmonic acid biosynthesis, *Nat. Plants*, 2019, **5**, 776–777.
51. J. Yang, G. Duan, C. Li, L. Liu, G. Han, Y. Zhang and C. Wang, The crosstalk between jasmonic acid and other plant hormone signaling highlight the involvement of jasmonic acid as a core component in plant response to biotic and abiotic stresses, *Front. Plant Sci.*, 2019, **10**, 1349.
52. M. Genva, F. O. Akong, M. X. Andersson, M. Deleu, L. Lins and M. Fauconnier, New insights into the biosynthesis of esterified oxylipins and their involvement in plant defense and developmental mechanisms, *Phytochem. Rev.*, 2019, **18**, 343–358.
53. R. Salwan and V. Sharma, Bioactive compounds of *Streptomyces*: Biosynthesis to applications, in *Studies in Natural Product Chemistry*, ed. Atta-Ur-Rahman, Elsevier BV, 2020, vol. 64, pp. 467–491.
54. E. J. N. Helfrich and J. Piel, Biosynthesis of polyketides by *trans*-AT polyketide synthases, *Nat. Prod. Rep.*, 2016, **33**, 120–121.
55. E. M. Musiol-Kroll and W. Wohlleben, Acetyltransferases as tools for polyketide synthase engineering, *Antibiotics*, 2018, **7**, 62.
56. A. Nivina, K. P. Yuet, J. Hsu and C. Kholsa, Evolution and diversity of assembly-line polyketide synthases, *Chem. Rev.*, 2019, **119**, 12524–12547.
57. M. L. Klejnstrup, R. J. N. Frandsen, D. K. Holm, M. T. Nielsen, U. H. Mortensen, T. O. Larsen and J. B. Nielsen, Genetics of polyketide metabolism in *Aspergillus nidulans*, *Metabolites*, 2012, **2**, 100–133.
58. J. Kim and G. Yi, PKMiner: a database for exploring type II polyketide synthases, *BMC Microbiol.*, 2012, **12**, 169.
59. G. P. Dinos, The macrolide antibiotic renaissance, *Br. J. Pharmacol.*, 2017, **174**, 2967–2983.
60. K. D. Lenz, K. E. Klosterman, H. Mukundan and J. Z. Kubicek-Sutherland, Macrolides: From toxins to therapeutics, *Toxins*, 2021, **13**, 347.
61. S. Lin, H. Liu, E. B. Svenninsen, M. Wollesen, K. M. Jacobsen, F. D. Andersen, J. Moyano-villameriel, C. N. Pedersen, P. Norby, T. Torring and T. B. Poulsen, Expanding the antibacterial selectivity of polyether ionophore antibiotics through diversity focused semisynthesis, *Nat. Chem.*, 2021, **13**, 47–55.

62. J. Rutkowski and B. Brezezinski, Structures and properties of naturally occurring polyether antibiotics, *BioMed Res. Int.*, 2013, 162513.
63. J. S. Parascandolo, J. Havemann, H. K. Potter, F. Huang, E. Riva, J. Connolly, I. Wilkening, L. Song, P. F. Leadlay and M. Tosin, Insights into 6-methylsalicylic acid bio-assembly by using chemical probes, *Angew. Chem., Int. Ed.*, 2016, **55**, 3463–3467.
64. O. Puel, P. Galtier and I. Oswald, Biosynthesis and toxicological effects of patulin, *Toxins*, 2010, **2**, 613–631.
65. E. Stocker-Worgotter, Metabolic diversity of lichen-forming ascomyceteous fungi: culturing, polyketide and shikimate metabolite production, and PKS genes, *Nat. Prod. Rep.*, 2008, **25**, 188–200.
66. D. C. S. Macedo, F. J. F. Almeida, M. S. O. Wanderley, M. S. Ferraz, N. P. S. Santos, A. M. Q. Lopez, N. S. Santos-Magalhaes and M. C. M. Lira-Nogueira, Usnic acid: from an ancient lichen derivative to promising biological and nanotechnology applications, *Phytochem. Rev.*, 2021, **20**, 609–930.
67. D. Saha, R. Fetzner, B. Burkhardt, J. Podlech, M. Metzler, H. Dang, C. Lawrence and R. Fischer, Identification of a polyketide synthase required for alternariol (AOH) and alternarion-9-methyl-ether (AME) formation in *Alternaria alternata*, *PLoS One*, 2012, **7**, e40564.
68. J. Liao, Z. He, Y. Xia, Y. Lei and B. Liao, A review on biosynthesis and genetic regulation of aflatoxin production by major *Aspergillus* fungi, *Oil Crop Sci.*, 2020, **5**, 166–173.
69. I. Caceres, A. A. Khoury, R. E. Khoury, S. Lorber, I. P. Oswald, A. E. Khoury, A. Atoui, O. Puel and J. Bailly, Aflatoxin biosynthesis and genetic regulation: a review, *Toxins*, 2020, **12**, 150.
70. R. A. Cacho, Y. Chooi, H. Zhou and Y. Tang, Complexity generation in fungal polyketide biosynthesis: a spirocycle-forming P450 in the concise pathway to the antifungal drug griseofulvin, *ACS Chem. Biol.*, 2013, **8**, 2322–2330.
71. X. Fan, J. P. Mattheis and R. G. Roberts, Biosynthesis of phytoalexin in carrot root requires ethylene action, *Physiol. Plant.*, 2000, **110**, 450–454.
72. R. Vrzal, K. Frauenstein, P. Proksch, J. Abel, Z. Dvorak and T. Haarmann-Stemmann, Khellin and visnagin differentially modulate AHR signaling and downstream CYP1A activity in human liver cells, *PLoS One*, 2013, **8**, e74917.
73. Q. Zhou, A. Brauer, H. Adihou and M. Schmalhofer, Molecular mechanism of polyketide shortening in anthraquinone biosynthesis of *Photorhabdus luminescens*, *Chem. Sci.*, 2019, **10**, 6341–6349.
74. Prateeksha, M. A. Yusuf, B. N. Singh, S. Sudheer, R. N. Kharwar, S. Siddiqui, A. M. Abdel-Azeem, L. F. Fraceto, K. Dashora and V. K. Gupta, Chrysophanol: a natural anthraquinone with multifaceted biotherapeutic potential, *Biomolecules*, 2019, **9**, 68.
75. A. Brauer, Q. Zhou, G. L. C. Grammbitter, M. Schmalhofer, M. Ruhl, V. R. I. Kaila, H. B. Bode and M. Groll, Structural snapshots of the minimal PKS system responsible for octaketide biosynthesis, *Nat. Chem.*, 2020, **12**, 9.

76. G. Raj, R. Kurup, A. A. Hussain and S. Baby, Distribution of napthaquinones, plumbagin, droserone, and 5-*O*-methyl droserone in chitin induced and uninduced *Nepenthes khasiana*: molecular events in prey capture, *J. Exp. Bot.*, 2011, **62**, 5429–5436.
77. K. Springob, S. Samappito, A. Jindaprasert, J. Schmidt, J. E. Page, W. De-Eknamkul and T. M. Kutchan, A polyketide synthase of Plumbago indica that catalyzes the formation of hexaketide pyrones, *FEBS J.*, 2007, **274**, 406–417.
78. S. J. Gagne, J. M. Stout, E. Liu, Z. Boubakir, S. M. Clark and J. E. Page, Identification of olivetolic acid cyclase from *Cannabis sativa* reveals a unique catalytic route to plant polyketides, *Proc. Natl. Acad. Sci. U. S. A.*, 2012, **109**, 12811–12816.
79. H. Hotti and H. Rischer, The killer of Socrates: Coniine and related alkaloids in the plant kingdom, *Molecules*, 2017, **22**, 1962.
80. C. Risdian, T. Mozef and J. Wink, Biosynthesis of polyketides in *Streptomyces*, *Microorganisms*, 2019, **7**, 124.

CHAPTER 4

Mevalonate and Methylerythritol Phosphate Pathways: Terpenoids and Steroids

Terpenoids (also known as isoprenoids) are a structurally diverse group of naturally occurring metabolites (containing more than 50 000 compounds) derived from 5-carbon building blocks known as isoprene units (C_5H_8). Based on the number of carbon atoms (C5 units) in the structure, terpenes may belong to different classes such as **hemiterpenes** (C5), **monoterpenes** (C10), **sesquiterpenes** (C15), **diterpenes** (C20), **sesterterpenes** (C25), **triterpenes** (C30), **tetraterpenes** (C40), and **polyterpenes** (>C40). The products of partial terpenoid origin (cytokinins and prenylated proteins) are known as **meroterpenes.**[1,2]

The terpenes are specialized metabolites essential for plants to interact with the environment: protecting plants against herbivores and pathogens, attracting pollinators and seed dispersing animals and acting as allelochemicals. These metabolites are also synthesized for primary functions such as respiration (ubiquinone), photosynthesis (chlorophylls, carotenoids, phylloquinones, and plastoquinone), growth regulation (brassinosteroids, cytokinins, gibberellins, and strigolactones), and as a part of the membrane structure (sterols). They are also important in medicine (taxol for cancer, artemisinin for malaria), agriculture, flavours, fragrances, cosmetics, nutrition, and other commercial uses (rubber).[3,4]

The active isoprene units involved in the biosynthesis of terpenoids are **isopentenyl diphosphate** (**IPP**) and its isomer **dimethylallyl diphosphate** (**DMAPP**). IPP and DMAPP may be produced *via* two pathways. These are

Chemical Diversity of Plant Specialized Metabolites: A Biosynthetic Approach
By Bratati De

Published by the Royal Society of Chemistry, www.rsc.org

the **mevalonate** (**MVA**) pathway and the **methylerythritol phosphate** (**MEP**) pathway. The MEP pathway is also known as the non-mevalonate pathway or **deoxyxylulose phosphate** (**DOXP** or **DXP**) pathway. But the name MEP pathway is most accepted.

In the case of the MVA pathway, three molecules of acetyl-coenzyme A give rise to 3-hydroxy-3-methylglutaryl-CoA (HMG-CoA) in two steps (Figure 4.1). First, two molecules of acetyl-coenzyme A are converted to acetoacetyl-CoA in a Claisen condensation. Another acetyl coenzyme A is incorporated

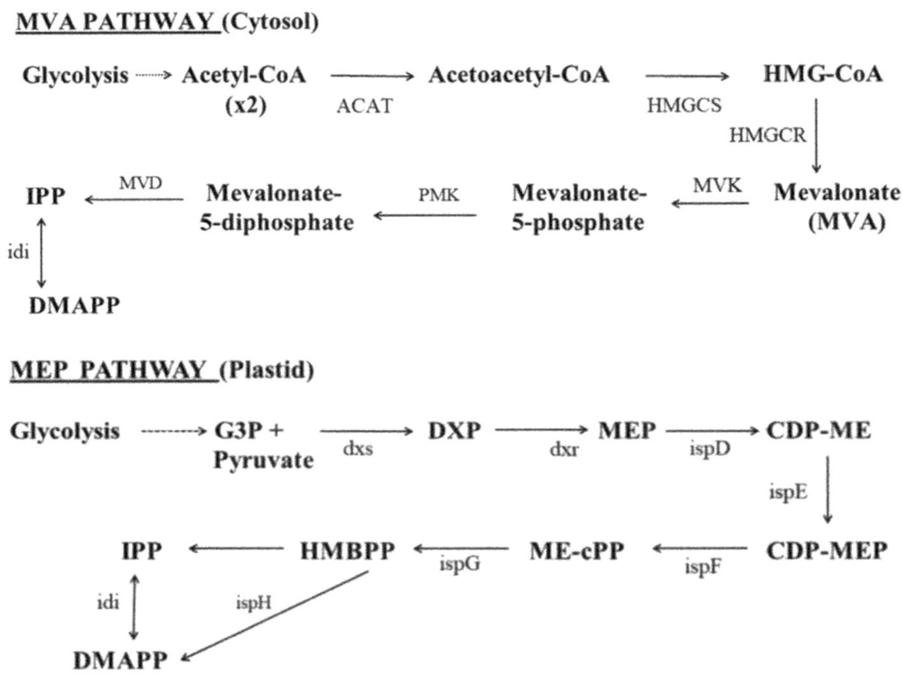

Figure 4.1 Biosynthesis of IPP and DMAPP. **MVA pathway enzymes and metabolites**: ACAT: acetoacetyl CoA thiolase; HMGCS: 3-hydroxy-3-methylglutaryl-CoA synthetase; HMG-CoA: 3-hydroxy-3-methylglutaryl-CoA; HMGCR: HMG-CoA reductase; MVK: mevalonic acid kinase; PMK: phosphomevalonate kinase; MVD: diphosphomevalonate decarboxylase; IPP: isopentenyl diphosphate; DMAPP: dimethylallyl diphosphate; idi: IPP delta-isomerase; **MEP pathway enzymes**: dxs: 1-deoxy-D-xylulose-5-phosphate synthase; dxr: 1-deoxy-D-xylulose-5-phosphate reductoisomerase; ispD: 2-*C*-methyl-D-erythritol 4-phosphate cytidylyltransferase; ispE: 4-diphosphocytidyl-methyl-D-erythritol kinase; ispF: 2-*C*-methyl-D-erythritol 2,4-cyclodiphosphate synthase; ispG: 4-hydroxy-3-methylbutenyl diphosphate synthase; ispH: 4-hydroxy-3-methylbutenyl diphosphate reductase; **MEP pathway metabolites**: G3P: glyceraldehyde 3-phosphate; DXP: 1-deoxy-D-xylulose 5-phosphate; MEP: 2-*C*-methyl-D-erythritol 4-phosphate; CDP-ME: 4-(cytidine 5′-diphospho)-2-*C*-methyl-D-erythritol; CDP-MEP: 2-phospho-4-(cytidine 5′-diphospho)-2-*C*-methyl-D-erythritol; ME-cPP: 2-*C*-methyl-D-erythritol 2,4-cyclodiphosphate; HMBPP: 1-hydroxy-2-methyl-2-butenyl 4 diphosphate.

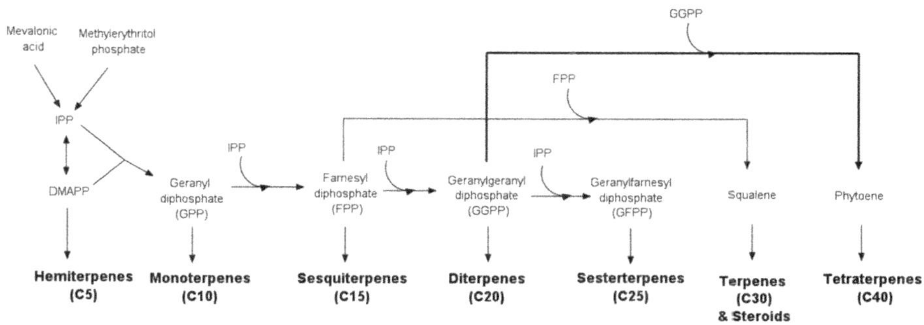

Figure 4.2 Schematic diagram for the biosynthesis of different groups of terpenes.

to produce HMG-CoA which is then converted to mevalonic acid (MVA). MVA produces IPP through formation of mevalonate 5-phosphate and mevalonate-5-diphosphate. IPP isomerase catalyzes interconversion of IPP and its isomer DMAPP. The MVA pathway occurs in animals, plants (cytosol), fungi, and archaea to form sesquiterpenes and triterpenes.[1,5–7]

In the MEP pathway, glyceraldehyde 3-phosphate (G3P) and pyruvate are condensed to 1-deoxy-D-xylulose-5-phosphate (DXP). DXP is reduced to form 2-*C*-methyl-D-erythritol 4-phosphate (MEP). Ultimately 4-hydroxy-3-methyl-but-2-enyl diphosphate (HMBPP) is formed through formation of 4-(cytidine 5′-diphospho)-2-*C*-methyl-D-erythritol (CDP-ME), CDP-ME 2-phosphate, and 2-*C*-methyl-D-erythritol 2,4-cyclodiphosphate (ME-cPP) (Figure 4.1). IPP and DMAPP are generated from HMBPP. The MEP pathway occurs in most bacteria, cyanobacteria, green microalgae, and plastids in plants to form monoterpenes, diterpenes and tetraterpenes.[1,5–7]

Formation of IPP and DMAPP are crucial for terpenoid biosynthesis as described schematically in Figure 4.2. Combination of DMAPP and IPP produces **geranyl diphosphate** (GPP) which is the precursor of **monoterpenes** (C10). Addition of one IPP unit to GPP yields **farnesyl diphosphate** (FPP) that can then lead to the formation of **sesquiterpenes** (C15). Further addition of an IPP molecule to FPP give rise to **geranylgeranyl diphosphate** (GGPP) which is the source of different **diterpenes** (C20). **Geranyfarnesyl diphosphate** (GFPP) generates the **sesterterpenes** (C25). Two molecules of FPP join tail-to-tail to yield **squalene** which generates **triterpenes** (C30) including the **steroids**. Tail-to-tail coupling of two molecules of GGPP produces tetraterpenes (C40). Details of biogenesis of terpenoids belonging to different classes based on number of carbon atoms will be discussed under corresponding headings.

Compartmentation of Terpenoid Biosynthesis Through MVA and MEP Pathways in Plants

The MVA pathway exists in the cytoplasm and peroxisome of eukaryotes. Most eubacteria use only MEP pathways for terpenoid biosynthesis. Archaebacteria, fungi, and animals use only the MVA pathway for production of terpenoids. Plants keep both MVA and MEP pathways.

In plants, the MVA and MEP pathways are compartmentalized as follows: the MVA pathway in the cytoplasm and the MEP pathway in the plastid. However, some evidence indicates that the common precursors of terpenoid biosynthesis such as IPP, GPP, FPP, and GGPP can be freely transported to different subcellular regions. The MVA pathway is generally involved in the supplies of precursors for the biosynthesis of sesquiterpenes, triterpenes and sterols, *cis*-polyisoprenoid alcohols, dolichol, brassinolides and the prenyl chains of ubiquinone. The MEP pathway generally produces precursors for the biosynthesis of hemiterpenes, monoterpenes, diterpenes, carotenoids, gibberellins, and side chains of plastoquinone and chlorophylls. Mitochondrial terpenoids are synthesized from MVA derived IPP imported from cytosol[3,8] (Figure 4.3).

4.1 Hemiterpenes

The terpene with a single isoprene unit is hemiterpene (C5) (Figure 4.4). The only hemiterpene is **isoprene** itself. Isoprene, a volatile compound, is released by many plants such as oaks, willows, poplars and spruce. IPP and

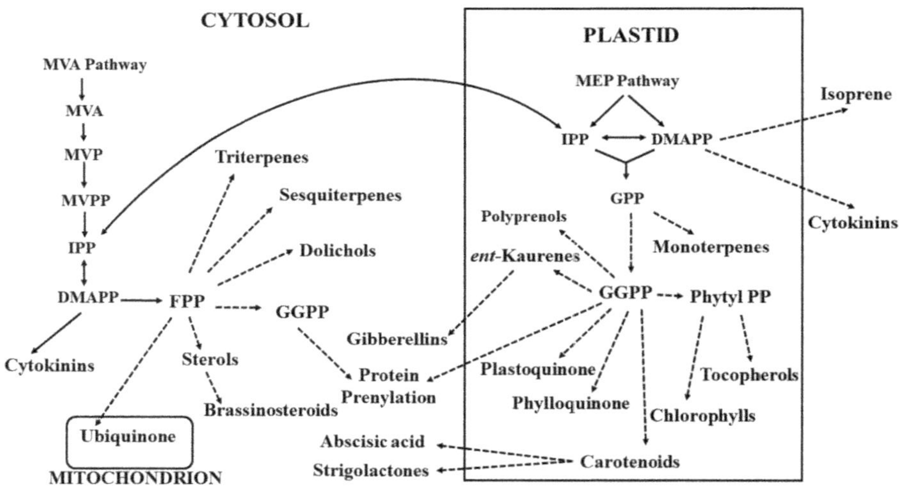

Figure 4.3 Compartmentation of terpenoid biosynthesis through MVA and MEP pathways in plants. MVP: mevalonate 5-phosphate; MVPP: mevalonate 5-diphosphate.

Isoprene 2-Methyl-3-buten-2-ol

Figure 4.4 Hemiterpenes.

DMAPP are reactive hemiterpene intermediates involved in terpenoid biosynthesis. **Methylbutenol** is produced by several species of pines.[7]

4.2 Monoterpenes

Monoterpenes are a large and diverse group of hydrocarbons. Around 1000 monoterpenes have been identified. They are mainly produced by plants including marine phytoplankton and mosses. Minor amounts of monoterpenes are also synthesized by insects and fungi. Basically, a monoterpene (C10 compound) consists of two isoprene units linked head to tail with each other (Figure 4.5).[9] Later they undergo cyclization and oxidization in a variety of ways. Due to their low molecular weight, monoterpenes are commonly found in essential oils. Many of the monoterpenes may exist as both enantiomers (+) and (−). **Iridoids** are cyclopentane pyran monoterpenes. In case of secoiridoids, the cyclopentane ring is opened. **Irregular monoterpenes** are synthesized by tail to middle binding of two isoprene units (Figure 4.5). Examples are lavandulol and pyrethrins.

Monoterpenes may exist as acyclic, monocyclic, and bicyclic compounds. **Acyclic monoterpene** hydrocarbons like myrcene, *Z*-β-ocimene, and *E*-β-ocimene are branched compounds with three double bonds. Oxygenated compounds are common in many essential oils and may exist as alcohols (geraniol, nerol, citronellol, linalool *etc.*) or aldehydes (geranial, neral, *Z*-isocitral, citronellal *etc.*) (Figure 4.6).[7,9–11]

Monocyclic compounds (menthane type) are exemplified by hydrocarbons (α-terpinene, γ-terpinene, α-phellandrene, β-phellandrene, limonene, terpinolene, *p*-cymene), hydroxylated derivatives (alcohols) (menthol, *trans*-piperitol, isopulegol, *trans*-pulegol, *trans*-carveol, thymol, carvacrol, terpinene-4-ol, α-terpineol), ketones (menthone, piperitone, pulegone, carvone), and an aldehyde (perilla aldehyde) (Figure 4.7).[7,9–11]

During biosynthesis of some of the **menthane type** structures, a second heterocyclic ring is formed after intramolecular reaction of oxygen containing groups. Examples are 1,8-cineol (eucalyptol) (with an ether bridge between C1 and C8), menthofuran and dill ether (five membered oxygen containing heterocycle, fused to the cyclohexane ring), *cis*- and *trans*-rose oxide (forming a tetrahydropyran ring) and ascaridol (with a peroxide structure) (Figure 4.8).[7,9–11]

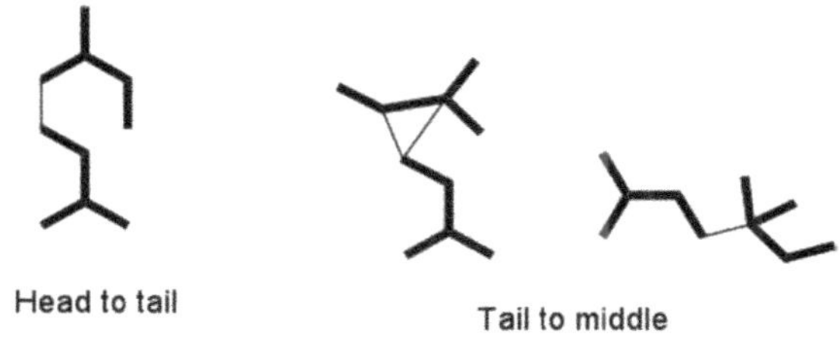

Figure 4.5 Linking of two isoprene units.

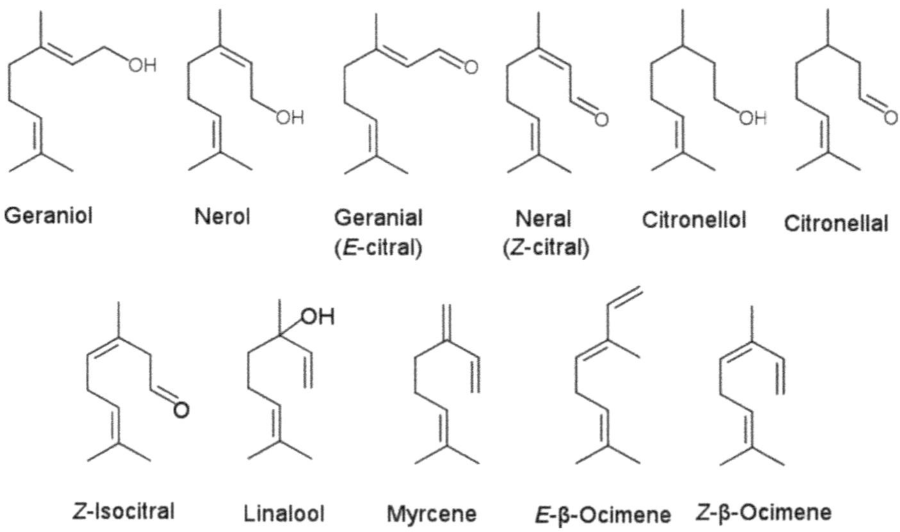

Figure 4.6 Acyclic monoterpenes.

Bicyclic monoterpenes may be of 6 structural types. These are pinane type, camphane/borane type, fenchane type, carane type, thujane type, and isocamphane type (Figure 4.9). In the carane type structure, an additional bond between C3 and C8 of the menthane structure give rise to a cyclopropane structure as found in Δ-3-carene. In the thujane type of monoterpenes, the cyclopropane occurs within the cyclohexane structure. These types of monoterpenes are found in nature as tertiary alcohols (*cis*- and *trans*-sabinene hydrate, sabinol) and ketones (*cis*-thujone, *trans*-thujone). The fenchane type of monoterpenes include fenchyl alcohol (fenchol), (+)-fenchone and (−)-fenchone. The camphene/bornane type is represented by (+)-borneol, (−)-borneol, bornyl acetate and camphor. The pinane type of bicyclic structure has a cyclobutene ring. The compounds are found as hydrocarbons (α-pinene, β-pinene), secondary alcohols (*cis*- and *trans*-verbenol, *cis*- and *trans*-chrysanthenol, *cis*- and *trans*-pinocarveol, and pinocampheol), ketones (pinocarvone, verbenone), the aldehyde myrtenal, and the corresponding primary alcohol myrtenol (Figure 4.10).[8–11]

Cleavage of geranyl diphosphate produces geraniol by geranyl diphosphate diphosphatase. Geraniol is oxidized by geraniol dehydrogenase to form geranial. Nerol, the *cis* isomer of geraniol, is generated from neryl diphosphate by monoterpenyl diphosphatase. Neral is produced from nerol. Geranyl diphosphate, after cyclization, is converted to (−)-limonene by the enzyme (−)-limonene synthase and to (+)-limonene by (+)-limonene synthase. Other menthane type monoterpenoids are then produced as described in Figure 4.11.[12]

Biosynthesis of bicyclic monoterpenes are depicted in Figure 4.12. (+)-Bornyl diphosphate synthase and (−)-bornyl diphosphate synthase

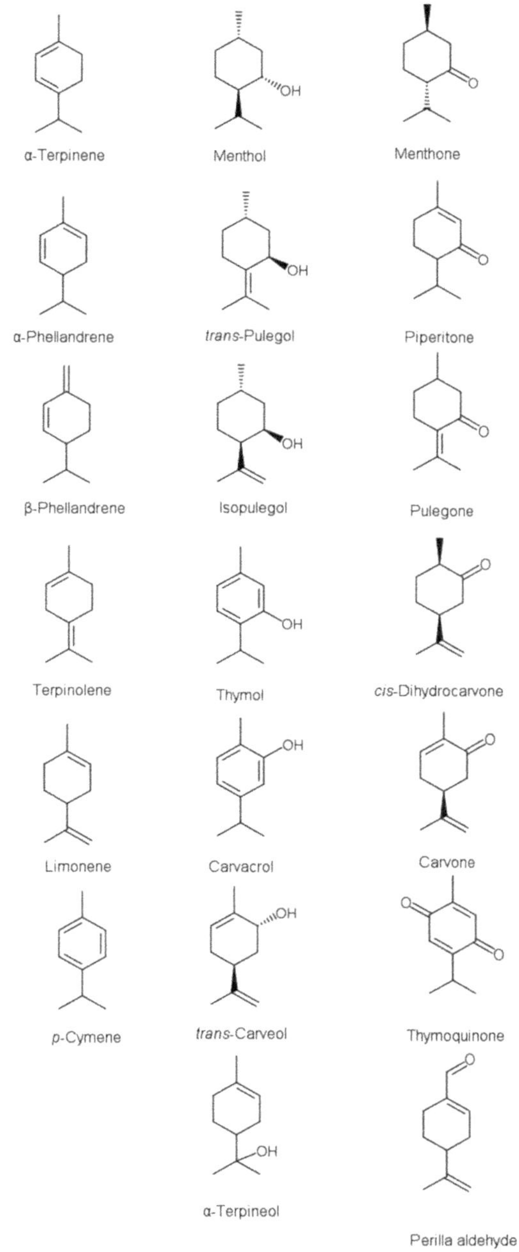

Figure 4.7 Monocyclic (menthane type) monoterpenes.

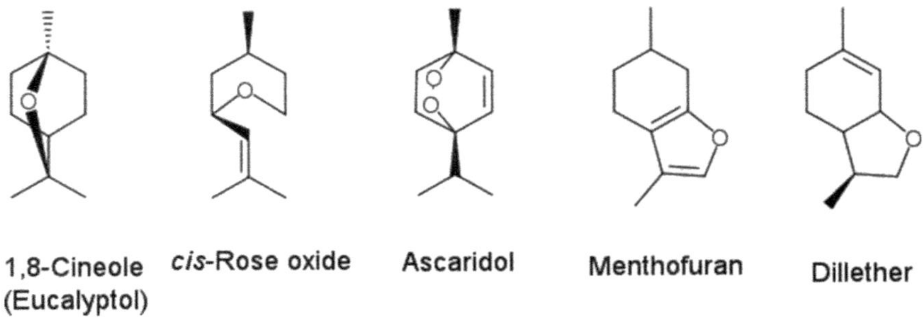

Figure 4.8 Menthane type monoterpenes forming a second heterocyclic ring.

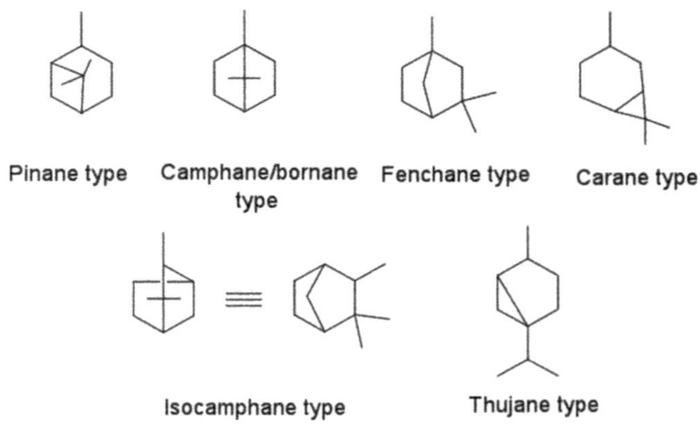

Figure 4.9 Bicyclic monoterpene types.

convert geranyl diphosphate to (+)-bornyl diphosphate and (−)-bornyl diphosphate, respectively. The diphosphate moieties are hydrolyzed by monoterpenyl diphosphatase to form borneol. Camphor is produced by respective dehydrogenase enzymes.[12]

4.2.1 Iridoids

Iridoids are cyclic monoterpenes containing a cyclopentane ring fused in many cases to a six membered dihydropyran ring. Secoiridoids have a cleaved cyclopentane ring where C7 and C8 are not connected. They are mostly found in the form of glycosides with glucose being combined at the unstable C1 hydroxyl group. Iridoids can be divided into four groups (Table 4.1). Dimers of iridoids are known as bis-iridoids.[13]

The biosynthetic pathway of iridoids has been studied in detail in *Catharanthus roseus*. Geraniol is converted into 8-hydroxygeraniol, and then to iridotrial *etc.* with ultimate formation of iridoid glucoside loganin and secoiridoid glucoside secologanin (Figure 4.13).[14]

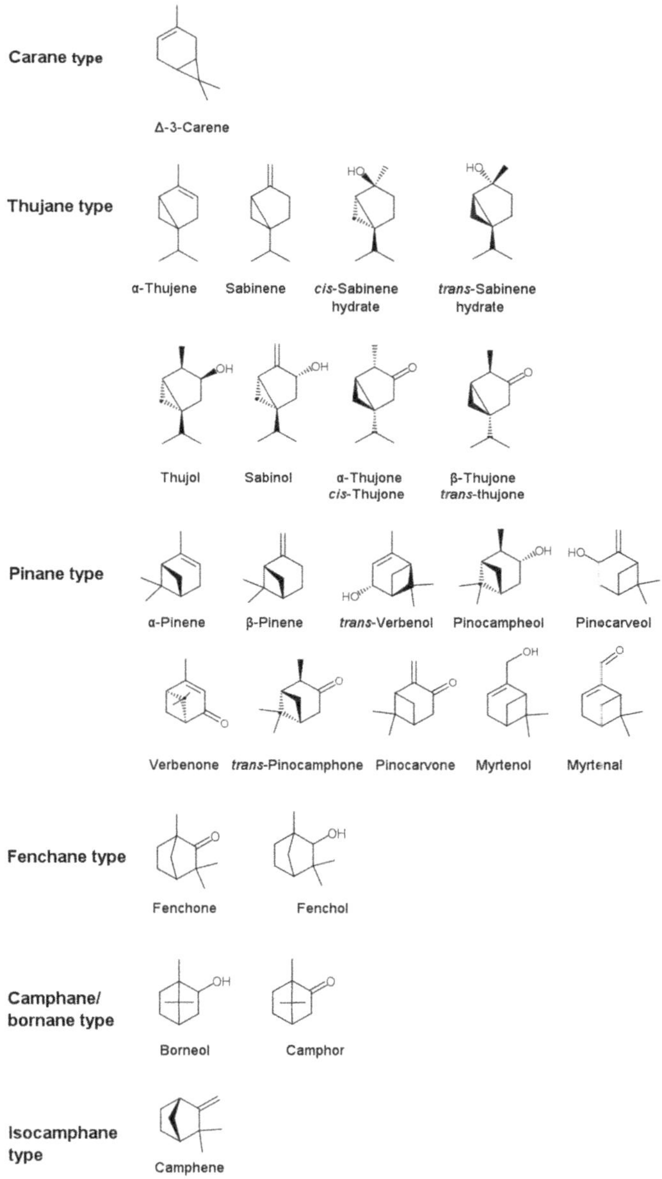

Figure 4.10 Examples of bicyclic monoterpenes.

There are many important medicinal plants which, in addition to other phytochemicals, contain iridoids (Figure 4.14). The *seco*-iridoid **gentiopicroside** (also known as gentiopicrin and gentiamarin) is the principal component of Gentian (*Gentiana lutea*, Gentianaceae) root. On hydrolysis this *seco*-iridoid yields the lactone **gentiogenin** and glucose. Gentiopicroside is bitter in taste.

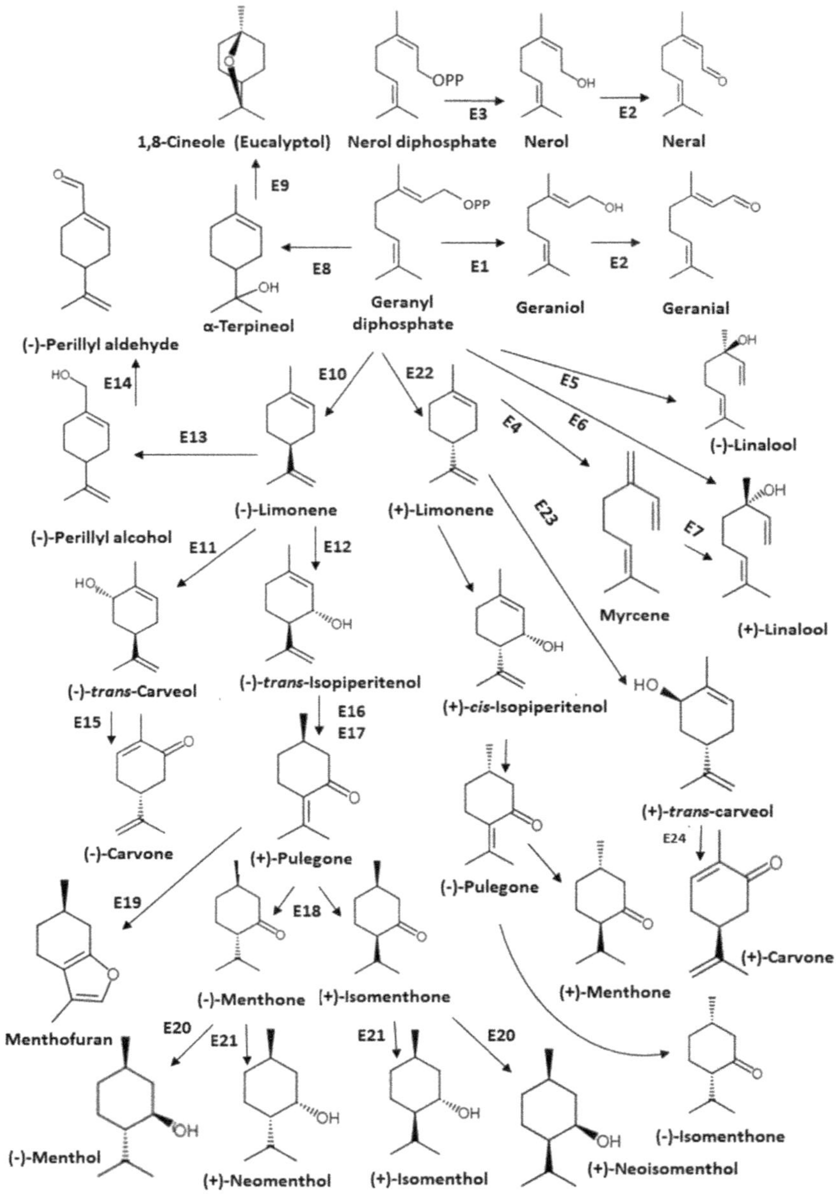

Figure 4.11 Biosynthesis of acyclic and menthane type monoterpenes. E1: geranyl diphosphate diphosphatase; E2: geraniol dehydrogenase; E3 monoterpenyl diphosphatase; E4: myrcene synthase; E5: *R*-linalool synthase; E6: *S*-linalool synthase; E7: linalool dehydratase; E8: (−)-α-terpineol synthase; E9: 1,8-cineole synthase; E10: (4*S*)-limonene synthase; E11: (−)-limonene-6-hydroxylase; E12: (−)-limonene-3-hydroxylase; E13: (−)-limonene hydroxylase: E14: perillyl-alcohol dehydrogenase; E15: carveol dehydrogenase; E16: isopiperitenol dehydrogenase; E17: (−)-isopiperitenone reductase; E18: (+)-pulegone reductase; E19: (+)-menthofuran synthase; E20: (−)-menthol dehydrogenase; E21: (+)-neomenthol dehydrogenase; E22: (+)-limonene synthase; E23: (+)-limonene-6-hydroxylase; E24: (+)-*trans*-carveol dehydrogenase.

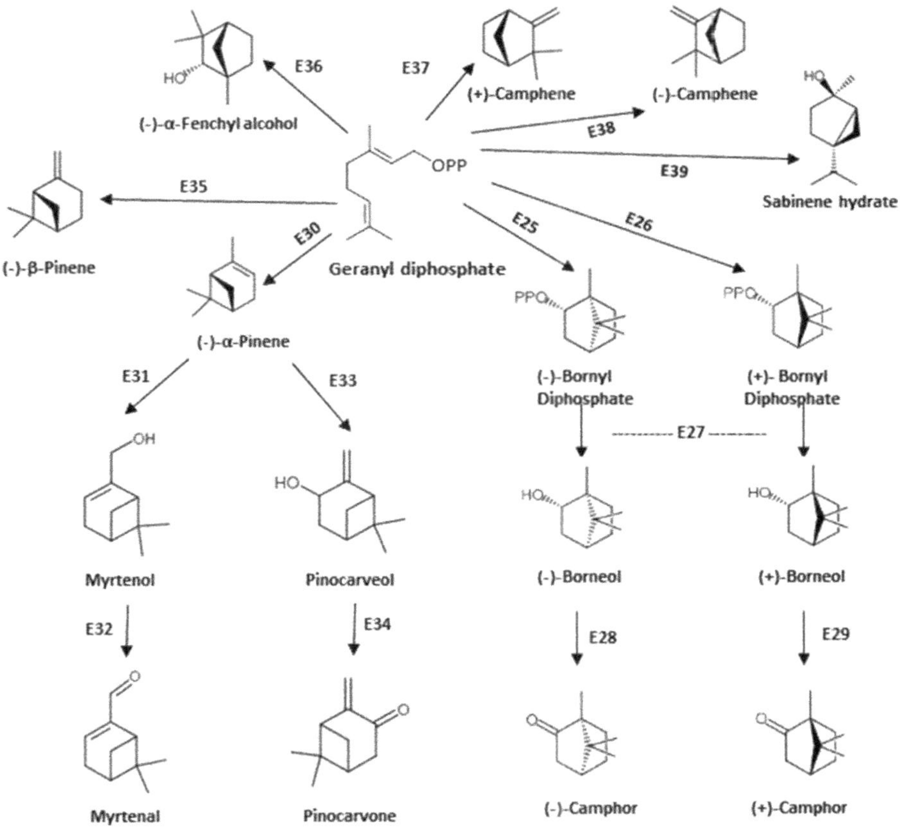

Figure 4.12 Biosynthesis of bicyclic monoterpenes. E25: (−)-bornyl diphosphate synthase; E26: (+)-bornyl diphosphate synthase; E27: monoterpenyl diphosphatase; E28: (−)-borneol dehydrogenase; E29: (+)-borneol dehydrogenase; E30: (−)-α-pinene synthase; E31: α-pinene dehydrogenase; E32: myrtenal dehydrogenase; E33: α-pinene monooxygenase; E34: pinocarveol dehydrogenase; E35: (−)-β-pinene synthase; E36: (−)-*endo*-fenchol synthase; E37: (+)-camphene synthase; E38: (−)-camphene synthase; E39: sabinene-hydrate synthase.

Amarogentin is the biphenolic acid ester of gentiopicroside. Bitterness value of amarogentin is 5000 times greater than that of gentiopicroside. Other iridoids are **sweroside** and **swertiamarin**. The important medicinal plant *Swertia chirayita* (Gentianaceae) is also reported to biosynthesize these bitter iridoids. Valerian root, rhizome and the stolon obtained from *Valeriana officinalis* (Caprifoliaceae) contain several non-glycosidal epoxy iridoid esters called **valepotriates** (Figure 4.14). The tranquilizing ability of *Valeriana officinalis* plant parts to relieve nervous tension is mainly due to the presence of the valepotriates. The principal valepotriate is **valtrate**. Others are **dihydrovaltrate, isovaltrate**, and **acevaltrate**. Other non-glycosidal iridoids are **baldrinal** and **homobaldrinal.**[8,15]

Table 4.1 Different types of iridoids.

Iridoid type	Examples	Occurrence
Iridoid glycosides	Acetylbarlerin, brasoside, geniposide, loganin	Scrophulariaceae, Lamiaceae, Gentianaceae, Ericaceae, Saxifragaceae, Hamamelidaceae
Secoiridoid glycosides	Secologanin, swertiamarin, gentiopicroside, sweroside, quinjiaoside A and B	Caprifoliaceae, Oleaceae, Gentianaceae, Apocynaceae, Loganiaceae, Cornaceae
Non-glycosidic iridoids	Valtrate, acevaltrate	Valerianaceae
Bis-iridoids	Cantleyoside, laciniatoside I and II, sylvestroside	Dipsacaceae

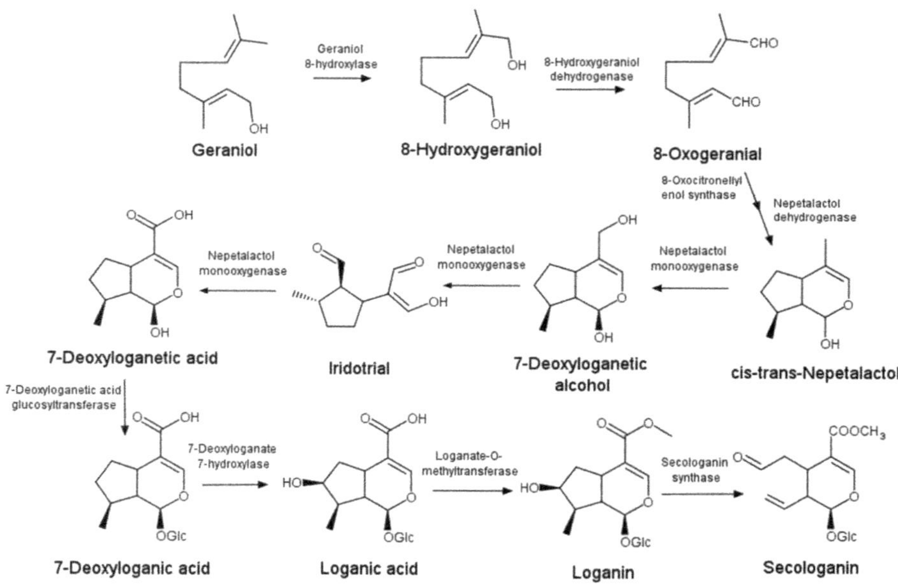

Figure 4.13 Biosynthesis of (*seco*)iridoids.

4.2.2 Irregular Monoterpenes

Lavender essential oil extracted from *Lavandula* spp. (Lamiaceae) contains some irregular monoterpenes, in addition to regular monoterpenes. Head to middle condensation (Figure 4.5) of isoprene units produces lavandulyl diphosphate by the enzyme lavandulyl diphosphate synthase. Lavandulol is formed from lavandulyl diphosphate by the action of the enzyme monoterpenyl diphosphatase. The ester derivative of **lavandulol** is **lavandulyl acetate** (Figure 4.15).[16]

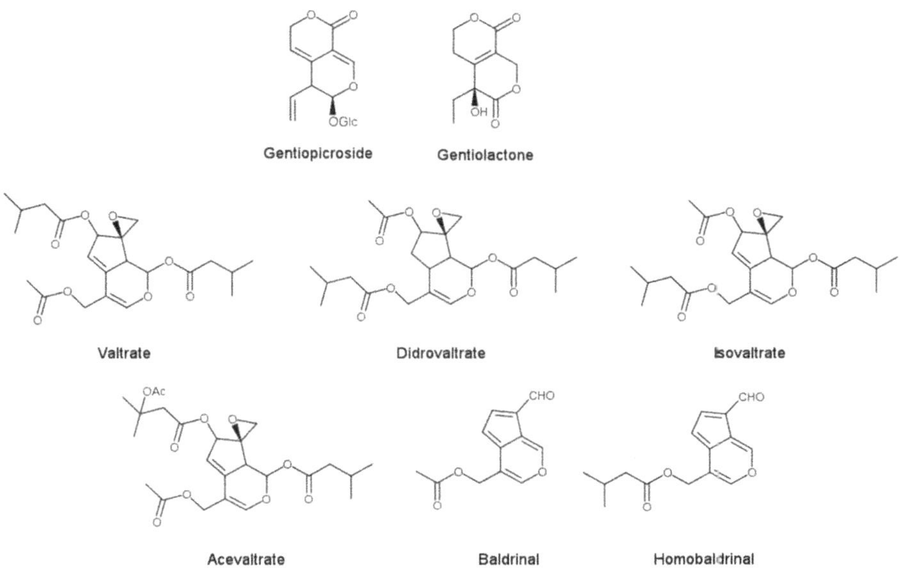

Figure 4.14 Iridoids.

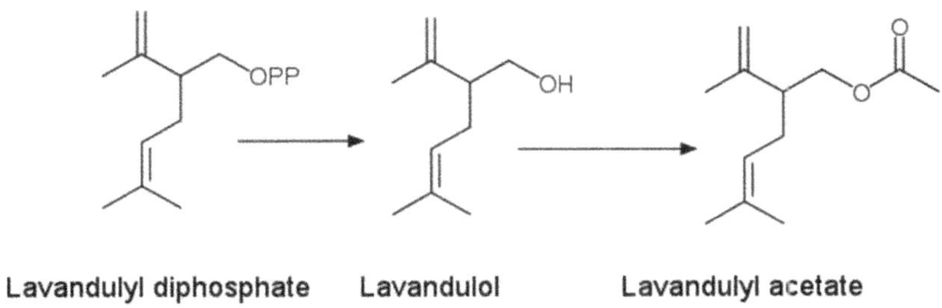

Figure 4.15 Irregular monoterpenes of lavender oil.

Pyrethrins are group of six chemicals mostly synthesized in the flowers of the pyrethrum plant (*Tanacetum cinerariifolium,* Asteraceae) and related *Tanacetum* species. Pyrethrins are esters of irregular monoterpene acids and alcohols derived from jasmonic acid. Pyrethrins belong to two types. **Type-I pyrethrins** are **jasmolin I**, **pyrethrin I**, and **cinerin I**. These are esters of chrysanthamic acid and the three alcohols jasmolone, pyrethrolone and cinerolone, respectively. **Type-II pyrethrins** are **jasmolin II, pyrethrin II** and **cinerin II**. They consist of esters of pyrethric acid and the same alcohols jasmolone, pyrethrolone and cinerolone respectively. The two irregular monoterpene acids (chrysanthemic acid and pyrethric acid) are produced *via* chrysanthemol by head to middle coupling of two molecules of dimethylallyl diphosphate followed by hydrolysis of the diphosphate group by phosphatases (Figure 4.16).[17]

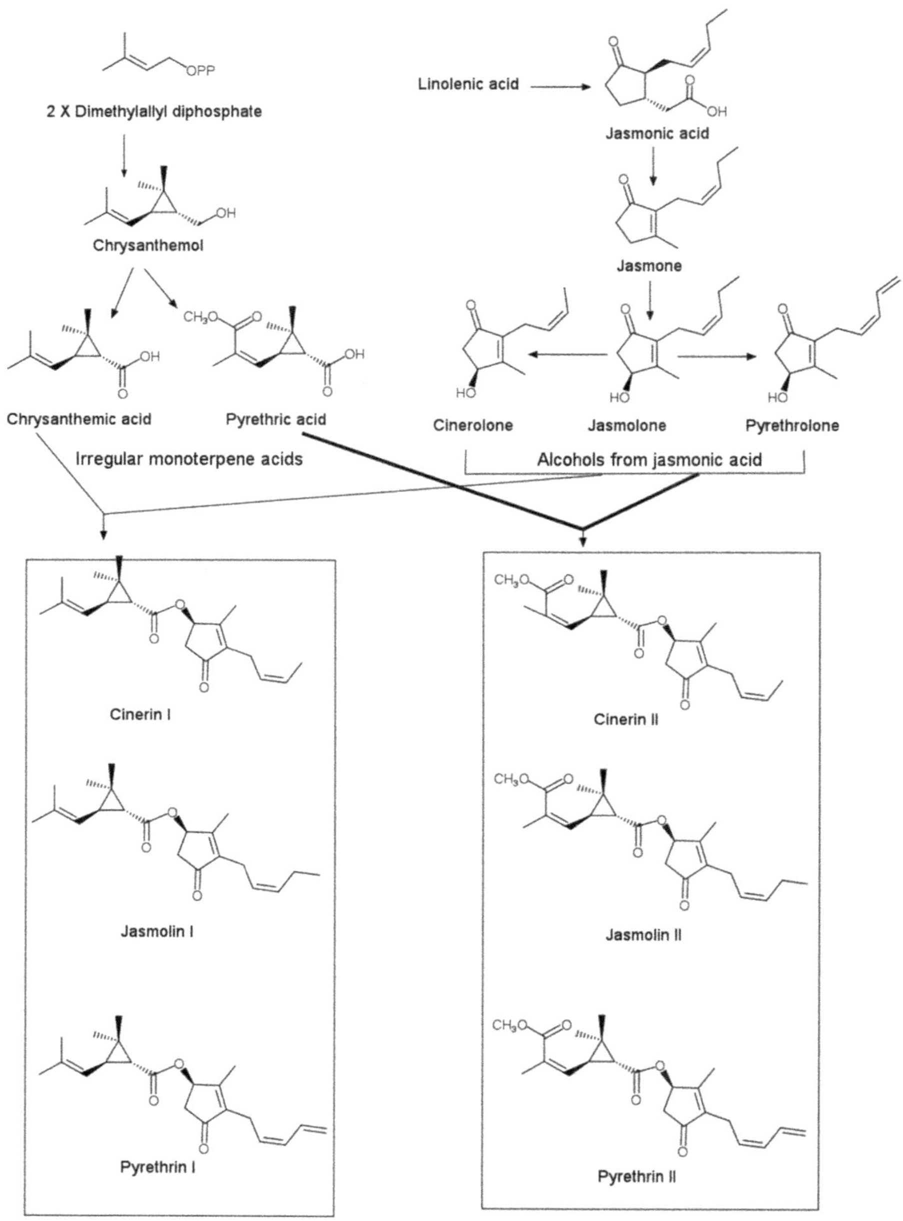

Figure 4.16 Biosynthesis of pyrethrins.

Pyrethrins are well known insecticides against insects which are immobilized and then killed. They have repellent properties for controlling mosquitoes. They are also biodegradable and photolabile. Pyrethrum flowers are the commercial source of these insecticides.

4.3 Sesquiterpenes

Sesquiterpenes consist of 15 carbon backbones and are diverse in structure. There are more than 11 000 sesquiterpenes of which about 8000 contain at least one γ-lactone ring. Sesquiterpenes may exist in acyclic (linear), mono-, bi-, or tricyclic forms. Most functional forms are cyclic.

Farnesane is the parent compound of the sesquiterpenes. **Farnesenes** *e.g.*, *E,E*-α-farnesene and *E*-β-farnesene are the acyclic sesquiterpenes widely detected in essential oils and contribute to the flavour of the oils. Oxidation products may exist as alcohol and aldehyde forms. The acyclic sesquiterpene alcohols **farnesol** and **nerolidol** are occasional essential oil constituents. The aldehydes **α-sinesal** and **β-sinesal** have extremely low odour threshold and are found in the essential oils of *Citrus* species (Figure 4.17).[9,18]

During biosynthesis, a great variety of cyclic skeletons (monocyclic, bicyclic, and tricyclic) are formed by the folding of the farnesyl chain. Some important skeletons such as **bisabolane, elemane, germacrane, humulane, bergamotane, cadinane, muurolane, amorphane, eudesmane, guaiane, himachalane, caryophyllane, patchoulane** and **aromadendrane** have been presented with examples in Figures 4.18 and 4.19. These sesquiterpenes are important constituents of different essential oils.

Gossypol is a dimeric sesquiterpene. δ-Cadinene generates hemigossypol which undergoes dimerization to yield gossypol (Figure 4.20). Gossypol is a phytoalexin found in the immature flower buds and seeds of cotton plant (*Gossypium* species, Malvaceae).[19]

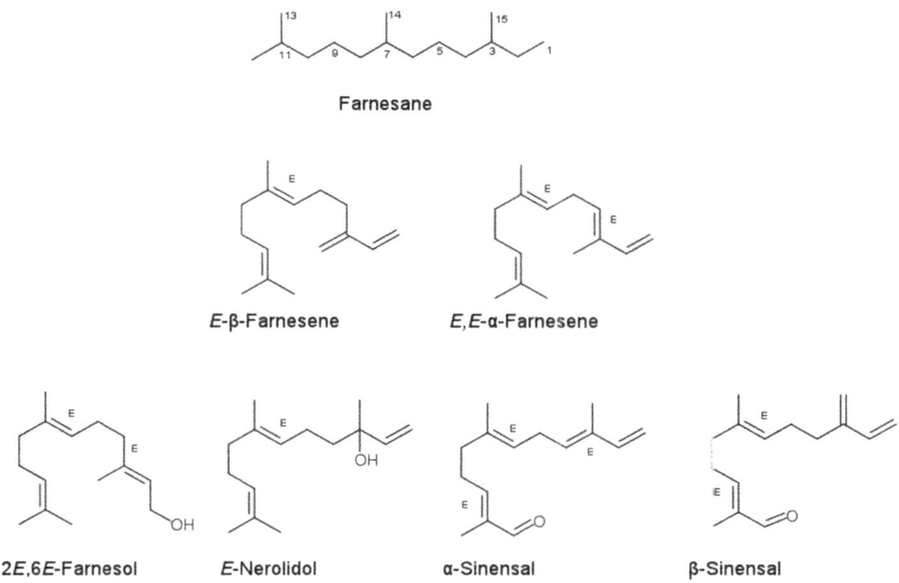

Figure 4.17 Acyclic sesquiterpenes.

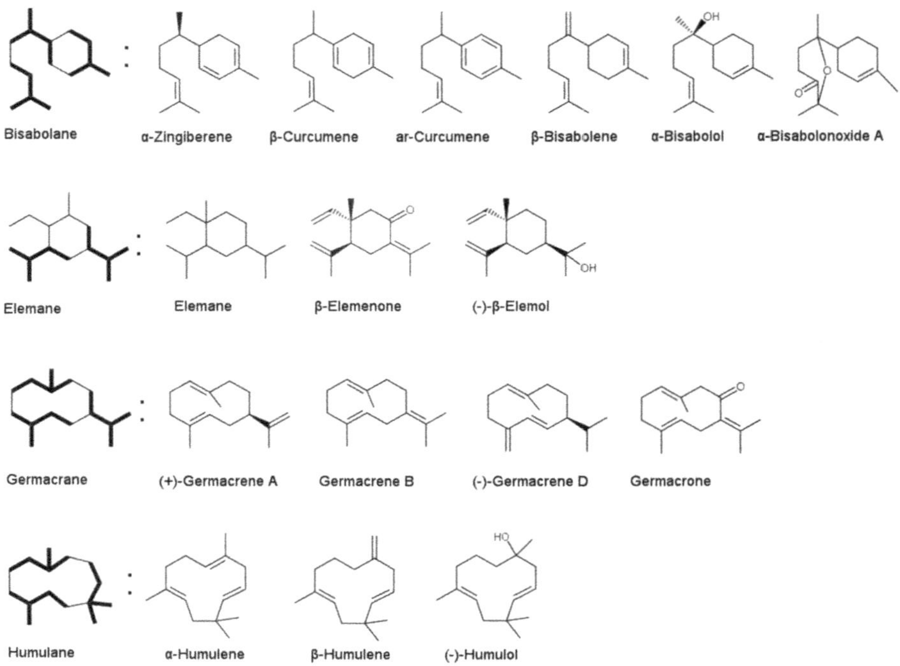

Figure 4.18 Monocyclic sesquiterpenes.

4.3.1 Sesquiterpene Lactones

Sesquiterpene lactones are C-15 compounds containing a γ-lactone ring and are structurally diverse due to numerous modifications. The majority of these compounds are cyclic in structure. In many cases sesquiterpene lactones contain exocyclic methylene groups conjugated to a carbonyl group. The exocyclic methylene is reduced in artemisinin, matricin, achillin, and santonin. There are various structural classes of the sesquiterpene lactones. Five important groups are **germacranolides** (have a 10 membered ring), **eudesmanolides** (have two fused six membered rings), **guaianolides** (have a 7-membered and a 5-membered ring and a methyl group at C-4.), **pseudoguaianolides** (have a 7-membered and a 5-membered ring and a methyl group at C-5), and **cadinanolides** (Figure 4.21). Some others are **elemanolides, ambrosanolides, xanthanolides** *etc.* All of them contain a fused 5-membered lactone group (γ-lactone). The suffix "olide" indicates the presence of a lactone group in the structure. The sesquiterpene lactones are generally bitter in taste and are non-volatile.[18,20,21]

Eudesmolides: Santonin obtained from *Artemisia santonica* (Asteraceae) is one important example of this group of compounds. Santonin was used as an ascaricide. But pharmaceutical use of this compound has been abandoned because of its toxic effect.

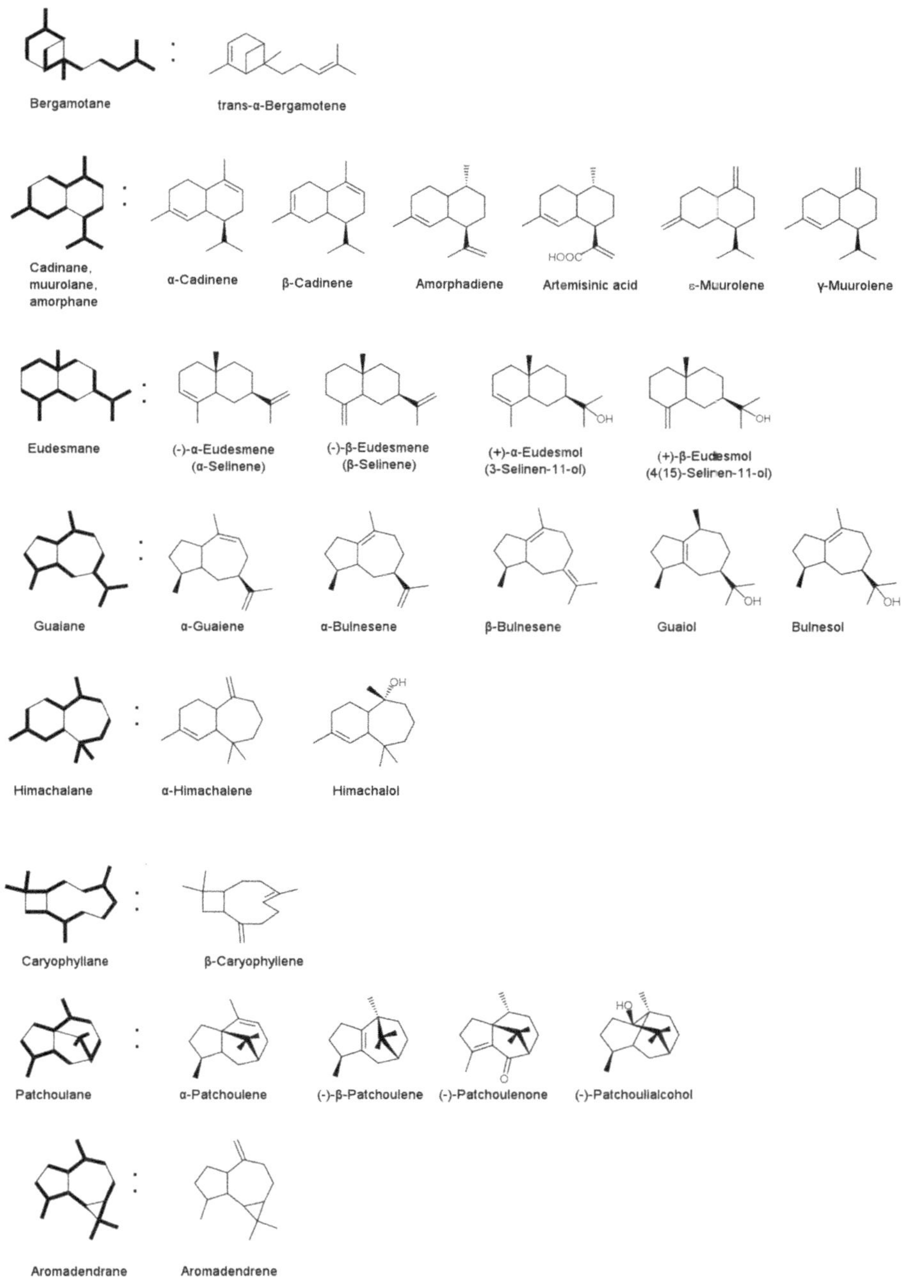

Figure 4.19 Bicyclic and tricyclic sesquiterpenes.

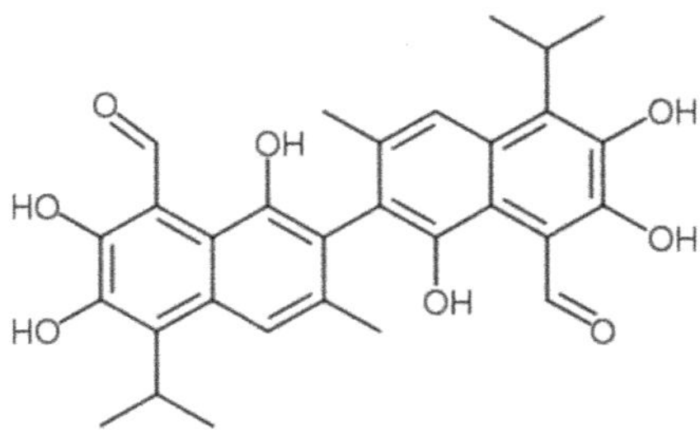

Gossypol

Figure 4.20 Dimeric sesquiterpene.

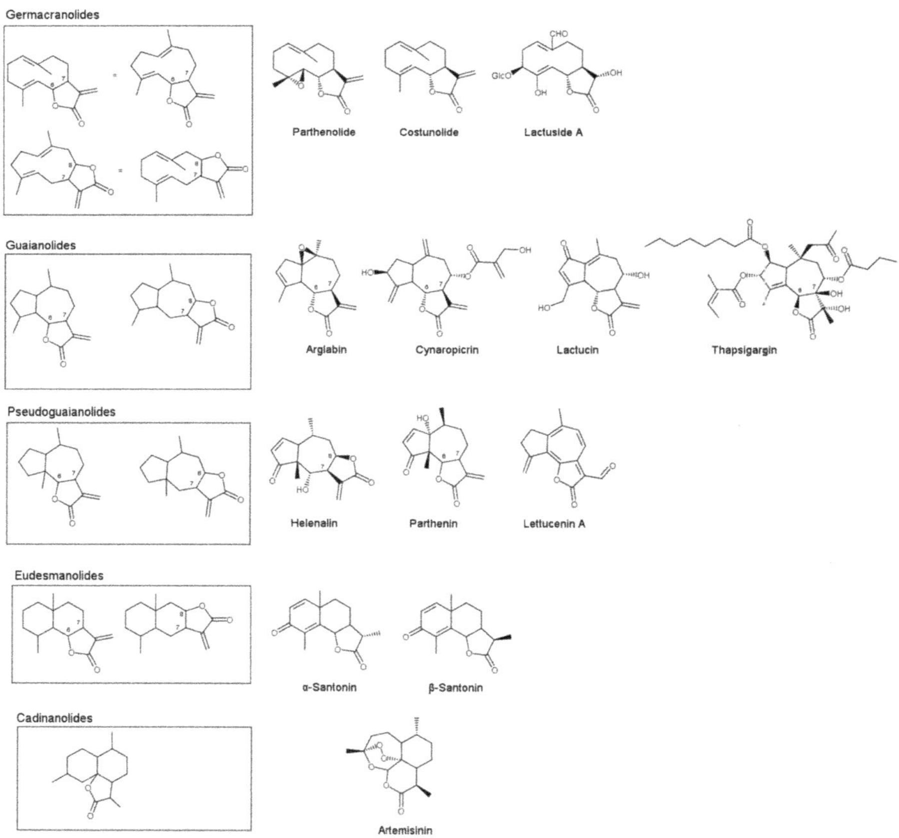

Figure 4.21 Sesquiterpene lactones.

Germacranolide: Costunolide was isolated from *Saussurea lappa* (Asteraceae) roots. This traditional Chinese medicinal plant has anticancer and anti-inflammatory properties. **Parthenolide** is obtained from the feverfew (*Tanacetum parthenium*, Asteraceae), the traditional herbal medicine used for the treatment of fever, arthritis, and migraines.

Guaianolide: Thapsigargin from *Thapsia garganica* (Apiaceae), **arglabin** from *Artemisia glabella* (Asteraceae), **lactucin** from *Lactuca sativa*, and **cynaropicrin** from *Cynara scolymus* (Asteraceae) and *Saussurea lappa* (Asteraceae) are examples of this group of compounds. Thapsigargin is a potent inhibitor of Ca^{2+}-ATPases and is capable of unbalancing cellular Ca^{2+} concentrations leading to disrupted cell function, growth, and apoptosis.

Pseudoguaianolides: Examples of this group of sesquiterpene lactones are **helenalin** isolated from *Arnica montana* (Asteraceae), **elephantopin** from *Elephantopus elatus* (Asteraceae). **Parthenin** from *Parthenium hysterophorus* (Asteraceae) is a highly alkylating agent. *Parthenium hysterophorus* is a bothersome weed in India because of its fast-spreading ability due to profuse seed production and its allelopathic effect on other plants.

Cadinanolides: Artemisinin has an endoperoxide ring in its structure. It has been isolated from *Artemisia annua* (Asteraceae), aerial parts of which have been used as a febrifuge in Chinese traditional medicine. Artemisinin, and its derivatives artemether and arteether are used for the treatment of malaria that is chloroquine resistant.

4.4 Diterpenes

Diterpenes are C-20 compounds with four isoprene units, biosynthetically derived from geranylgeranyl diphosphate. Geranylgeranyl diphosphate is produced predominantly by the MEP pathway but also the mevalonate pathway. Diterpenes are widespread in the plant kingdom. There are more than 2500 diterpenes distributed in many plant families. According to the nature of cyclization, the diterpenoids are divided into more than 45 skeletal types classified as acyclic, monocyclic, bicyclic, tricyclic, tetracyclic, pentacyclic, and macrocyclic ring. Commonly, the diterpenes are found as polyoxygenated compounds with keto and hydroxyl groups. The hydroxyl groups are often esterified by aliphatic or aromatic acids.[22–25] Diterpenes are often present in plants as lactones. Diterpenes are identified as bioactive natural products. Some of them are produced commercially for uses in medicine (a few of them being clinically effective), food additives and agriculture. Diterpenes are also components of resin and oleoresins. Several diterpene phytoalexins such as momilactones A and B, oryzalexins (A–F) and phytocassanes are produced by different parts of rice plants.[26,27]

The diterpenes are biosynthesized from geranylgeranyl diphosphate (GGPP). Cyclization of GGPP proceeds *via* four different conformational folds. GGPP in the "chair–chair" normal conformation yields copalyl diphosphate. GGPP in the chair–chair antipodal conformation produces *ent*-copalyl diphosphate. GGPP in the chair–boat antipodal conformation generates

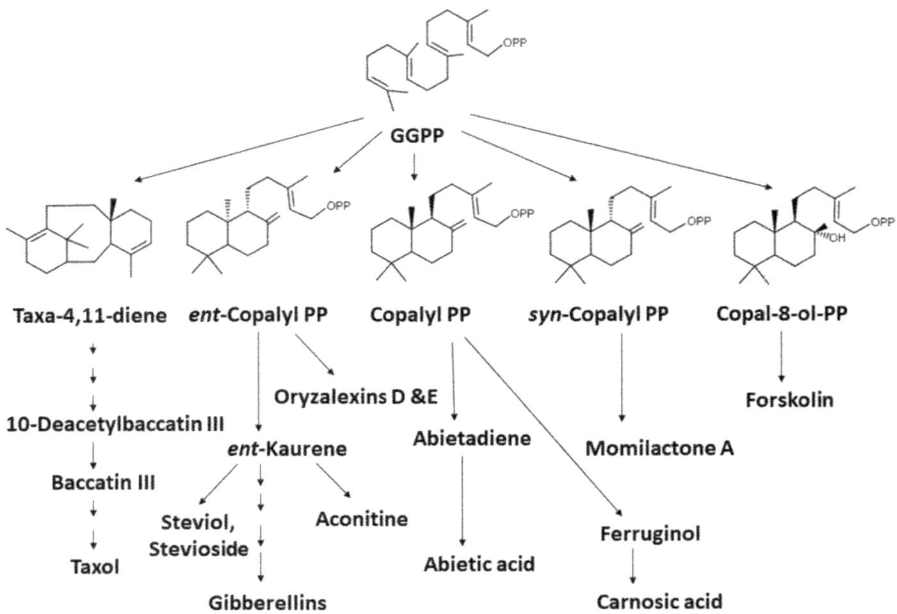

Figure 4.22 Schematic pathway for biosynthesis of diterpenes.

syn-ent-copalyl diphosphate. Chair–boat normal conformation of GGPP produces *syn*-copalyl diphosphate.[23] Biosynthetic origin of some important diterpenes is shown in Figure 4.22.

4.4.1 Acyclic Diterpenes

Phytol (skeletal type phytane) is an example of a linear or acyclic diterpene. Phytol constitutes the lipophilic side chain of chlorophyll a and the side chain of phylloquinone (vitamin K_1). Phytylation of homogentisic acid forms vitamin E (tocopherol)[8] (Figure 4.23).

4.4.2 Bicyclic Diterpenes

Bicyclic diterpenes may belong to different skeletal types (labdanes, halimanes, clerodanes). One important diterpene belonging to labdane type is **forskolin** (synonyms: coleonol, colforsin) obtained from *Coleus forskohlii* of family Lamiaceae (Figure 4.24) which is used in Ayurvedic traditional system of medicine. This compound contains a third ring. Forskolin has potential to treat hypertension, heart failure, and alleviate glaucoma. It has platelet aggregation inhibitory activity, anti-HIV, and anti-tumour properties. The efficacy of forskolin depends on activation of the enzyme adenylate cyclase and increased cAMP. It is biosynthesized from copal-8-ol diphosphate.[28]

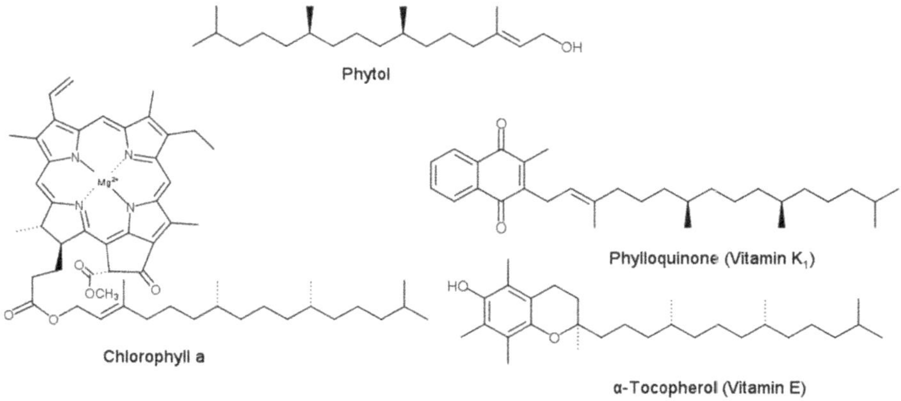

Figure 4.23 Acyclic diterpenes.

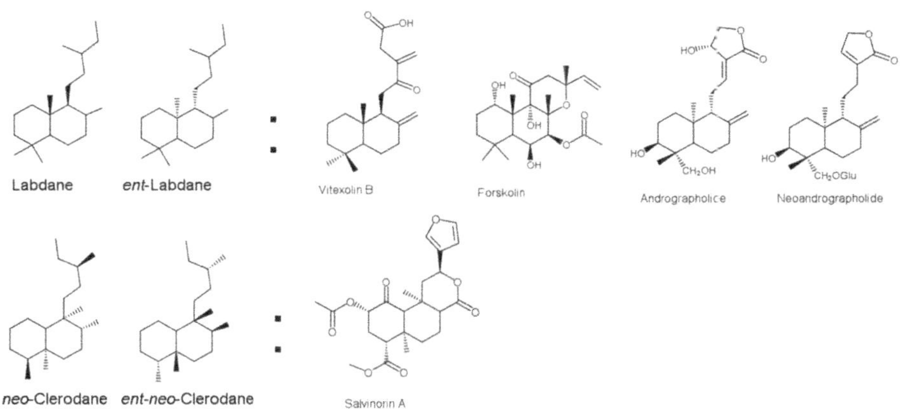

Figure 4.24 Bicyclic diterpenes.

Colforsin daropate, a water soluble forskolin derivative, has been introduced as a drug. **Andrographolide** and **neoandrographolide** (Figure 4.24), the two constituents of *Andrographis paniculata* (Acanthaceae), are two *ent*-labdanes with γ-lactone between carbons C16 and C15.Andrographolide shows anti-inflammatory properties by inhibiting NF-κB/MAPK activation. Andrographolide is also a potent hepatoprotective diterpene.[29–31] **Salvinorin A** (Figure 4.24) is a neo-clerodane diterpene isolated from the Mexican plant *Salvia divinorum* (Lamiaceae). The leaves of the plant were used for centuries by the Mazatecan culture as a psychoactive ingredient. Salvinorin A has been shown to possess hallucinogenic activity, anti-nociceptive property, and depressant-like effects.[32]

4.4.3 Tricyclic Diterpenes

The skeletal types of tricyclic diterpenes are primanes, abietanes, cassanes, rosanes, vouacapanes, podocarpanes *etc.* Many important plant metabolites of abietane type have been reported. **Abietic acid** (synonym: abietinic acid, silvic acid) is a part of resin obtained from several species of *Pinus* (Pinaceae).[33] **Dehydroabietic acid** and **ferruginol** are aromatic abietanes. Aromatic abietanes are known as chemical defense agents. Ferruginol and commercially available abietane **carnosic acid**, reported from Lamiaceae, are antioxidants.[34] **Margocin**, **margocinin** and **nimosone** are also abietane type diterpenes reported from *Azadirachta indica* (Meliaceae)[35,36] (Figure 4.25).

4.4.4 Tetracyclic Diterpenes

The tetracyclic diterpenes have been divided into several skeletal types such as trachylobanes, kauranes, stemodanes, stemaranes, atisanes, gibberellanes and triglianes. **Gibberellins** (GAs) are a large group of tetracyclic diterpenes biosynthesized from *ent*-kaurene *via ent*-kaurenoic acid. *ent*-Kaurene is produced *via ent*-copalyl diphosphate as an intermediate. Gibberellanes possess mostly *ent*-gibberellane (C20) or *ent*-20-norgibberellane (C19) skeletons. There are 136 different GAs, named GA_1 to GA_{136}, reported from higher plants, fungi, and bacteria. Not all the GAs are bioactive. GA_1 and GA_4 are the major biologically active GAs in angiosperms having phytohormonal activity. They play significant roles in plant growth and development, starting from seed germination to vegetative and reproductive development, through stimulation of cell division and elongation. GA_4 is more active than GA_1. GA_7 is also biologically active. GAs with phytohormonal activity possess C19

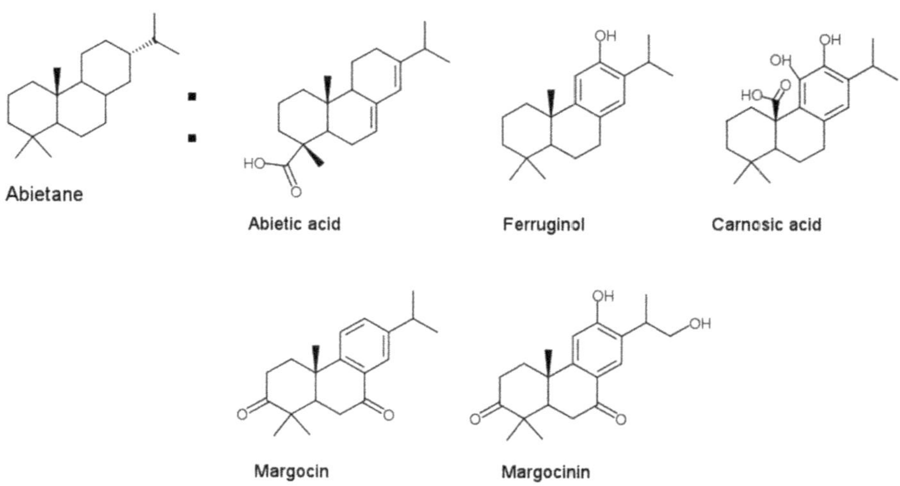

Figure 4.25 Tricyclic diterpenes.

skeletons with a C-19–C-20 lactone, a 7-carboxylic acid group and a 3β-hydroxyl group (Figure 4.26). GA_3 (also known as gibberellic acid), first characterized in *Gibberella fujikuroi* reclassified as *Fusarium fujikuroi,* is produced in commercial scale in fungal cultures for agricultural applications, particularly for production of seedless grapes. Ferns of family Schizaeaceae use the methyl esters of GA_9, GA_{20} and GA_{73} as regulators of antheridium development. These methyl esters of GA are not active in angiosperms. The methyl ester of GA_9, a fern antheridiogen, regulates protonema differentiation of the moss *Physcomitrella patens.*[37–40]

Biosynthesis of gibberellins involves three intracellular compartments: plastids, endoplasmic reticulum, and cytoplasm. In the plastid, GGPP is converted into *ent*-kaurene. *ent*-Kaurenol is converted to GA_{12} and GA_{53} in the endoplasmic reticulum. Other gibberellins are biosynthesized in the cytoplasm[40] (Figure 4.27).

Other important tetracyclic diterpenes are the aglycone **steviol**, the glycosides **stevioside**, **rebaudioside A–F**, and **others** biosynthesized *via ent*-kaurene. The steviol glycosides are found in the leaves of the shrub *Stevia rebaudiana* (Family Asteraceae), native to the Paraguay and Brazil, now

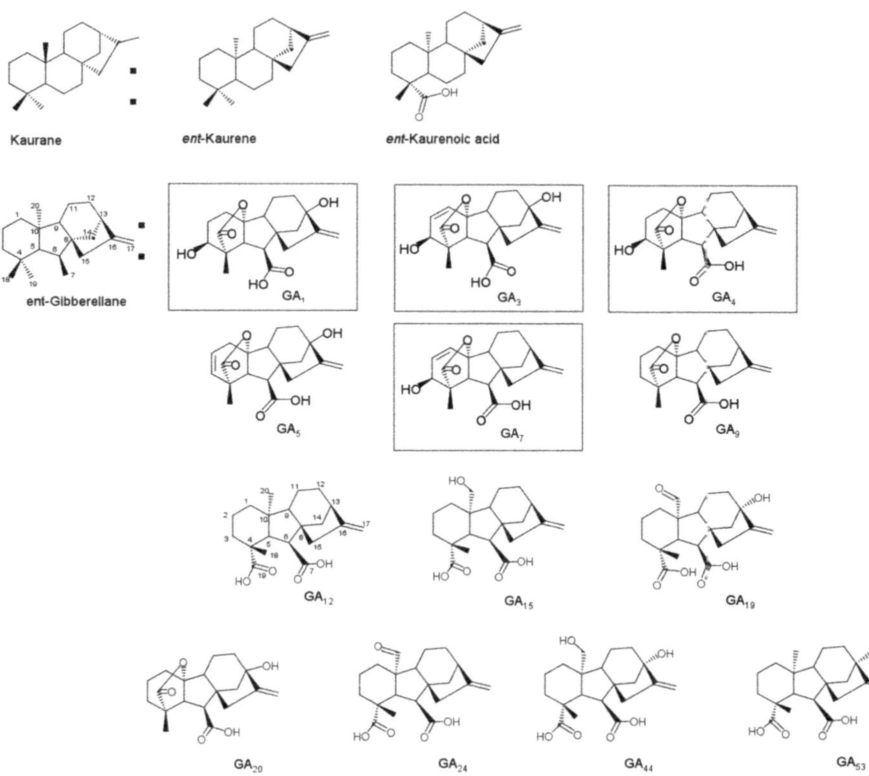

Figure 4.26 Different gibberellins.

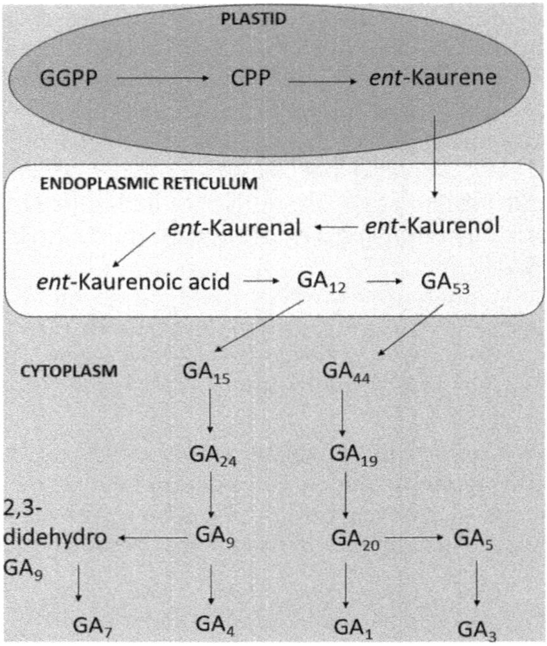

Figure 4.27 Biosynthesis of gibberellins showing intracellular compartmentation between plastids, endoplasmic reticulum, and cytoplasm. GGPP: geranyl geranyl diphosphate; CPP: copalyl diphosphate.

cultivated in regions of Asia, Canada, and Europe. The plant was used for centuries as a sweetener. The aglycone steviol is *ent*-hydroxykaur16-en-18-oic acid. The two glycosides stevioside and rebaudioside A (Figure 4.28) are 200 to 300 times sweeter than sucrose. Stevioside and rebaudioside A enhance glucose-induced insulin secretion. The two non-caloric glycosides are now approved by European Union and United States as a sweetening food additive.[41–44]

Metabolites belonging to tetracyclic diterpenes with a tigliane skeleton are **phorbol** and **phorbol esters** (PEs) (Figure 4.29). PEs are mainly obtained from the genera *Croton* (*Croton sparsiflorus*, *Croton tiglium*, *Croton ciliatoglandulifer*) and *Euphorbia* (*Euphorbia tirucalli*) of Euphorbiaceae. Other plants reporting the presence of PEs are *Sapium indicum*, *Sapium japonicum*, *Jatropha curcas*, *Excoecaria agallocha*, and *Homalanthus nutans* of Euphorbiaceae. The phorbol contains five hydroxyl groups. Ring A is *trans* linked to the ring B (7-membered). Ring B is *trans* linked to ring C (six-membered). Ring C is *cis* linked to the ring D. In case of PEs, two hydroxyl groups on neighbouring carbon atoms of the polycyclic phorbol are esterified to fatty acids. The active phorbols belong to the β-series with *trans* (A/B fusion). α-Series phorbols with *cis* A/B fusion are inactive. Non-esterified phorbol is inactive. Esterification at C12 and C13 is essential for activity. One of the most studied phorbols is

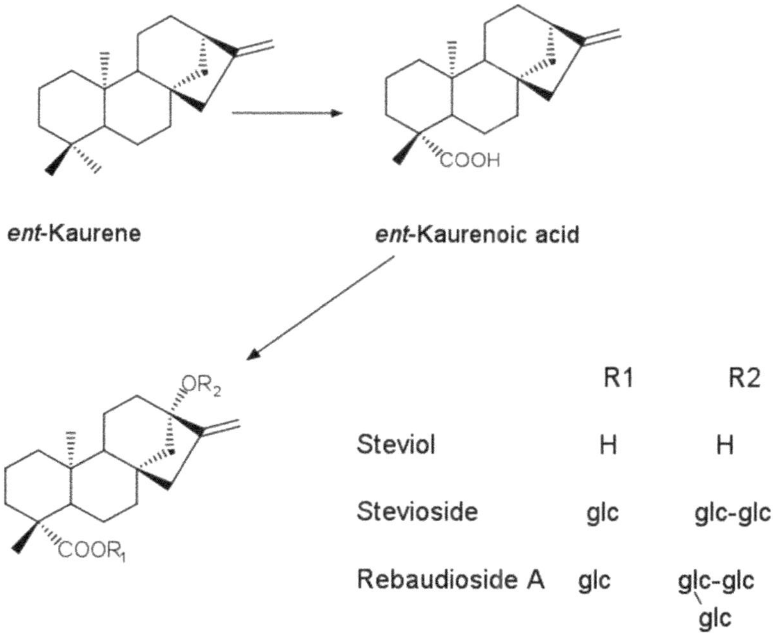

Figure 4.28 Tetracyclic diterpenes of *Stevia rebaudiana*.

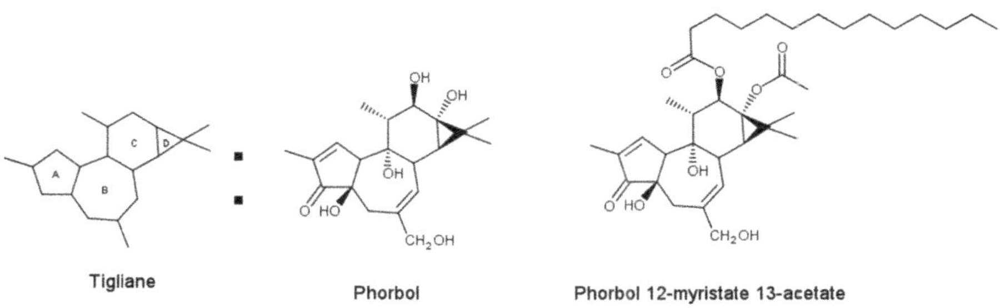

Figure 4.29 Diterpenes with tigliane skeletons.

phorbol 12-myristate 13-acetate. PEs are known for their tumour promoting activities and activate protein kinase C. After more extended incubation period, PEs cause down regulation of the enzyme.[45,46]

The ingenane skeleton of tetracyclic diterpene skeleton is represented by ingenol. Ingenol-3-angelate (Figure 4.30) was obtained from *Euphorbia peplus* (Euphorbiaceae). This diterpenoid is an irritant toxin. Ingenol-3-angelate (ingenol mebutate) has been granted approval by the FDA in 2012 to be used clinically in the treatment of pre-cancerous skin disease actinic keratosis.[47,48]

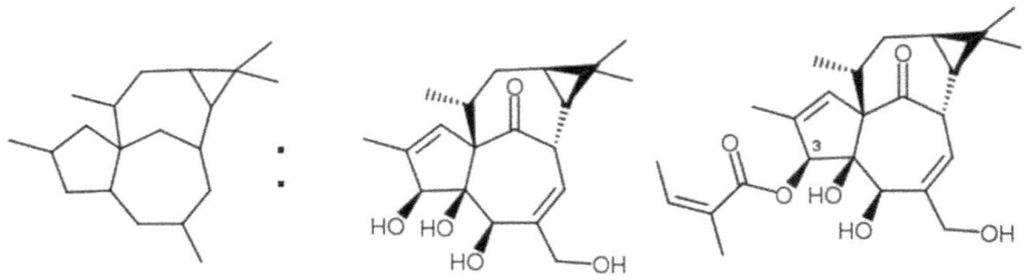

Figure 4.30 Diterpenes with ingenane skeletons.

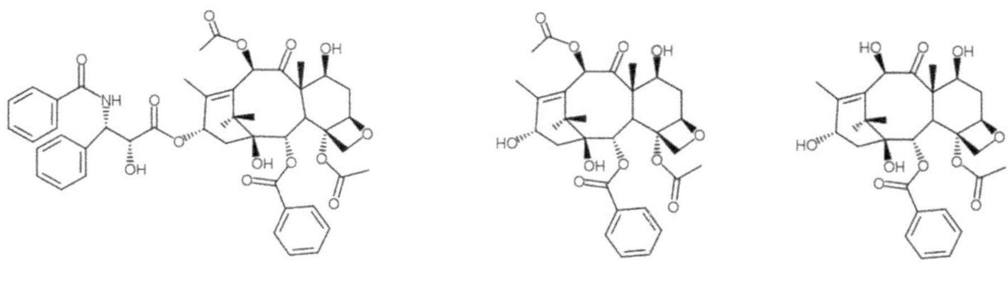

Figure 4.31 Macrocyclic diterpenes.

4.4.5 Macrocyclic Diterpenes

Taxol is one of the most important compounds belonging to the taxane skeleton of diterpenes. The highly oxygenated diterpene taxol (generic name paclitaxel), originally isolated from the bark of *Taxus brevifolia* (Taxaceae), is a highly effective and widely used anti-cancer agent. Taxol was later identified from many other species of *Taxus* such as *Taxus baccata*, *Taxus cuspidata etc.* Paclitaxel and its semisynthetic analogue docetaxel are used in the treatment of ovarian cancer, metastatic breast cancer, lung cancer, gastroesophageal, cervical, prostrate, head and neck cancers. Taxol is difficult and expensive to isolate from the plants and is a threat to the environment. Isolation of 1 kg taxol requires barks from 2000–3000 trees. The synthetic process is also complex and not economic. At present paclitaxel is produced in a semisynthetic way from baccatin III which is again prepared from deacetyl baccatin III (Figure 4.31). Deacetyl baccatin III can be isolated from the renewable needles of *Taxus baccata* in significant quantity (1 g kg^{-1}). The endophytic fungi *Taxomyces andreanae* isolated from inner bark of *Taxus brevifolia* and *Pestalotiopsis microspora* isolated from the inner bark of *Taxus wallichiana* produce taxol and other taxanes *de novo*. But the yield is insignificant. Taxol is sometimes classified as a diterpene alkaloid. However, the nitrogen is not incorporated into the diterpene skeleton. The side chains of taxol are derived from the amino acid phenylalanine which contain the nitrogen atom.[8,49,50]

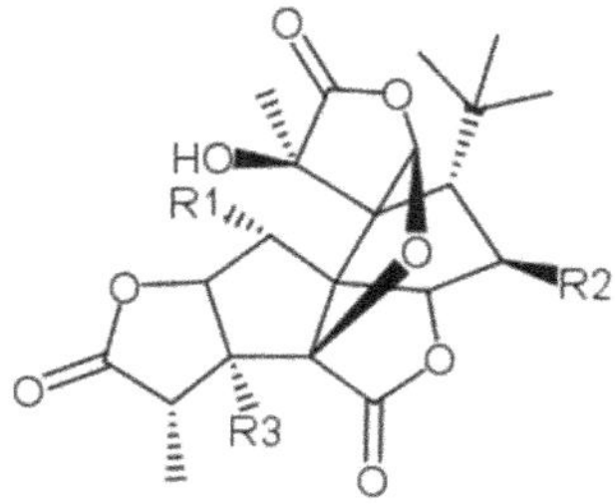

Ginkgolides

	R1	R2	R3
GA	H	H	OH
GB	OH	H	OH
GC	OH	OH	OH
GJ	H	OH	OH
GM	OH	OH	H
GK	OH		H

Figure 4.32 Ginkgolides.

4.4.6 Other Diterpenes

Ginkgolides (Figure 4.32) were isolated from the root bark and leaves of *Ginkgo biloba* (Ginkgoaceae). Ginkgolides are diterpene trilactones consisting of six 5-membered rings (one spiro-nonane carbocyclic ring, three lactones, and a tetrafuran ring) and a tertiary butyl group that is distinctive and rarely found in natural products. Ten ginkgolides (ginkgolide A–Q) have been isolated and identified from *Ginkgo biloba.* All are pharmacologically active. Ginkgolides are active for inhibition of platelet activating factor (PAF) receptors. Ginkgolide B (GB) is the most potent. PAF initiates inflammation and is an important factor in neural injury such as ischemia/reperfusion injury, stroke and multiple sclerosis. Ginkgo extracts are also applied clinically as herbal medicine for treatment of early-stage Alzheimer's disease.[51–55]

Diterpenoid alkaloids are characteristics in the genera of Ranunculaceae specially *Aconitum*, *Delphinium*, *Thalictrum* and *Consolida*. Over 1000 diterpenoid alkaloids have been identified. These alkaloids are very toxic. Diterpenoid alkaloids are divided into three categories C-18, C-19, and C-20.Biosynthetically these alkaloids are derived from GGPP that is cyclized to produce *ent*-copalyl diphosphste (*ent*-CPP). *ent*-CPP leads to the *ent*-kaurene and *ent*-atisane types of diterpenes that are the two major precursors of different categories of diterpenoid alkaloids. The nitrogen atoms are introduced to the structure in the end by transamination. L-Serine is the source of nitrogen. Examples are aconitine from the dried roots of *Aconitum napellus* and atisine from *Delphinium* species (Figure 4.33). Aconitine is extremely toxic. Atisine is less toxic.[56,57]

4.5 Sesterterpenes

Sesterterpenes are pentaprenyl terpenoids whose complex polycyclic structures are derived from the linear precursor geranylfarnesyl diphosphate. These are widely distributed in higher plants. They have also been isolated from fungi, lichens, insects and marine sponges. More than 1400 sesterterpenes have been isolated and identified through 2019. The sesterterpenes

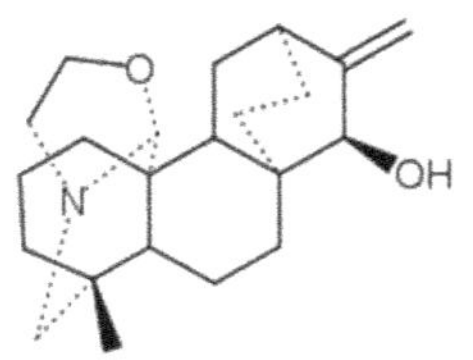

Atisine

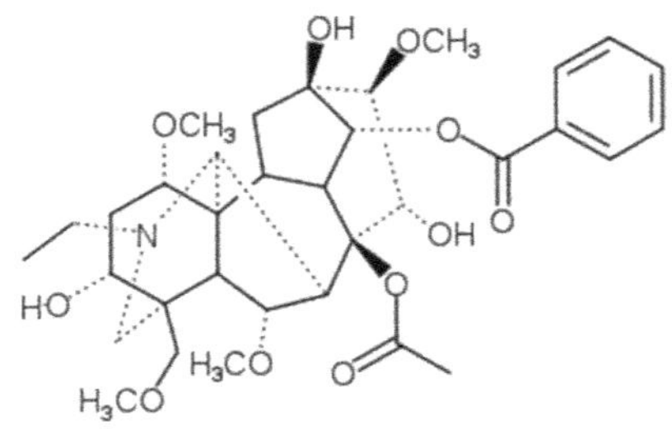

Aconitine

Figure 4.33 Diterpenoid alkaloids.

have potential bioactivities such as inhibition of enzymes, suppression of cancer cell growth, antimicrobial activities, deterrence of herbivorous insects *etc.*[58]

The sesterterpenes may be divided into linear, monocarbocyclic, bicarbocyclic, tricarbocyclic, tetracarbocylic and pentacabocyclic classes. Linear sesterterpenes lack carbocyclic rings. But they may contain furan, pyran or pyrrolidine rings.[58] A few examples of this group of terpenoids isolated from plants and fungi are presented in Table 4.2 and Figure 4.34.

4.6 Triterpenes and Steroids

The triterpenes are one of the largest groups of plant metabolites. The group includes pentacyclic triterpenes and tetracyclic triterpenes (steroids), limonoids, and quassinoids. Other organisms like bacteria synthesize the simple triterpene hopene. Triterpenes occur widely as C-30 aglycones or as esters or glycosides. Over 20 000 such compounds with approximately 200 structural skeletons are recorded. Simple triterpenes are present as a component of the leaf surface waxes and membranes. Most of the complex triterpenoids are produced as defensive chemicals to protect the plants against pathogens and herbivores. Humans use many of these metabolites as pharmaceuticals, food, cosmetics, and agrochemicals.[59,60]

Triterpenes are biosynthesized after cyclization of 2,3-oxidosqualene, derived from squalene, by the enzymes oxidosqualene cyclases. Oxidosqualene cyclases are usually divided into two categories based on the structure formed during substrate binding and folding. The P group forms a protosteryl cation through a chair–boat–chair configuration and generates a tetracyclic triterpene (6-6-6-5) skeleton of steroids. The D group of oxidosqualene cyclases forms a dammarenyl cation through chair–chair–chair configuration. Dammarenyl type tetracyclic triterpenes ginsenosides are produced from the dammarenyl cation. The dammarenyl cation by ring expansion generates the baccharenyl cation. The pentacyclic triterpene skeletons (6-6-6-6-5 or 6-6-6-6-6) arise from the baccharenyl cation. Further elaboration and modification of the triterpene skeleton, such as acylation and/or glycosylation take place by the cytochrome P450 enzymes (P450s). Hopanes are cyclized from squalene by

Table 4.2 Different classes of sesterterpenes.

Sesterterpenoids	Organisms
Linear	
Coleifolides A and B	Parts of the herb *Scutellaria coleifolia* (Lamiaceae)
Acetylated sesterterpene	Straw of *Oryza sativa* (Poaceae)
Monocarbocyclic	Mostly from sponges
Bicarbocyclic	
Leucosceptroids E–N	Flowers of *Leucosceptrum canum* (Lamiaceae)
Colquhounoids A–C	From peltate glandular trichomes of *Colquhounia coccinea* var. *mollis* (Lamiaceae)
Lachnocalyxolides A–C	Aerial parts of *Salvia lachnocalyx* (Lamiaceae)
Salvidominicolides A and B	From *Salvia dominica* (Lamiaceae)
Leucosceptroid O	From the flowers of *Leucosceptrum canum* (Lamiaceae)
Goniocarpic acid	Leaves of *Serjania goniocarpa* (Sapindaceae)
Emericellenes A–E	Endophytic fungal strain *Emericella* sp. AST0036 from leaf tissue of *Astragalus Lentiginosus* (Fabaceae)
24-α-D-Glucosyl-(−)-terpestacin	Endophytic fungus *Alternaria alternate* obtained from roots of *Ceratostigma griffithii* (Plumbaginaceae)
Tricarbocyclic	
Ophiobolin O and 16-*epi*-ophiobolin O	Marine fungus *Aspergillus* sp.
Ophiobolin A	Phytopathogenic fungus *Bipolaris oryzae*
Ophiobolin U	A fungal strain of *Aspergillus insuetus*
Asperophiobolins A–D	Endophytic strain of *Aspergillus* sp. isolated from the mangrove plant *Kandelia candel* (Rhizophoraceae)
Bipolarolides E–G (nitrogenous ophiobolin analogues)	Fungus *Bipolaris* sp.
Tetracarbocylic	
Nambiscalarane	Bioluminescent fungus *Neonothopanus nambi*
Sesteralterin	*Alternaria alternata* strain (k21-1) isolated from the surface of marine red alga *Lomentaria hakodatensis*
Pentacarbocyclic	
Asperterpenoid A	Mangrove endophytic fungus *Aspergillus* sp.
Aspterpenacids A	Mangrove endophytic fungus *Aspergillus terreus* H010
Aspterpenacid C	*Swertia bimaculata* (Gentianaceae)

the enzyme squalene cyclase. The *seco*-triterpenoid α-onocerin is biosynthesized from bis-oxidosqualene, instead of 2,3-oxidosqualene (Figure 4.35).[59–62]

4.6.1 Hopanes

Fossil hopanes of cyanobacterial origin are recorded. Hopanoids with skeletal similarity with that of hopane have been considered as "molecular fossils" for ancient bacteria. 2-Methylhopane hydrocarbons have been detected not

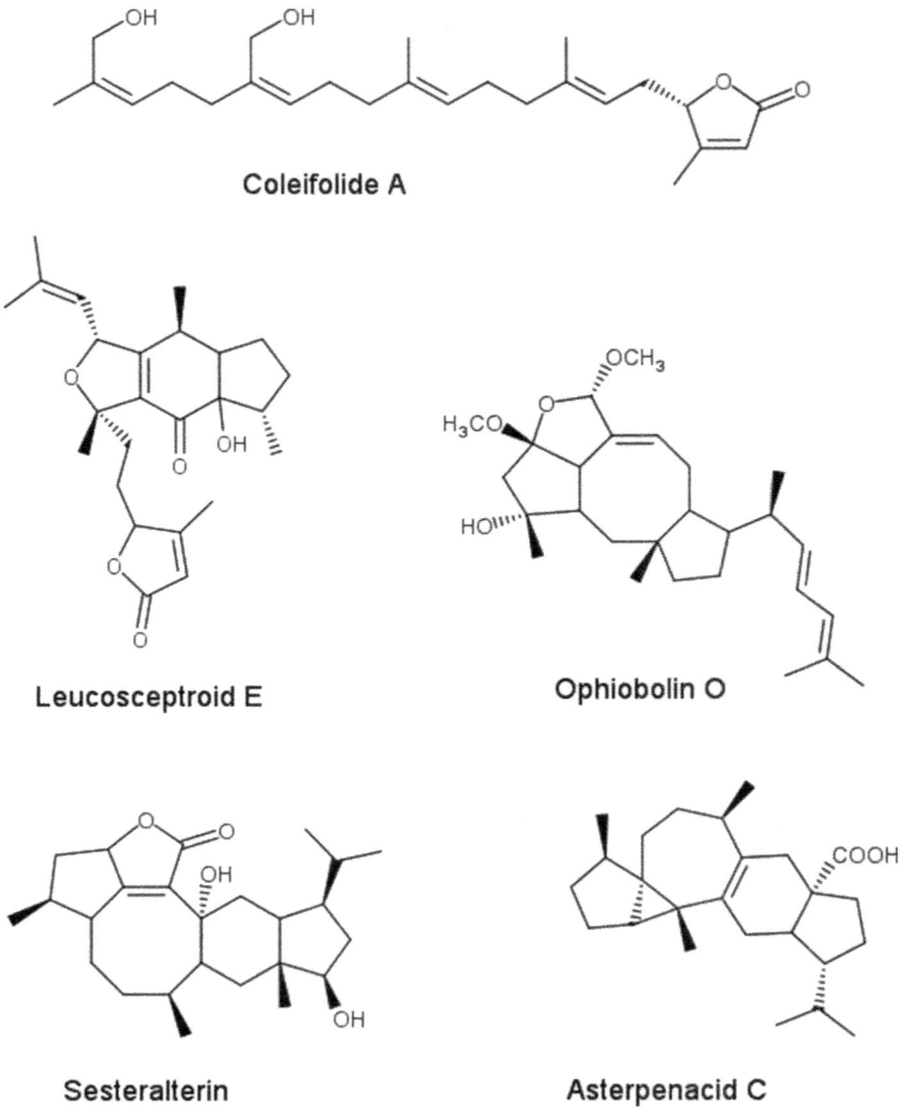

Figure 4.34 Sesterterpenes.

only in ancient marine and non-marine sedimentary rocks, but also in cyanobacteria and in the anoxygenic phototroph *Rhodopseudomonas palustris* in the modern environment. Hopanoids are thought to play a role in membrane integrity and stress tolerance in bacteria.[63]

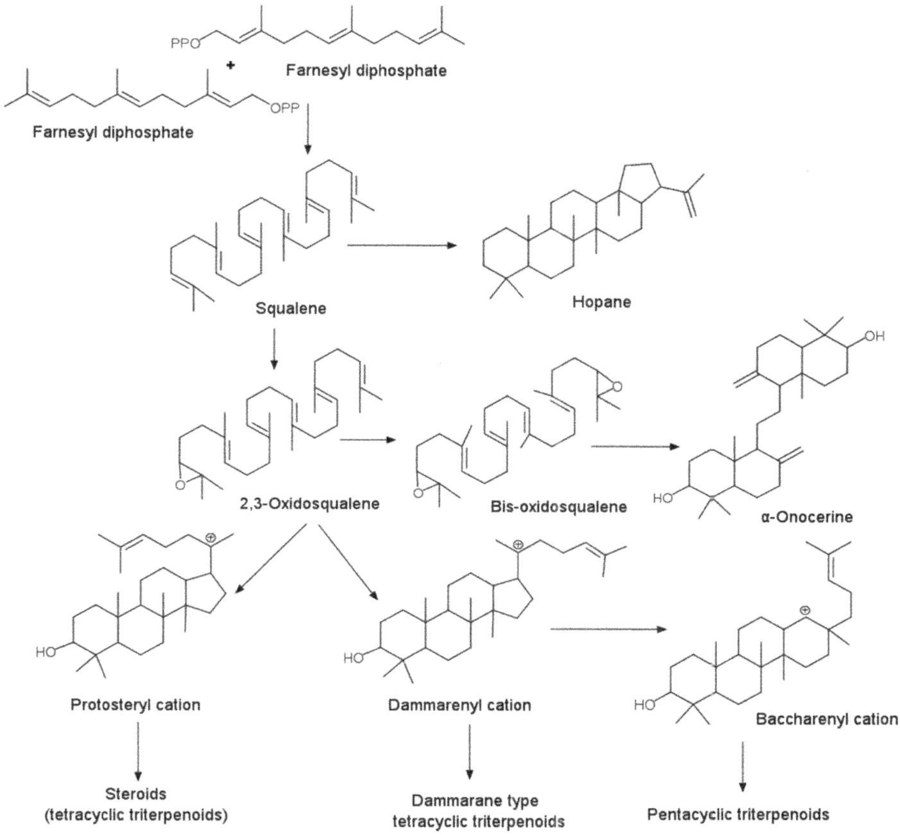

Figure 4.35 Biosynthetic origin of triterpenes.

4.6.2 Tetracyclic Dammarane Type Triterpenes

Asian **ginseng** is a medicinal herb used in China for thousands of years. Ginseng is derived from the roots and rhizomes of *Panax ginseng* (Family Araliaceae) and is used as energy booster. Ginseng has many other biological activities such as antioxidant, antitumour and anti-inflammatory. It acts against stress and has potential neuroprotective properties. More than 100 ginsenosides have been identified. The triterpenoid **saponins ginsenosides** were identified as the main constituents of ginseng. Ginsenosides are divided into three types belonging to the dammarane type and the oleanane type of triterpenoids. Tetracyclic dammarane type can be divided into protopanaxadiol group (*e.g.*, Rb1, Rb2, Rb3, Rc, Rd, Rh2, Rg3 *etc.*) and protopanaxatriol group (*e.g.*, Re, Rg1, Rg2, Rh1) (Figure 4.36). Oleanolic acid group (*e.g.*, Ro) belongs to the pentacyclic triterpenoid oleanane type.

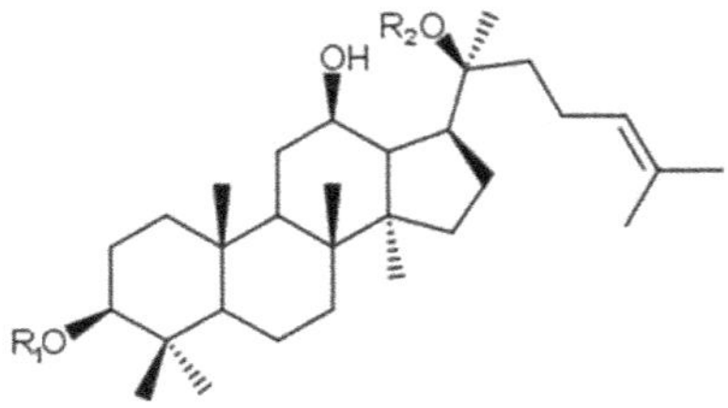

	R_1	R_2
Protopanaxadiol	H	H
Ginsenosides		
Rb1	Glc(2→1)Glc	Glc(6→1)Glc
Rb2	Glc(2→1)Glc	Glc(6→1)Ara(p)
Rc	Glc(2→1)Glc	Glc(6→1)Ara(f)
Rd	Glc(2→1)Glc	Glc
Rg3	Glc(2→1)Glc	H
Rh2	Glc	H

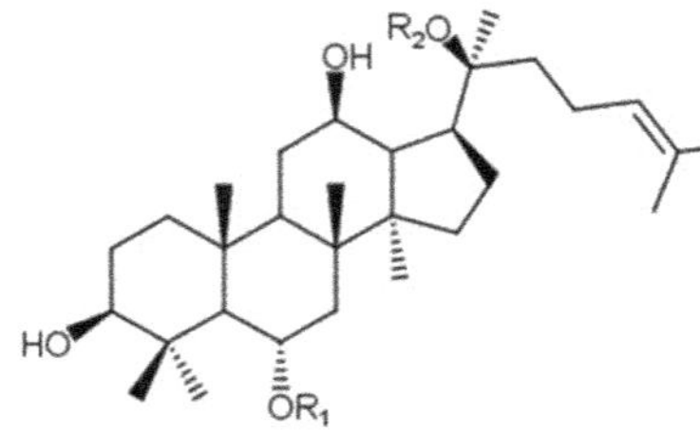

	R_1	R_2
Protopanaxatriol	H	H
Ginsenosides		
Re	Glc(2→1)Rha	Glc
Rg1	Glc	Glc
Rg2	Glc(2→1)Rha	H
Rh1	Glc	H

Figure 4.36 Dammarane type saponins: ginsenosides.

Dammarane type ginsenosides dominate over oleanane type in amount and structural diversities. Many of the ginsenosides are reported to have immunomodulatory and anticancer properties. Ginsenoside Rg3 is the major active component in an anticancer drug in China. Both protopanaxadiol type and protopanaxatriol type ginsenosides are derived from dammarenediol-II. Dammarenidiols are also found in Dammar resin from *Balanocarpus heimii* (Dipterocarpaceae).

American ginseng (*Panax quinquefolius*), growing in North America, possesses different properties in clinical applications as well as in many of the ginsenosides identified.[64–66]

Dammarane type saponins *e.g.*, **bacoside A** and **bacopaside X** (Figure 4.37) are the major active principles of *Bacopa monnieri* (Plantaginaceae) traditionally used as neural tonic and memory enhancer in Ayurvedic medicine. The aglycone is **jujubogenin**.[67]

4.6.3 Pentacyclic Triterpenes and Saponins

There are different types of pentacyclic triterpenes produced from the baccharenyl cation. Biosynthetic origins of some of them are mentioned in Figure 4.38. Pentacyclic triterpenes are found in fruit peels of apple, mango, green pepper, strawberries, guava, and in some aromatic herbs like basil and rosemary.[59,68–71]

4.6.3.1 Lupane-type Triterpenes

Lupeol (Figure 4.38) is found in plants in resins and waxes and is a precursor of specialized triterpenes involved in plant defense and development. Lupeol is involved in nodule formation in *Lotus japonicus* (Fabaceae). **Betulinic acid**

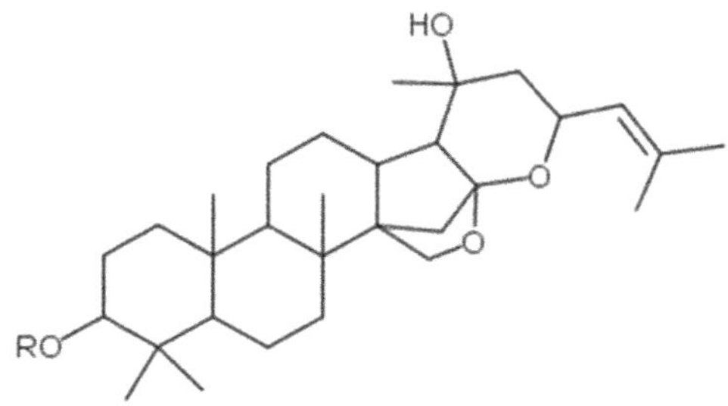

	R
Jujubogenin	H
Bacoside A3	α-L-Arabinofuranosyl(1→2)-[β-D-glucopyranosyl-(1→3)]-β-D-glucopyranosyl
Bacopaside X	α-L-Arabinofuranosyl(1→2)-[β-D-glucopyranosyl-(1→3)-α-L-arabinopyranosyl

Figure 4.37 Dammarane type saponins: bacoside and bacopaside.

is a lupane type pentacyclic triterpene and an oxidized form of lupeol. The alcohol **betulin** is a major component of the bark of *Betula alba* (Betulaceae), commonly known as birch. Both betulin and betulinic acid (Figure 4.38) are of interest in anti-HIV therapy. These triterpenes inhibit viral growth in a way different from other HIV drugs. Bevirimat is a semisynthetic ester and a promising anti-HIV component undergoing clinical trial.[72] Betulin and betulinic acid have potential anticancer properties.[73]

4.6.3.2 Oleanane-type Triterpenes

β-Amyrin, erythrodiol, oleanolic acid and **maslinic acid** are few common examples of oleanane type pentacyclic triterpenes. β-Amyrin (Figure 4.38) is also found in plants in resins and waxes. Maslinic acid (Figure 4.38) obtained from *Olea europaea* (Oleaceae) exhibited promising antidiabetic effect. Specialized triterpenoid saponins of glycyrrhetic acid, quillaic acid and soyasapogenol are derived from β-amyrin. **Glycyrrhetic acid** or glycyrrhetinic acid is the aglycone of the saponin **glycyrrhizic acid** (Figure 4.39) found in the roots and stolons of the medicinal plant liquorice. It is obtained from several species of *Glycyrrhiza* such as *Glycyrrhiza glabra*, *Glycyrrhiza inflata*, *Glycyrrhiza uralensis* (Fabaceae). **Glycyrrhizin** is a mixture of potassium and calcium salts of glycyrrhizic acid. The sweet taste of liquorice is due to glycyrrhizin. Glycyrrhizic acid, glycyrrhetinic acid, and 18-β-glycyrrhetic acid were developed in China and Japan as drugs for liver diseases.[15]

Soyasapogenol (Figure 4.39) has been isolated from *Glycine max* (Fabaceae), commonly known as soyabean. Soyabean sapogenins are divided into

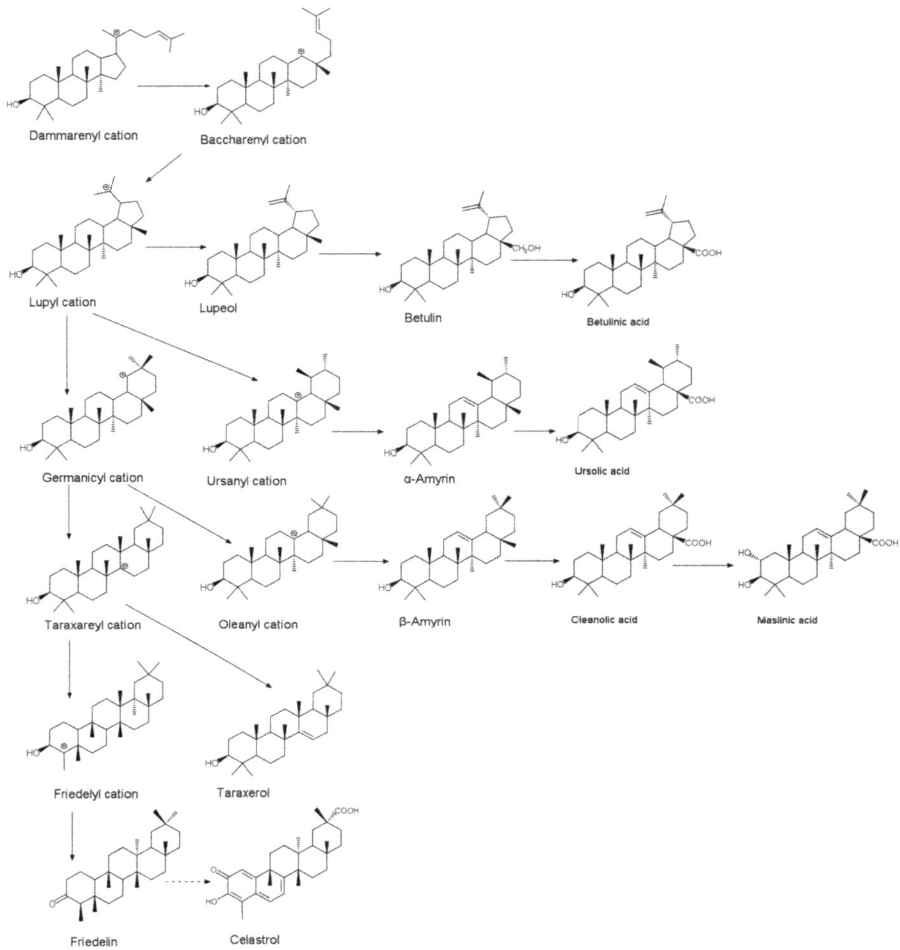

Figure 4.38 Biosynthesis of pentacyclic triterpenes.

two groups: group A and group B. Group A acetyl saponins contain **soyasapogenol A** as the aglycone. The undesirable bitter and astringent tastes of foods prepared from soybean are due to Group A saponins. Group B saponins have **soyasapogenol B** as the aglycone. These two soyasapogenols are produced from β-amyrin by a series of oxidative reactions. The saponin is soyasaponin.[74]

β-Amyrin is also the precursor of **hederagenin glycosides**. The aglycone hederagenin (Figure 4.39) has insect feeding deterrent activity.[75] Hederagenin monoglucoside, obtained from *Medicago sativa* (Fabaceae), acts as plant growth inhibitor. *Gymnema sylvestre* (Apocynaceae) is an Ayurvedic medicine used to treat diabetes. **Gymnemic acids** (Figure 4.39) improve glucose tolerance and decrease blood glucose level in diabetic patients.[76] Gymnemic acids

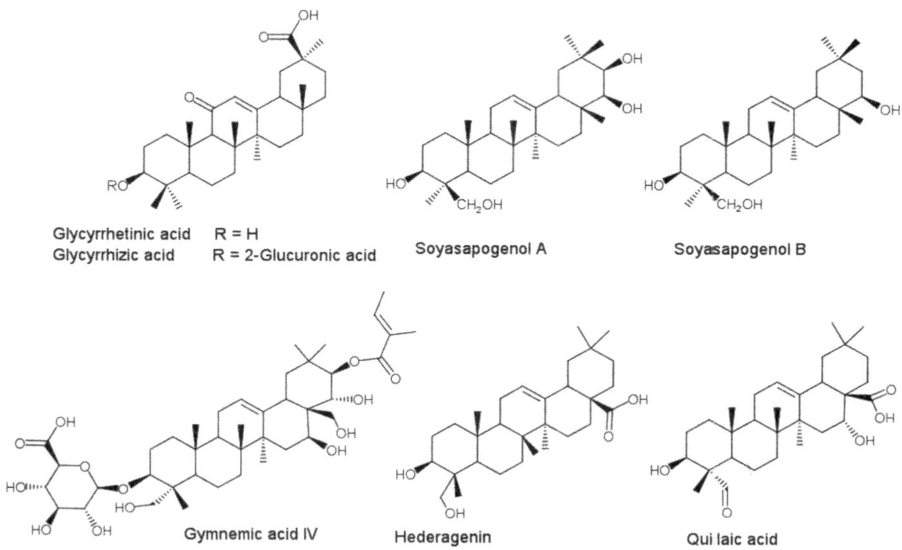

Figure 4.39 Oleanane-type triterpenes.

suppress taste responses to various sweet substances in humans.[77] Gymnemic acids are also oleanane type pentacyclic triterpenoid saponins.

Quillaja saponaria (Quillajaceae) saponins are important for industrial applications. Nearly 100 saponins have been identified from *Quillaja saponaria*. In response to different antigens, the saponins alone or incorporated into immunostimulating complexes modulate immune system by increasing antigen uptake, stimulating cytotoxic T lymphocyte production and cytokines. Saponin natural product QS-21 (Figure 2.13), as immunological adjuvant, helps stimulate the immune response to co-administered antigens. However clinical use has certain drawbacks.[78–80] **Quillaic acid** (Figure 4.39) and the saponins belong to β-amyrin related group.

4.6.3.3 *Ursane-type Triterpenes*

The plant *Centella asiatica* (Family Apiaceae), a traditional Ayurvedic medicine, contains the triterpene glycosides **madecassoside** and **asiaticoside** (Figure 2.13) and their aglycone **asiatic acid** (Figure 4.40). The triterpenes suppress inflammation and reduce blood glucose levels. The plant also contains ursolic acid.[81] Resins from *Boswellia serrata* and other species of *Boswellia* (Family Burseraceae) are used in the perfumery industry. **Boswellic acids** generally occur as α-isomers (oleanane type) and β-isomers (ursane type). The constituents **boswellic acid** and 3-*O*-**acetyl-11-keto-β-boswellic acid** (Figure 4.40) have potential anti-inflammatory and anti-tumour properties.[82,83]

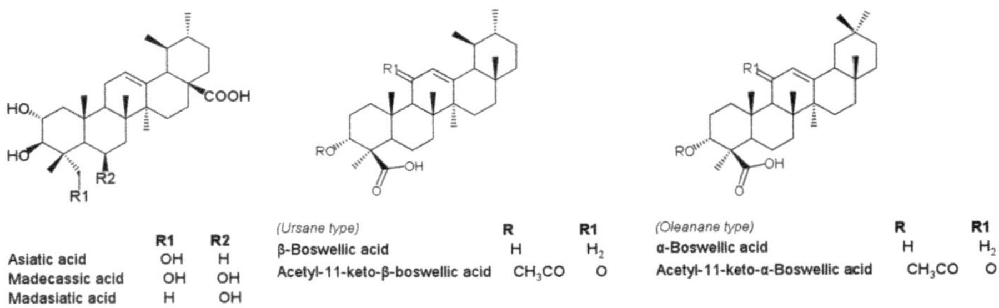

Figure 4.40 Ursane type triterpenes.

4.6.3.4 *Friedelane-type Triterpene*

Celastrol (also named tripterine), a friedelane type pentacyclic triterpene isolated from the plant *Tripterygium wilfordii* (Celastraceae), is one of the potential therapeutic agents effective against liver, lung, breast, and ovarian cancer.[84,85]

4.6.4 *seco*-Terpenes

The polycyclic triterpenes can lose their endocyclic C–C bond due to modification of the cyclization cascade or due to post-cyclization ring cleavage by late-stage oxidative modifications, resulting in the formation of *seco*-terpenes (Figure 4.41). One example is **α-onocerin** (ring C 8,14 bond missing) which was detected from two unrelated sources, the roots of the medicinal plant *Ononis spinosa* (Fabaceae) and the club moss *Lycopodium clavatum* (Lycopodiaceae). **Lansic acid**, a bicyclic triterpene reported from the fruit peels of *Lansium domesticum* (Meliaceae) is another example. Here carbons 3,4 in ring A and 21,22 in ring D are cleaved. Fragmentation of carbons 8,14 is associated with the formation of **β-*seco*-amyrin** and **α-*seco*-amyrin.** In case of **camelliol A** from *Camellia sasanqua* (Theaceae), cleavages of the 8,14 bond and 9,10 bond of C-13 result in formation of an oleanane cation. Other examples are **sootepin A** (3,4-*seco*-cycloartane carboxylates) from apical buds of *Gardenia sootepensis* (Rubiaceae); **dammarenolic acid** (a 3,4-*seco*-dammarane), a major acid reported from various trees of Dipterocarpaceae, plants of Meliaceae such as *Cabralea eichleriana* and *Aglaia ignea* and wax of galls of *Aglaia ignea*; and ***seco*-moronic acid** (a 3,4-*seco*-oleanane) from *Phoradendron reichenbachianum* (Santalaceae) *etc.*[86]

4.6.5 *nor-seco*-Terpenoids

Ring opening associated with carbon loss and fragmentation of the polycyclic structure of triterpenes results in the formation of *nor-seco*-triterpenoids. Limonoids and quassinoids are examples of *nor-seco*-triterpenoids.[86]

Limonoids are confined mainly to the families Meliaceae and Rutaceae. Limonoids are highly oxygenated modified terpenoids. They contain or are

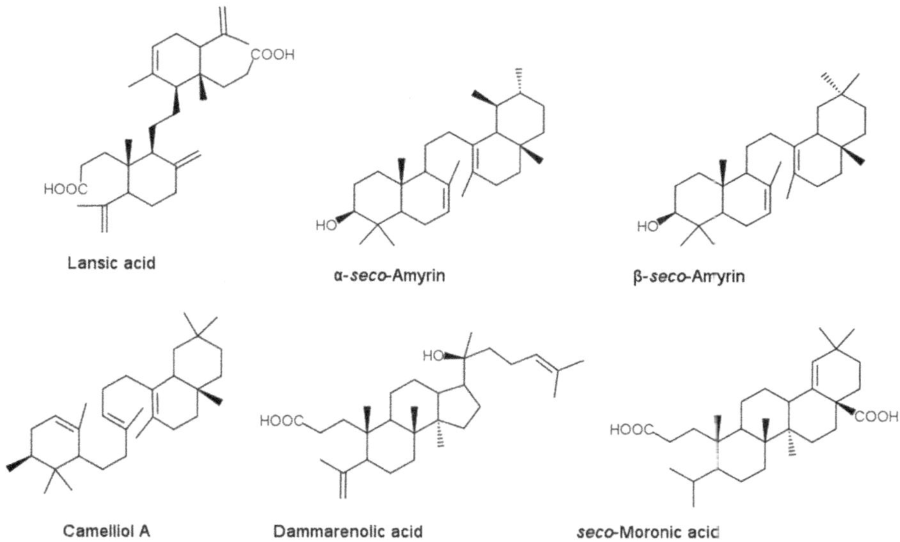

Figure 4.41 *seco*-Terpenes.

derived from a precursor 4,4,8-trimethyl-17-furanylsteroid skeleton. The citrus limonoids contain a furan ring attached to the D-ring at C-17. They also contain oxygen containing functional groups at C-3, C-4, C-7, C-16, and C-17. In Meliaceae the structures are more complex with remarkably high degree of oxidation. Limonoids from Meliaceae have potential activity against insect pests without harming the environment.[87] **Azadirachtin** (Figure 4.42) found in the seeds of *Azadirachta indica* (Meliaceae) is a powerful feeding deterrent for several phytophagus insects. The compound, at parts per million concentrations, inhibits growth and development of all insects ingesting it. It has activities against many other arthropods, annelids, and nematodes. But, it has little or no toxicity for vertebrates. **Nimbin** (Figure 4.42) has almost no antifeedant activity. The activity of **salannin** (Figure 4.42) is small in comparison to that of azadirachtin. **Marrangin** (Figure 4.42) is the insecticidal component of *Azadirachta excelsa*. Meliacarpin from *Melia azedarach* (Meliaceae) has similar structure. But the extracts are toxic to mammals.[88]

Quassinoids are bitter constituents of different species of Simaroubaceae. These are chemically modified triterpenes generated due to extensive oxidative biodegradation. According to their basic skeletons, quassinoids are classed into five types C-18, C-19, C-20, C-22, and C-25. Quassia, the wood of *Picrasma excelsa* (Simaroubaceae) is used as a bitter tonic, as an insecticide, and as an enema for expulsion of worms. *Quassia amara* (Simaroubaceae) is traditionally used in South America for its antiamoebic, antimalarial activity. Both these plants contain **quassin**, and **neoquassin**[89–91] (Figure 4.43).

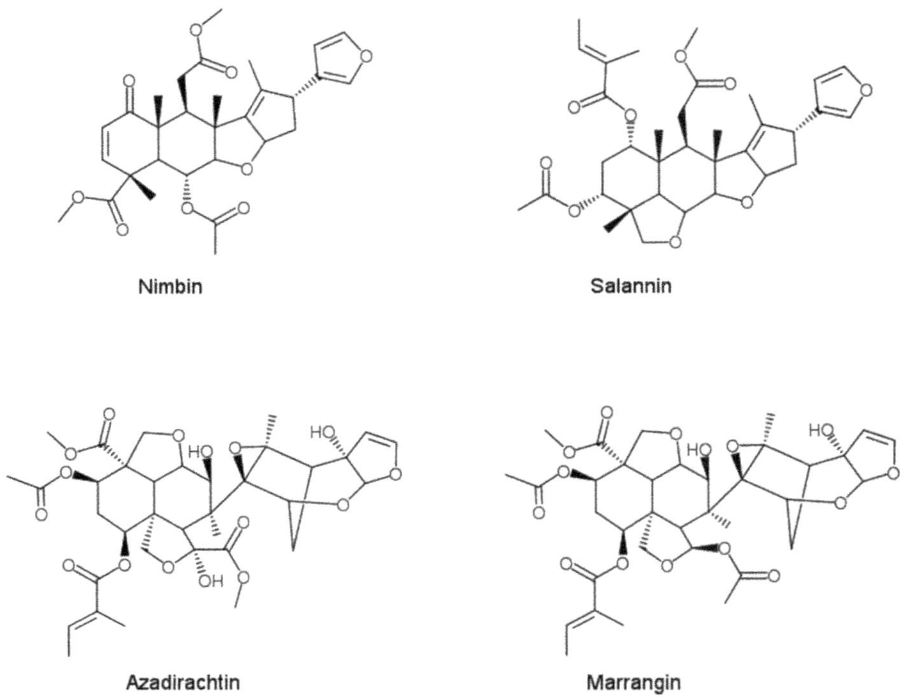

Figure 4.42 Limonoids.

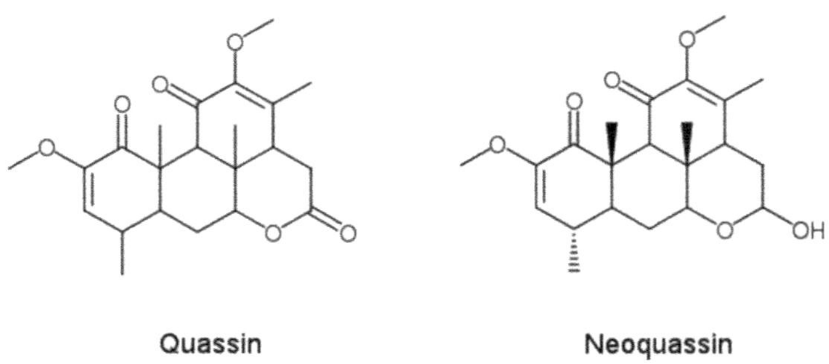

Figure 4.43 Quassinoids.

4.6.6 Steroids

Steroids are tetracyclic hydrocarbons containing a cyclopentanoperhydrophenanthrene nucleus.[92] These natural products are widely distributed in both plant and animal kingdoms. The steroids that naturally occur in animals can also be found naturally in plants. The simplest steroid nucleus consists of three six membered cyclohexane rings (designated as rings A, B, C) and a five membered cyclopentane ring (designated as ring D). Methyl groups are normally present at C-10 and C-13 and a hydroxyl group at C-3.An alkyl side chain may be present at C-17. Based on the number of hydrocarbons, different steroid nomenclature is used namely oestrane, androstane, pregnane, cholane, cholestane, ergostane, campestane, stigmastane, poriferastane, lanostane and cycloartane[8,15] (Figure 4.44).

In plants steroids belong to different classes such as phytosterols, steroidal sapogenins, steroidal alkaloids, cardiac glycosides, withanolides, cucurbitacins, brassinolides and phytoecdysteroids. In addition, progesterone, testosterone, and other testosterone derivatives as well as oestrogens have been reported from many plants.

Steroids are synthesized *via* the mevalonic acid pathway. Tail to tail linkage of two molecules of farnesyl diphosphate produces squalene. The cyclization of 2,3-oxidosqualene, formed from squalene, leads to the formation of steroidal products. Squalene is converted to 2,3-oxidosqualene by the enzyme squalene monooxygenase or epoxidase. Cyclization of 2,3-oxidosqualene is catalyzed by the enzyme known as oxidosqualene cyclase. 2,3-Oxidosqualene,

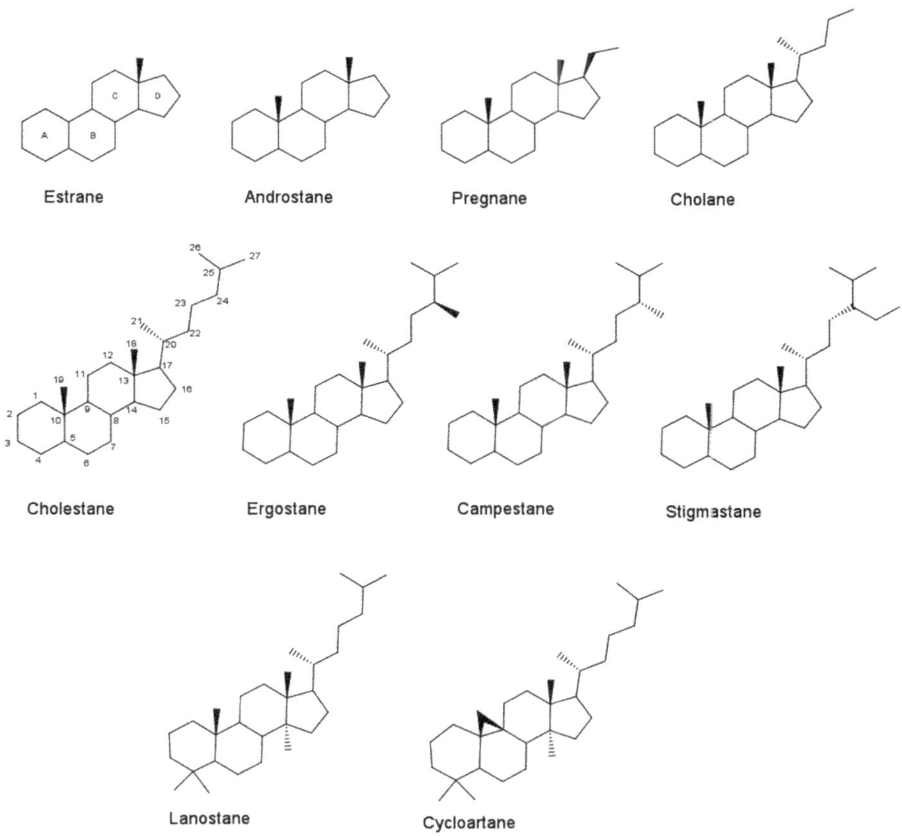

Figure 4.44 Steroid nomenclature based on the number of hydrocarbons.

via the chair–boat–chair (CBC) conformation (protosteryl cation), generates **lanosterol** (in fungi and animals), **cycloartenol** (in plants), and **cucurbitadienol** (in plants) by the enzymes lanosterol synthase, cycloartenol synthase and cucurbitadienol synthase, respectively.[59] Other steroids are produced following diverse steps (Figure 4.45).

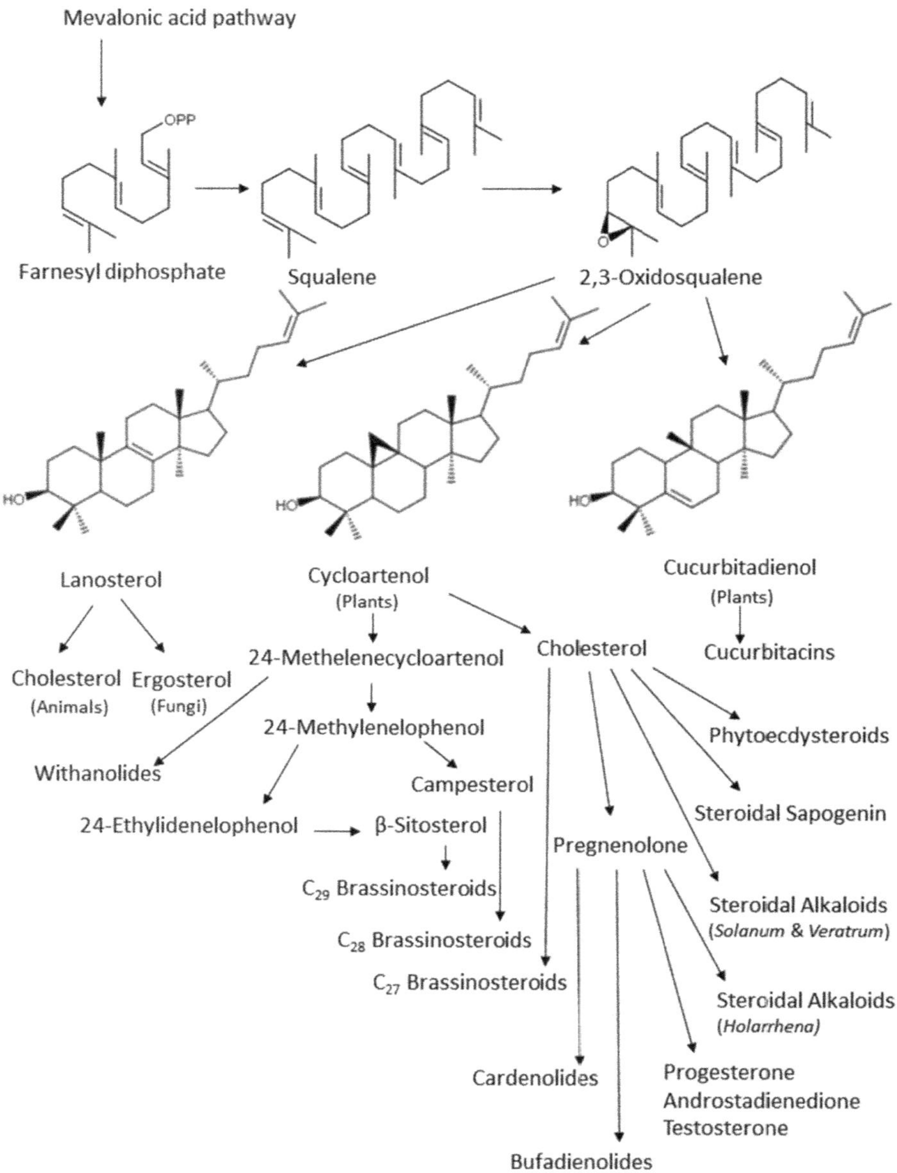

Figure 4.45 Biosynthetic origin of different steroids.

4.6.6.1 Sterols

Plant sterols, also known as phytosterols, are steroid alcohols. Sterols are ubiquitously present in eukaryotic organisms. Over 250 different sterols have been reported from different plants and marine sources. They have 4 steroid rings (A, B, C, D) (*trans* ring structures) with two methyl groups (C-18, C-19) and a long flexible side chain at the C-17 position. The free sterols have a 3-hydroxyl group.[93] Sterols in plants are biosynthesized (Figure 4.46) from cycloartenol, the first polycyclic steroid intermediate. The fungal sterol **ergosterol** is produced from lanosterol. Sterol side chain reductase 2 (SSR2) converts cycloartenol to the cholesterol branch. The methyl and ethyl substituents at the C-24 position are not derived from acetate mevalonate. These alkyl groups are added to cycloartenol by sterol methyl transferase 1 (SMT1) to produce plant sterols.[94,95]

Based on the structure and biosynthesis, phytosterols may be divided into 3 groups. These are (1) 4-desmethyl sterols, (2) 4α-monomethyl sterols, and (3) 4,4-dimethyl sterols. Depending on the position and number of double bonds in B-ring, 4-desmethyl sterols may be Δ^5 sterols (examples: **campesterol, sitosterol**), Δ^7 sterols (example: **avenasterol**), and $\Delta^{5,7}$ sterols

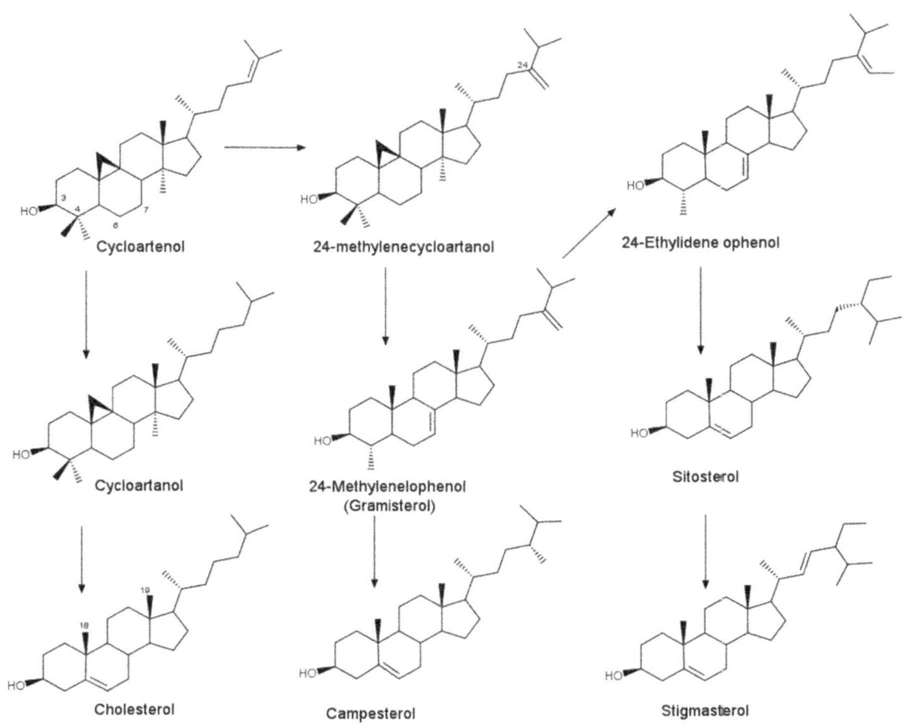

Figure 4.46 Biosynthesis of sterols.

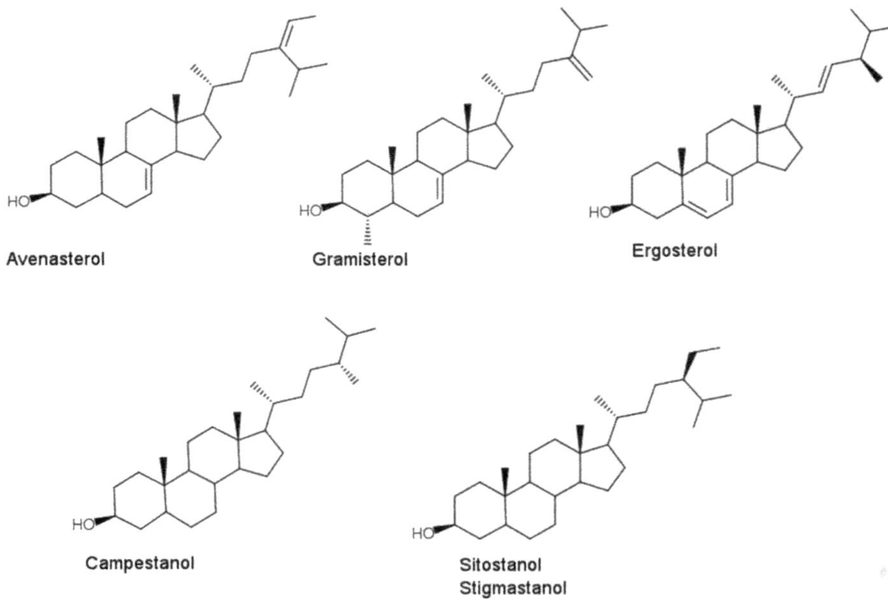

Figure 4.47 Other sterols and stanols.

(example: **ergosterol**). **Gramisterol** is an example of a 4α-monomethyl sterol. Cycloartenol is a 4,4-dimethyl sterol (Figures 4.46 and 4.47) that produces most of the phytosterols. Phytosterols can be converted into **phytostanols** (C-5 saturated) by chemical hydrogenation. Small amounts of such stanols such as **sitostanol, stigmastanol** and **campestanol** occur in wheat and rye.[93]

The free sterols may be esterified with fatty acids or phenolic acids at the 3- position to form steryl esters. It may also be β-linked to the 1′-position of a carbohydrate to form ethyl steryl glycosides.[93]

Sterols are structural components of the membrane lipid bilayers in plants and contribute to maintaining membrane fluidity and permeability. Sterols also affect the activity of membrane proteins such as enzymes, ion channels, receptors, and components of signal transduction pathways like ATPases. Sterols also regulate growth and development to stress resistance, such as UV radiation, cold and drought stress.[93,94]

Dietary phytosterols and stanols inhibit intestinal cholesterol absorption and lower low-density lipoprotein (LDL) cholesterol level.[93] Ergosterol in fungi (produced *via* lanosterol) and 7-dehydrocholesterol (produced *via* lanosterol in animals and *via* cycloartenol in plants) are converted into vitamins D_2 (ergocalciferol) and D_3 (cholecalciferol) (Figure 4.48), respectively, under UV radiation followed by human metabolism. So ergosterol and 7-dehydrocholesterol are known as pro-vitamin Ds.[96]

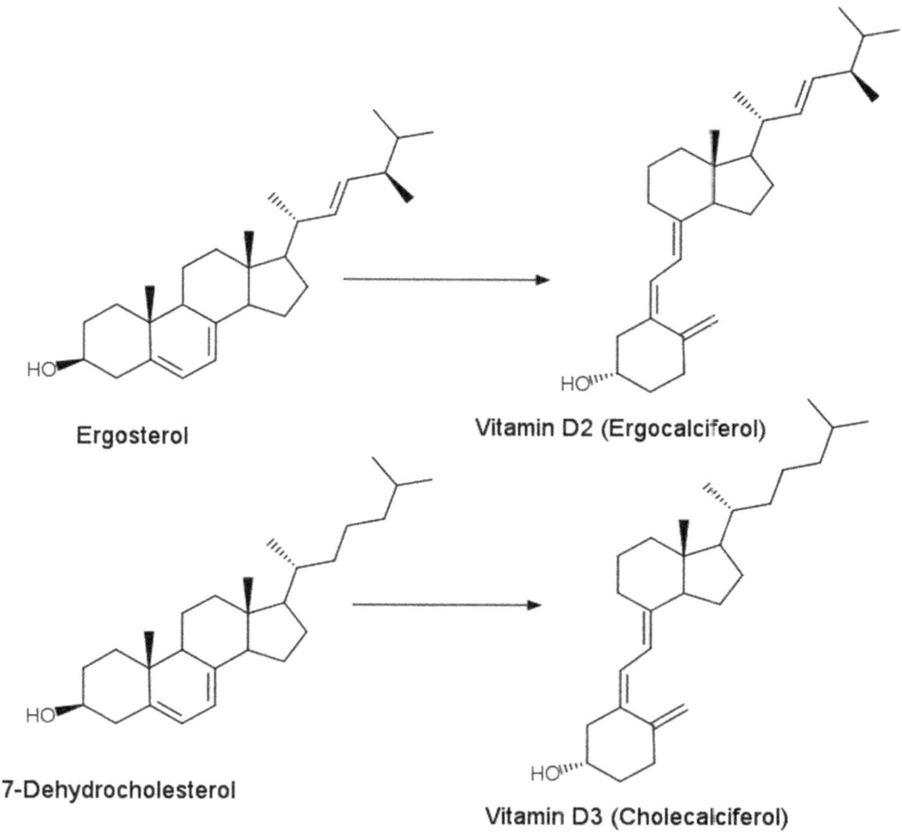

Figure 4.48 Two forms of Vitamin D.

4.6.6.2 Steroidal Sapogenins

Steroidal sapogenins are found in plants in the form of glycosides known as saponins. The steroidal saponins have spiroketal configuration at C-22. Steroidal sapogenins (steroidal aglycones) occur in different plants (Table 4.3). The sapogenins may exist either as 3-β-hydroxy or 2,3-β-dihydroxy forms. The sugar moiety is usually attached at position 3. Again, there may be non-ketonic or a few 12-keto forms. Diosgenin, one of the commercially important steroidal sapogenins, exists as the glycoside dioscin in plants. Acid hydrolysis of dioscin yields the aglycone sapogenin diosgenin. Yamogenin is a C-25 isomer of diosgenin. Different steroidal sapogenins, based on the number and position of hydroxyl groups and the presence of carbonyl group,[97,98] are shown in Figure 4.49 and Table 4.3.

The steroidal sapogenins are biosynthesized from cholesterol. The spiroketal function is derived from the side chain of cholesterol by oxygenation

Table 4.3 Different types of steroidal aglycones.

Monohydroxy, non- ketonic		
Diosgenin	Δ^5-3-β-Hydroxy	Different species of *Dioscorea* (Dioscoreaceae) *e.g.*, *Dioscorea collettii, Dioscorea composita, Dioscorea deltoidea, Dioscorea floribunda, Dioscorea mexicana, Dioscorea nipponica, Dioscorea prazeri, Dioscorea tokoro, Balanites aegyptiaca* (seed kernel) (Zygophyllaceae), *Costus speciosus* (Costaceae), *Kallstroemia pubescens* (Zygophyllaceae), *Solanum nigrum* (Solanaceae), *Trigonella foenum-graecum* (Fabaceae), *Trillium kamtschaticum* (Melanthiaceae)
Yamogenin	Δ^5-3-β-Hydroxy	*Dioscorea* spp. (Dioscoreaceae), *Balanites aegyptiaca* (Zygophyllaceae), *Trigonella foenum-graecum* (Fabaceae)
Sarsasapogenin	3-β-Hydroxy	*Yucca* spp. (Asparagaceae), *Smilax* spp. (Smilaceae)
Smilagenin	3-β-Hydroxy	Roots of *Smilax* spp. (commonly known as Sarsaparilla) *e.g.*, *Smilax aristolochiifolia, Smilax regelii, Smilax febrifuga*
Tigogenin	3-β-Hydroxy	*Digitalis purpurea* (Plantaginaceae)
Neotigogenin	3-β-Hydroxy	*Digitalis purpurea*
Monohydroxy, ketonic		
Hecogenin	12-Keto-3-β-hydroxy	*Agave sisalana* (Asparagaceae), *Agave americana*
Dihydroxy, non-ketonic		
Yuccagenin	Δ^5-2,3-β-Dihydroxy	*Allium fistulosum* (Amaryllidaceae), *Agave* spp.
Gitogenin	2,3-β-Dihydroxy	*Agave* spp.
Dihydroxy, ketonic		
Manogenin	12-Keto-2,3-β-dihydroxy	*Agave* spp.
Kammogenin	Δ^5-12-Keto-2,3-β-dihydroxy	*Agave* spp.

reactions and producing a ketone structure at C-22. Whole genome sequencing of *Dioscorea zingiberensis* (Dioscoreaceae) revealed the diosgenin biosynthetic pathway[99] (Figure 4.50). Diosgenin is one of the steroidal sapogenins used commercially to produce different steroidal drugs.

4.6.6.3 Steroidal Alkaloids

Steroidal alkaloids (considered as pseudoalkaloids) and their glycosides are nitrogen-containing metabolites of plants occurring mainly in the families Solanaceae, Liliaceae, Apocynaceae and Buxaceae. Based on the structure and biosynthesis there are mainly three types of steroidal alkaloids: *Solanum* alkaloids (C27 group), *Veratrum* alkaloids (C27 group) and alkaloids of *Holarrhena* (C21 group).[15,100,101]

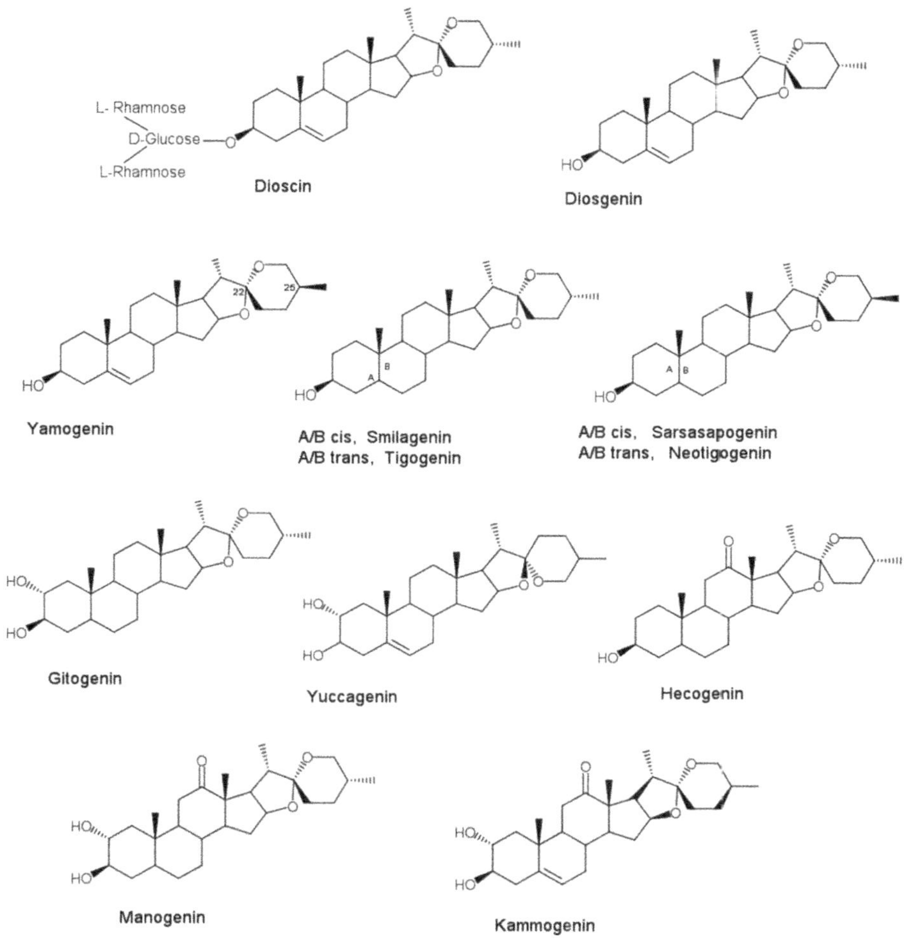

Figure 4.49 Different types of steroidal sapogenins.

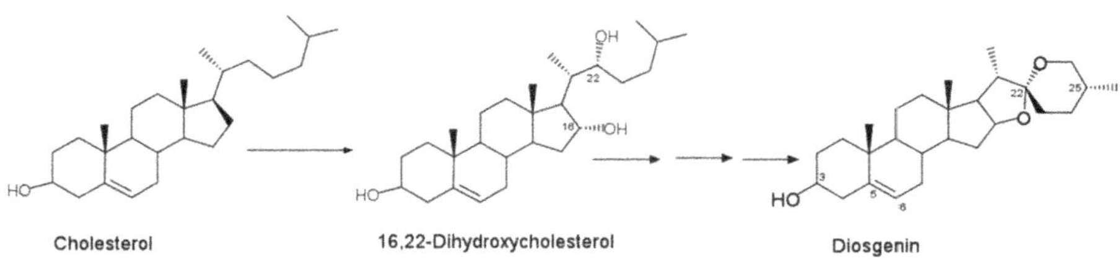

Figure 4.50 Biosynthesis of diosgenin from cholesterol.

***Solanum* alkaloids**: In Solanaceae, several steroidal alkaloids are present as glycosides and have the haemolytic and toxic properties of saponins. The oligosaccharide attached to the aglycone may vary from monosaccharide to pentasaccharide, and includes D-glucose, D-galactose, L-rhamnose, D-xylose

Table 4.4 Steroidal alkaloids.

Glycoside	Aglycone part	Sugar part	Sources
Solasonine	Solasodine	—D-Galactose—D-Glucose | L-Rhamnose	*Solanum* species
Tomatine	Tomatidine	—D-Galactose—D-Glucose—D-Xylose | D-Glucose	*Solanum lycopersicum* (tomato)
α-Solanine	Solanidine	—D-Galactose—D-Glucose | L-Rhamnose	*Solanum* species

and rarely L-arabinose. The sugar is usually linked by a β-glycosidic bond at C-3.[102] The nitrogen is associated with the steroidal part. An amino acid is the donor of nitrogen atom which is introduced by a transamination reaction. Examples of such metabolites are shown in Table 4.4.

Biosynthetically the *Solanum* steroidal alkaloids are derived from cholesterol produced *via* a cytosolic mevalonic acid pathway. The side chain of cholesterol undergoes modification through hydroxylation, oxidation, and transamination. L-Arginine is the source of incorporated nitrogen in the steroidal skeleton. The pathway includes several cytochrome P450 enzymes, a transaminase, and a 2-oxoglutarate-dependent dioxygenase. Cholesterol is putatively hydroxylated first at C-22 and then at C-26 to produce 22,26-dihydroxycholesterol. 22,26-Dihydroxycholesterol is then again hydroxylated at C-16, oxidized at C-22, followed by E-ring closure. The 26-aldehyde intermediate, produced in the process, undergoes transamination. Nucleophilic attack of the amino nitrogen results in the formation of **tomatidenol** which produces **tomatidine**. A divergent pathway from 22,26-dihydroxycholesterol has been proposed in potato. Hydroxylation at C-16 to produce 16,22,26-trihydroxycholesterol, followed by oxidation and transamination at C-26 produce **solanidine** after two nucleophilic substitutions[103] (Figure 4.51).

***Veratrum* alkaloids**: The steroidal alkaloids produced by different species of *Veratrum* (Family Melanthiaceae) are divided into three subgroups. These are the **jervanine**, the **veratranine**, and the **cevanine** subgroups.[100] Veratrum alkaloids have skeletons known as **C-nor-D-homosteroids** formed due to certain changes in the normal steroid skeleton resulting in a five membered C ring and a six-membered D ring (6-6-5-6 ring).[104] The ABCD system is spiro-connected to a five membered E ring that is again fused to a piperidine F-ring in the jervanine subgroup. In the veratranine subgroup, the alkaloids have no E ring. Examples of the three subgroups are given in Figure 4.52.

Pregnant sheep, grazing in the fields of *Veratrum californicum* and other species of *Veratrum*, were found frequently to give birth to young animals with a single eye. Cyclopamine, obtained from *Veratrum californicum*, was identified as the cause of the teratogenic effect producing cyclopia. The teratogenic effect of cyclopamine is associated with inhibition of Hedgehog signaling pathway. Further research has found this alkaloid to be a promising

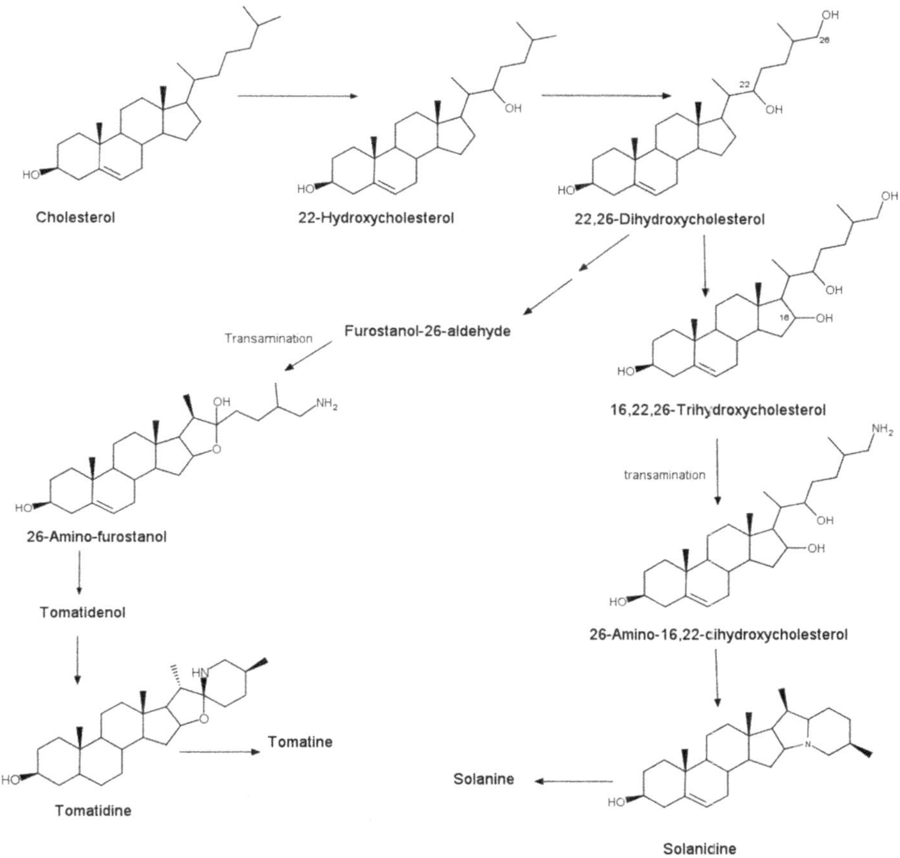

Figure 4.51 Biosynthesis of steroidal alkaloids in *Solanum*.

candidate with antineoplastic activities against several cancers caused by Hedgehog signaling malfunction.[105]

The cyclopamine biosynthesis begins with cholesterol which is a common precursor of most of the steroidal alkaloids (Figure 4.53). L-Arginine is the source of nitrogen atom.[105]

***Holarrhenna* alkaloids**: Different species of *Holarrhena* (Family Apocynaceae) such as *Holarrhena antidysenterica* (Synonym: *Holarrhena pubescens*) and *Holarrhena floribunda* are reported to contain several steroidal alkaloids such as **holarrhimine, conarrhimine, conessine, isoconessine, conessidine, conimine, kurchicine, kurchine, kurchinine, holaphyllamine, holaphylline** *etc.* Pregnenolone, produced from cholesterol, is involved in the biosynthesis of alkaloids of this group (Figure 4.54). In India, the bark of *Holarrhena antidysenterica* is used in traditional system of medicine for the treatment of amoebic dysentery.[106]

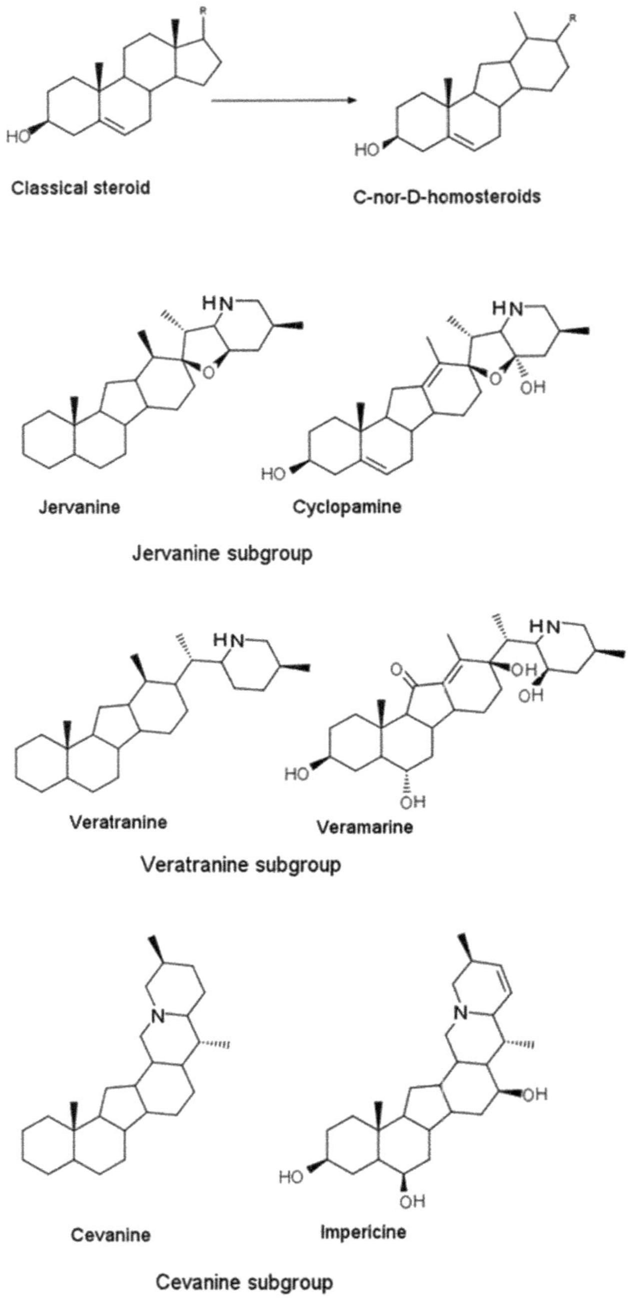

Figure 4.52 Different *Veratrum* alkaloids.

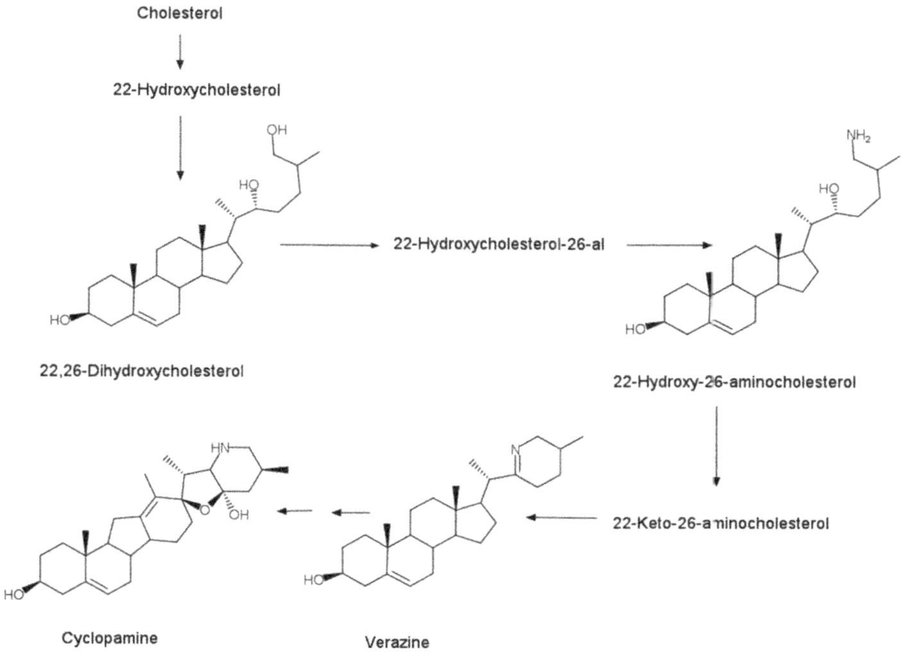

Figure 4.53 Biosynthesis of cyclopamine.

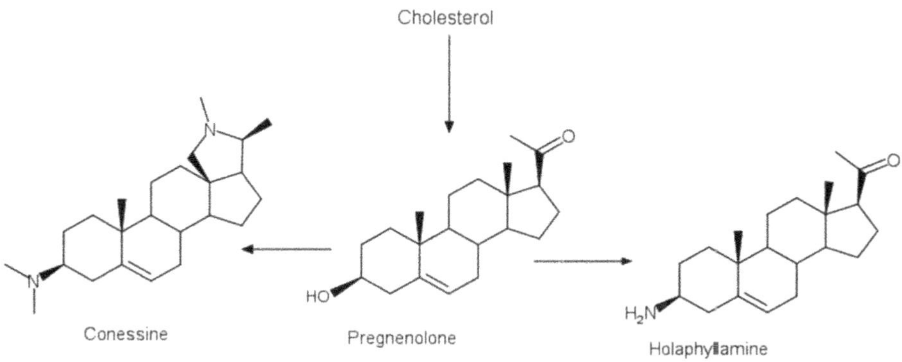

Figure 4.54 *Holarrhena* alkaloids.

4.6.6.4 Cardiac Glycosides

Cardiac glycosides are naturally occurring plant metabolites. More than one hundred cardiac glycosides have been reported from plants confined to Angiosperms. Cardiac glycosides may have up to five sugars attached to a steroidal aglycone (genin). The genin has an unsaturated steroidal lactone ring attached at the 17 β position. Cardiac glycosides are characterized by a 14-hydroxy group. The sugar moieties are linked at C-3 of the aglycone by β-linkages. The sugars may include D-glucose, L-rhamnose, L-arabinose, D-xylose, and other deoxy sugars that are naturally reported only in cardiac glycosides. The deoxy sugars 6-deoxy or 2,6-dideoxyhexoses and their methyl ethers such as D-digitoxose, D-digitalose, D-cymarose, D-fucose, L-oleandrose, D-sarmentose, L-talomethylose, D-gulomethylose are common. The aglycones of cardiac glycosides have A/B and C/D *cis* and B/C *trans* ring junctions. The cardiac glycosides are more abundant than their aglycones in nature.

Two main classes of cardiac glycosides (Figure 4.55) are **cardenolides** (C23 steroids) and **bufadienolides** (C24 steroids). They differ from each other by the structure of the respective aglycones. Cardenolides have a α,β-unsaturated five membered γ-lactone ring (furanone) at the C17 position. The bufadienolides have doubly unsaturated six-membered lactone (α-pyrone) ring at the C17.[107]

4.6.6.4.1 Cardenolides. Cardenolides are abundantly found in genera of Apocynaceae (*Adenium*, *Apocynum*, *Carissa*, *Cerbera*, *Nerium*, *Stropanthus*, *Thevetia*, *Urechites*, *Asclepias*, *Calotropis*, *Cryptostegia*, *Gomphocarpus*, *Menabea* and *Periploca*). Cardenolides were also reported from the genera of Celastaceae (*Euonymus*, *Lophopetalum*), Brassicaceae (*Cheiranthus*, *Erysimum*), Fabaceae (*Coronilla*), Moraceae (*Antiaris*, *Castilla*, *Naucleopsis*, *Maquira*), Plantaginaceae (*Digitalis*, *Isoplexis*) and Malvaceae (*Mansonia*, *Corchorus*). Some of the most important plant sources of cardenolides are *Digitalis purpurea*,

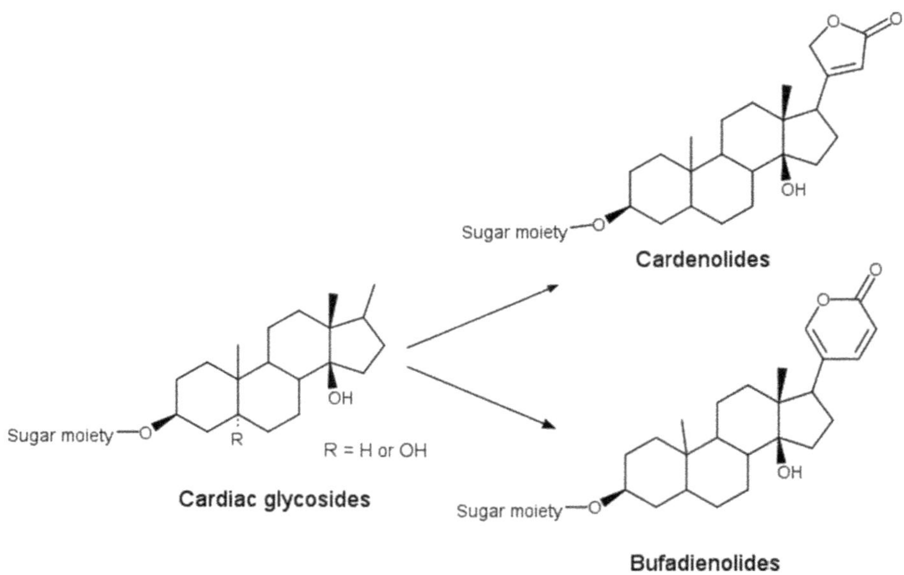

Figure 4.55 Cardiac glycosides.

Digitalis lanata, *Strophanthus gratus*, *Strophanthus kombe*, *Nerium oleander* and *Thevetia peruviana* (Synonym: *Cascabela thevetia*).[15,108] The cardenolides of *Digitalis* species can be grouped according to their aglycone genins (Figure 4.56) *e.g.*, **digitoxigenin, digoxigenin, gitoxigenin, gitaloxigenin** and **diginatigenin.** Some of the cardenolides reported from *Digitalis purpurea* (Table 4.5) and *Digitalis lanata* (Table 4.6) are presented.[15]

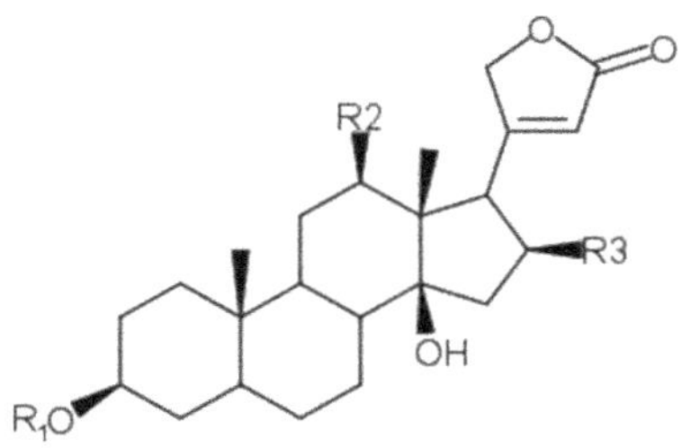

R1	R2	R3	
H	H	H	Digitoxigenin
H	OH	H	Digoxigenin
H	OH	OH	Diginatigenin
H	H	OH	Gitoxigenin
H	H	OCHO	Gitaloxigenin

Figure 4.56 Genins of *Digitalis* cardenolides.

Table 4.5 Cardenolides of *Digitalis purpurea*.

Sugar components	Aglycone	Glycosides
Glucose-(digitoxose)3-	Digitoxigenin	Purpurea glycoside A
(Digitoxose)3-	Digitoxigenin	Digitoxin
Glucose-(digitoxose)2-acetyldigitoxose-	Digitoxigenin	Purlanoside A
Digitalose	Digitoxigenin	Odoroside H
Glucose-(digitoxose)3-	Gitoxigenin	Purpurea glycoside B
(Digitoxose)3-	Gitoxigenin	Gitoxin
Glucose-(digitoxose)2-acetyldigitoxose-	Gitoxigenin	Purlanoside B
Glucose-digitalose-	Gitoxigenin	Digitalinum verum
Glucose-(digitoxose)3-	Gitaloxigenin	Glucogitaloxin
(Digitoxose)3-	Gitaloxigenin	Gitaloxin
Digitalose-	Gitaloxigenin	Verodoxin

Table 4.6 Cardenolides of *Digitalis lanata*.

Sugar components	Aglycone	Glycosides
Glucose-acetyldigitoxose-(digitoxose)2-	Digitoxigenin	Lanaloside A
Acetyldigitoxose-(digitoxose)2-	Digitoxigenin	Acetyldigitoxin
(Digitoxose)3-	Digitoxigenin	Digitoxin
Glucose-acetyldigitoxose-(digitoxose)2-	Gitoxigenin	Lanatoside B
Glucose-digitoxose-	Gitoxigenin	Glucogitoroside
Glucose-digitalose-	Gitoxigenin	Digitalinum verum
Glucose-acetyldigitoxose-(digitoxose)2-	Digoxigenin	Lanatoside C
Acetyldigitoxose-(digitoxose)2-	Digoxigenin	Acetyldigoxin
(Digitoxose)3-	Digoxigenin	Digoxin
Glucose-acetyldigitoxose-(digitoxose)2-	Diginatigenin	Lanatoside D
Glucose-acetyldigitoxose-(digitoxose)2-	Gitaloxigenin	Lanatoside E

The primary glycosides detected in the fresh leaves of *Digitalis purpurea* were purpurea glycoside A, purpurea glycoside B and glucogitaloxin. All these three glycosides have three digitoxose sugar moieties with a terminal glucose. During the drying process, terminal glucose is lost due to enzyme degradation resulting in the formation of digitoxin, gitoxin and gitaloxin respectively. Further hydrolysis may take place with complete loss of activity if the leaves are stored under poor storage conditions.[15]

Cardiac glycosides, particularly cardenolides have long been used in folk medicine as heart tonics. They have also been used as arrow poisons. At present, *Digitalis* cardenolides have no major role in cardiovascular drug market except digoxin. Digoxin is produced from the dried leaves of *Digitalis lanata* due to lack of economic viability in chemical synthesis. *Digitalis* glycosides increase myocardial contraction and were used in congestive heart failure. However, digitoxin and some other cardenolides failed to show clinical benefit. In addition, digoxin intoxication was also observed. In recent years, digoxin showed beneficial effects in patients who remained symptomatic after being treated with diuretics and/or angiotensin-converting enzyme inhibitors. The role of some of the cardenolides such as digitoxin, digoxin, ouabain, and some other cardenolides have been evaluated in recent times as antineoplastic agents. Promising data show these cardenolides to inhibit cancer cell proliferation and block tumour growth by inducing immunogenic cell death.[109–113]

In the pregnane pathway of cardenolide formation, it was hypothesized that pregnenolone may be derived from cholesterol. Cholesterol side chain cleaving enzyme (SCCE) cleaves the cholesterol side chain between C-20 and C-22. NAD:3β-hydroxysteroid dehydrogenase/Δ^5-Δ^6 ketosteroid isomerase (3β-HSD/KSI) converts pregnenolone to progesterone first by oxidation of the 3-hydroxy group and then by shifting the double bond from 5 to 4 position. After a few more steps, progesterone is converted to 5β-pregnane-3,14,21-triol-20-one. Carbon atoms C-22 and C-23 of the cardenolides are derived from a malonyl CoA. An intra- or intermolecular attack of malonyl CoA at C-20 carbonyl of pregnane-20-one has been proposed for attaching C-22 and C-23 to the pregnane skeleton. Malonyl-coenzyme A:21-hydroxypregnane 21-*O*-malonyl transferase (21MaT) is involved in the reaction to produce 21-*O*-malonyl-5β-pregnane-3β,14β-diol-20-one which is then converted into digitoxigenin (Figure 4.57).[114,115]

Strophanthus kombe and *Strophanthus gratus*, growing in equatorial Africa, contain cardiac glycosides (Figure 4.58) *e.g.*, **k-strophanthin-β**, and **cimarin**. Purified total cardiac glycosides from *Strophanthus kombe* seeds were often used as an injection under the name **strophanthin k** for the treatment of cardiac deficiencies. *Strophanthus gratus* also contains **ouabain** (the rhamnoside of ouabagenin). This cardiotonic steroid ouabain was later described as an endogenous mammalian substance. Ouabain was found in bovine adrenal glands, human plasma *etc.* Besides the effect on the cardiovascular system and blood pressure control, evidence indicates that in mammals ouabain regulates several immune functions.[111,115]

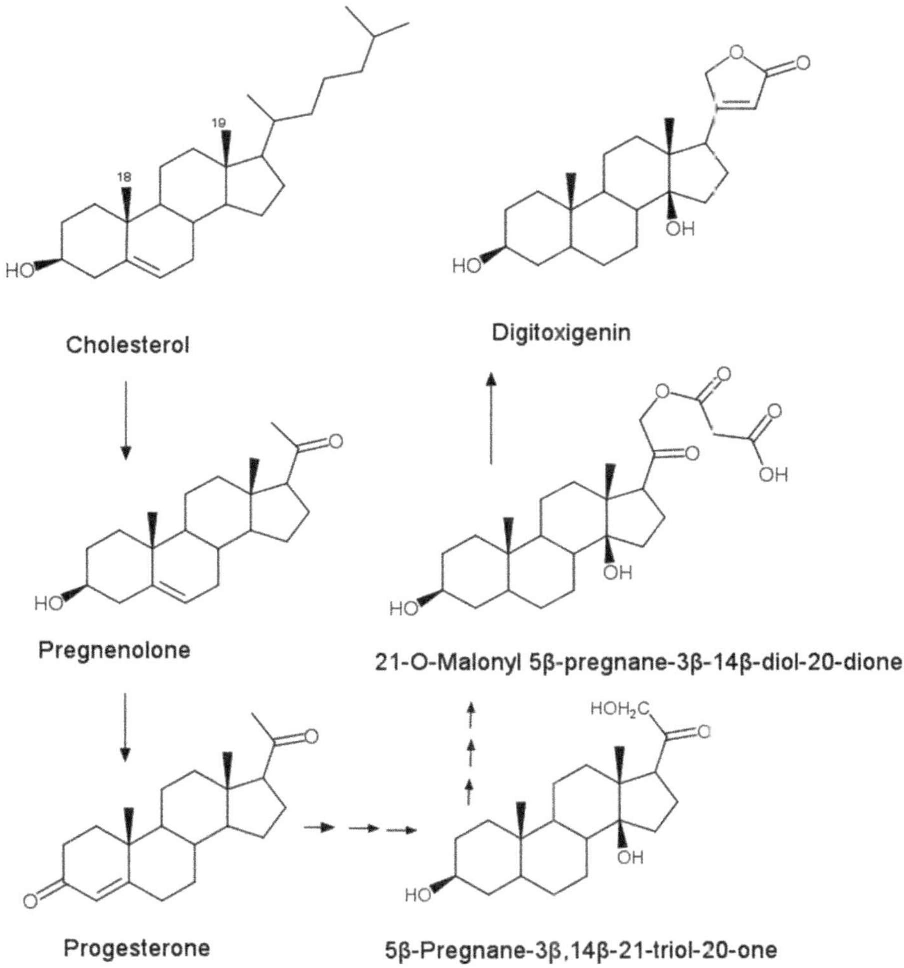

Figure 4.57 Biosynthetic pathway of cardenolide formation in *Digitalis.*

Strophanthidin .... R = H
(Aglycone)
Cymarin .... R = D-cymarose
K-Strophanthin- β ...... R = D-glucose-D-cymarose
K-Strophanthoside ..... R = D-glucose-D-glucose-D-cymarose

Ouabagenin .... R = H
Ouabain .... R = L-rhamnose

Figure 4.58 Cardiac glycosides of *Stropanthus* spp.

Nerium oleander (Family Apocynaceae) contains cardiac glycosides such as **oleandrin, adynerin** and **digitoxigenin**. Reports indicate acute oleander toxicity due to presence of cardiac glycosides in all parts. *Thevetia peruviana* (Apocynaceae) contains a few cardenolides *e.g.*, **thevetin A, thevetin B, peruvoside, thevetoxin** *etc.* These cardiac glycosides are toxic in nature.[15]

4.6.6.4.2 Bufadienolides. Bufalin (the name originated from *Bufo* – toad), a bufadienolide structure, was originally isolated from the skin of toads (*Bufo* spp.). Then many bufadienolides were identified from different plant sources[116] (Table 4.7 and Figure 4.59).

Bufadienolides detected in Crassulaceae induce symptoms of cardiac poisoning in small animals, particularly sheep. *Urginea maritima* (Asphodelaceae), also known as squill, contains the bufadienolides scillaren A, and proscillaridin A. However, the glycosides have a short duration of actions. So squill is not used for its properties on the heart. Squill is used for its expectorant action in some preparations. The red variety of squill (*Urginea maritima*) containing glucoscilliroside and scilliroside has rodenticide properties. This variety should not be present in medicinal squill.[8] Pregnenolone, derived from cholesterol is the precursor of hellebrigenin,[116] a bufadienolide found in genera of *Urginea, Helleborus* (Ranunculaceae) and *Kalanchoe* (Crassulaceae) (Figure 4.60).

Table 4.7 Bufadienolides.

Family	Genera	Bufadienolides
Crassulaceae	*Cotyledon orbiculata*	Orbicuside A, B, and C
	Tylecodon wallichii	Cotyledoside
	Tylecodon grandiflorus	Tyledoside A, tyledoside C
	Kalanchoe lanceolata	Lanceotoxin A (5-*O*-acetyl-hellebrigenin), lanceotoxin B (5-*O*-acetyl-3-*O*-α-L-rhamnosylhellebrigenin)
	Kalanchoe tomentosa	Kalanchoside
	Kalanchoe tubiflora	Bryotoxin A, bryotoxin B, and bryotoxin C
Asphodelaceae	*Urginea maritima*	Scillarenin, proscillaridin A, scillaren A, scillirosidine, scilliroside, glucoscilliroside and related bufadienolides
Asparagaceae	*Bowiea volubilis*	Bovoside A
Iridaceae	*Homeria glauca*	Epoxyscillirosidine, epoxy-12-hydroxyscillirosidine
	Moraea polystachya *Morea graminicola*	Hydroxybovogenin A, formyloxybovogenin A
Francoaceae	*Melianthus comosus*	Melianthugenin
	Bersama abyssinica	Abyssinin, abyssinol A–C
Ranunculaceae	Species of *Helleborus*	All derivatives of hellebrigenin *e.g.*, hellebrin, desglucohellebrin, hydroxydesglucohellebrin
Santalaceae	*Thesium lineatum*	Thesiuside (5-*O*-acetyl-3-*O*-β-D-glucopyranosylhellebrigenin)

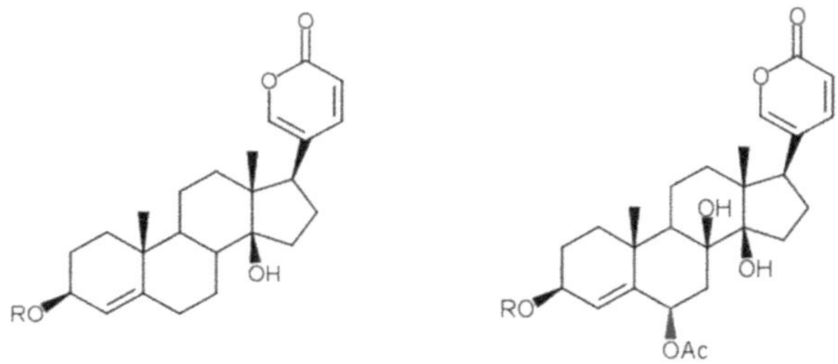

Scillarenin .... R = H
(Aglycone)
Proscillaridin A .... R = L-rhamnose
Scillaren A .... R = L-Rhamnose-D-glucose

Scillirosidin .... R = H
(Aglycone)
Scilliroside .... R = D-glucose
Glucoscilliroside .... R = D-glucose-D-glucose

Figure 4.59 Bufadienolides of *Urginea maritima.*

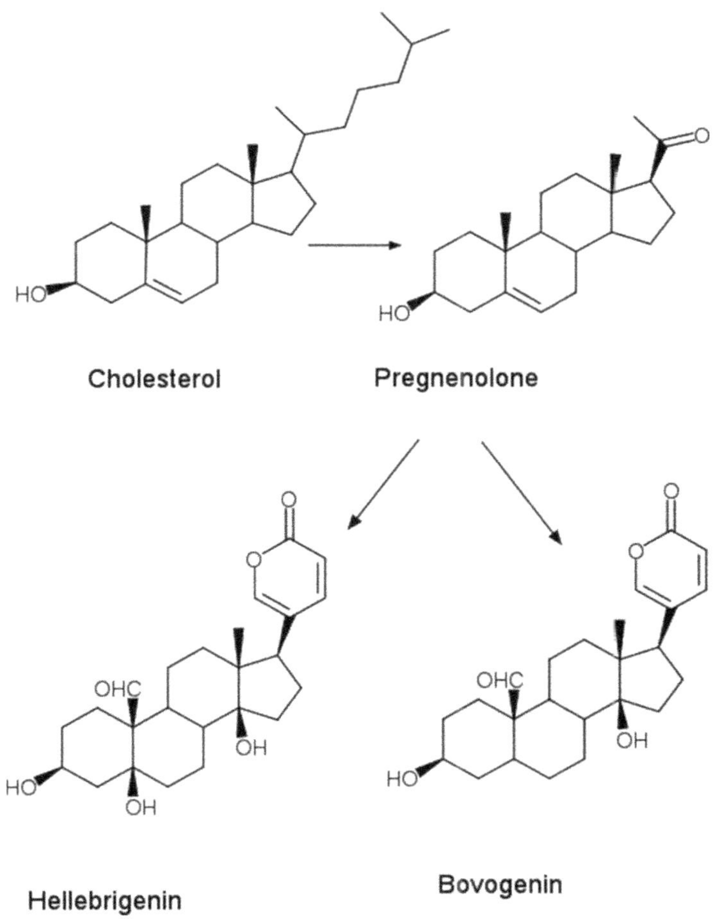

Figure 4.60 Origin of a few bufadienolides.

4.6.6.5 Withanolides

Withanolides are polyoxygenated steroids based on a C28 ergostane skeleton in which C-26 and C-22, or C-26 and C-23 are oxidized to form a δ-lactone (type A) or γ-lactone (type B).[117,118] Withanolides are present in the form of aglycones (Figure 4.61). A few of the withanolides have been found in the form of glycosides with the sugar moiety at C3 and C27.

24-Methylenecholesterol is a proposed precursor of withanolide biosynthesis. However, conversion of 24-methylenecholesterol to withanolides is not yet fully understood. Using phylogenetics, metabolomics and gene expression analysis along with heterologous expression and virus-induced gene silencing showed conversion of 24-methylenecholesterol to 24-methyldesmosterol to be the committing step in withanolide biosynthesis[119,120] (Figure 4.62).

Withanolides are found in Solanaceae (*Acnistus*, *Datura*, *Deprea*, *Discopodium*, *Dunalia*, *Iochroma*, *Jaborosa*, *Lycium*, *Larnax*, *Nicandra*, *Mandragora*, *Physalis*, *Solanum*, *Withania* and *Witheringia*), and in families like Fabaceae, Lamiaceae, Myrtaceae and Dioscoreaceae. Approximately 900 withanolides have been identified.[121,122]

Withania somnifera, commonly known as Ashwagandha, has been used in Ayurveda, Siddha and Unani medicine and other indigenous systems of medicines for over 3000 years. In Ayurveda, *Withania* is used for its potent aphrodisiac, rejuvenating, and life prolonging properties. The plant also enhances memory function and cures geriatric problems, promotes vigour and strength and counteracts fatigue, weakness, bone weakness,

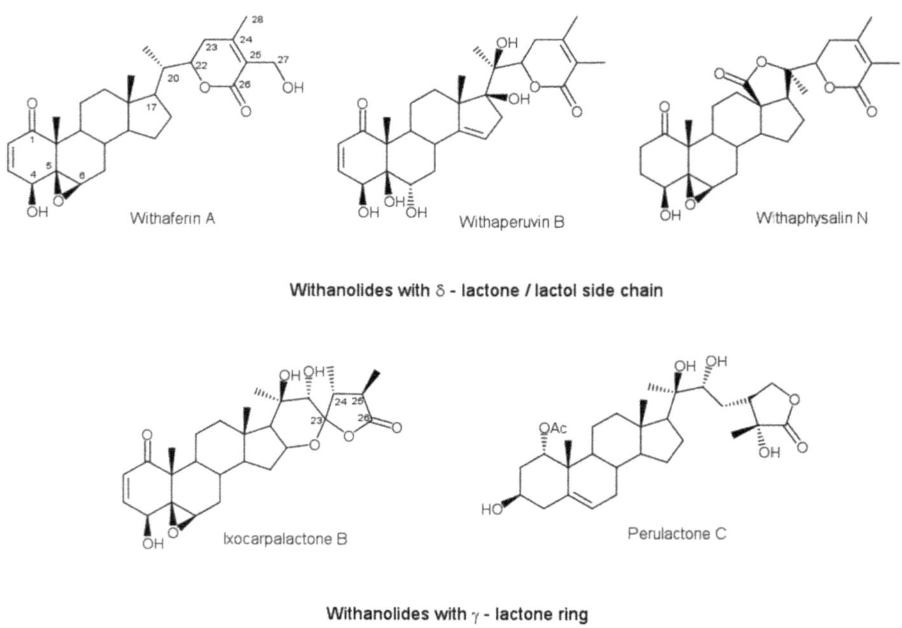

Figure 4.61 Withanolides.

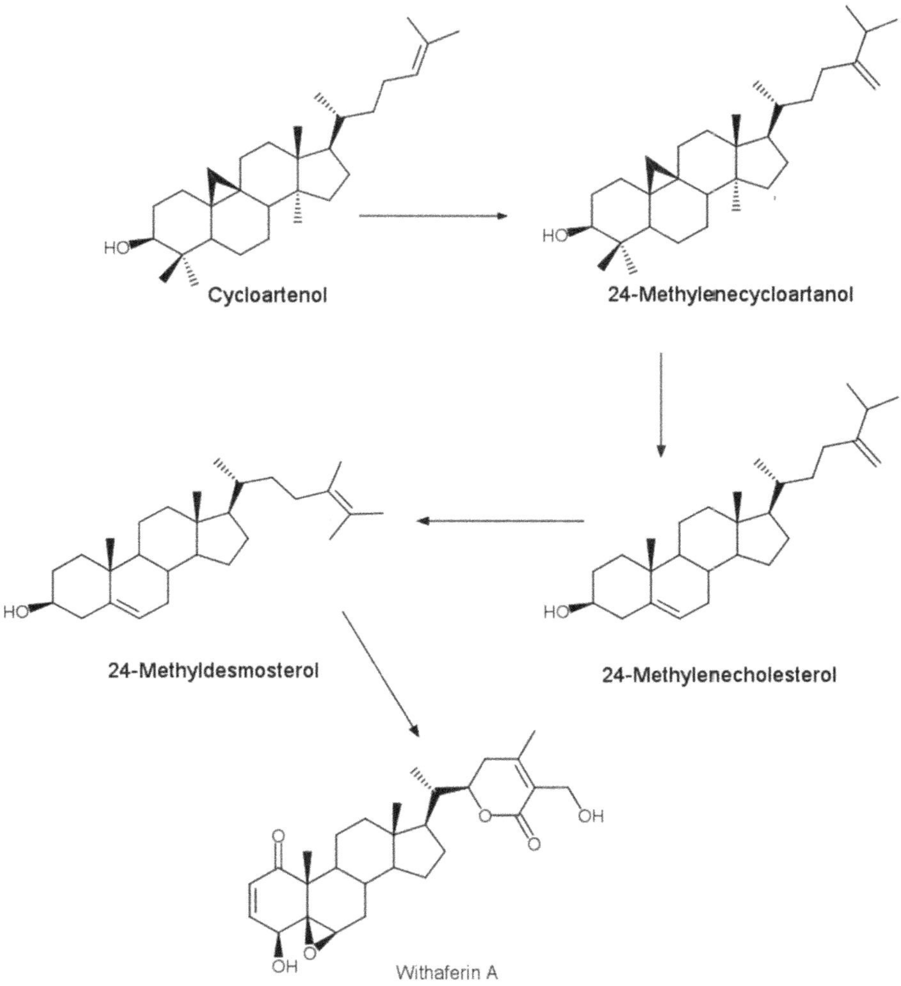

Figure 4.62 Biosynthesis of withaferin A.

impotency, premature aging, debility and muscle tension. Recent studies exhibited diverse pharmacological activities of withanolides such as antimicrobial, anticancer, anti-inflammatory and immunoregulatory activities. The withanolides were found to have roles to prevent osteoporosis and rheumatoid arthritis, inflammatory bowel disease and neurodegenerative diseases.[123,124]

4.6.6.6 Cucurbitacins

Cucurbitacins (Figure 4.63) are characterized by tetracyclic cucurbitane steroid skeleton. They have a 5,6-double bond, and a methyl group at C-9. They may exist in the form of glycosides. There are about 40 cucurbitacins

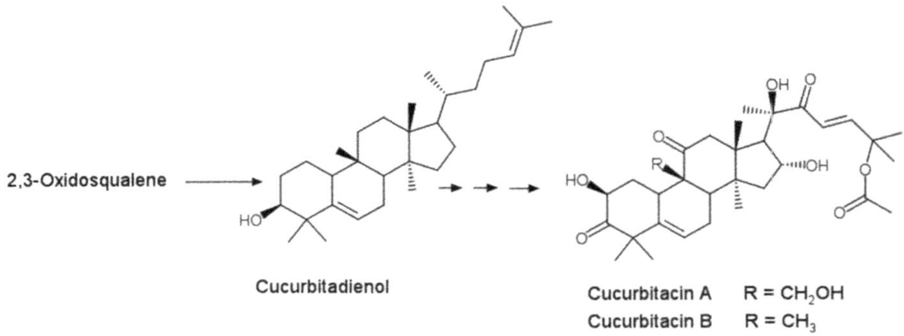

Figure 4.63 Cucurbitacins.

divided into 12 groups *e.g.*, cucurbitacins A, B, C, D, E, I, H, Q, R and dihydrocucurbitacin B. Cucurbitacins confer bitter taste in cucurbits.[125] Cyclization of 2,3-oxidosqualene produces cucurbitadienol. Subsequent hydroxylation, acetylation, glucosylation produce different types of cucurbitacins.[59,126]

Cucurbitacins are found in many Cucurbitaceae plants (*Bryonia*, *Cucumis*, *Cucurbita*, *Luffa*, *Lagenaria*, *Citrullus*), and in members of Begoniaceae, Brassicaceae, Elaeocarpaceae, Primulaceae, Rosaceae, Rubiaceae, Scrophulariaceae *etc.* Cucurbitacins have also been reported from some genera of mushrooms *e.g.*, *Russula* and *Hebeloma*.[127]

Cucurbitacins are highly toxic metabolites. But they have pharmacological potential against inflammation, atherosclerosis, diabetes, and cancer. However, the cytotoxicity of cucurbitacins hinders their application in medicine. Cucurbitacin B is a promising anticancer compound against various types of cancers. It acts through various molecular signaling pathways.[125]

4.6.6.7 *Brassinosteroids*

Brassinosteroids are polyhydroxylated steroidal phytohormones. Brassinolide was first discovered in 1979. There are about 70 such compounds occurring in nature either as free form or conjugated with glucose or fatty acids. Brassinosteroids have been reported from lower to higher plants in different parts like roots, stems, leaves, flowers, seeds and grains.[128,129]

Based on the number of carbons, brassinosteroids can be divided into C27, C28, C29 compounds (Figure 4.64). The brassinosteroid biosynthetic pathway is dependent on cycloartenol and cycloartanol. C27 Brassinosteroids are biosynthesized from cholesterol. C28 Brassinosteroids are biosynthesized from 24-methylenecholesterol or campesterol. The C29 Brassinosteroids are derived from sitosterol or avenasterol[128,129] (Figure 4.65).

Brassinosteroids regulate plant growth and development of shoots and roots, vascular differentiation, fertility, seed germination, immunity, and responses to environmental stresses.[128]

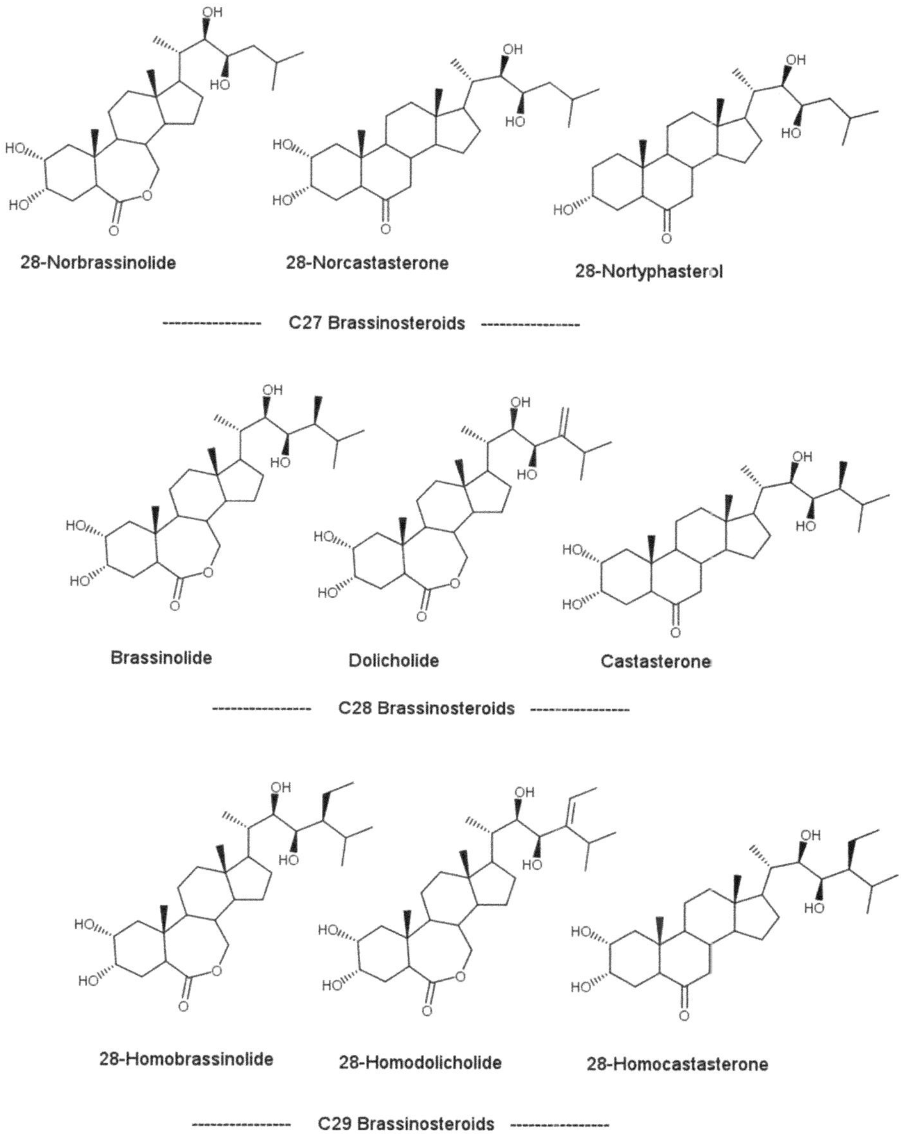

Figure 4.64 Different types of brassinosteroids.

4.6.6.8 Phytoecdysteroids

Phytoecdysteroids are ecdysteroids produced by plants as defensive chemicals against herbivore insects. More than 300 phytoecdysteroids have been isolated from plants. 20-Hydroxyecdysone (Figure 4.66) is the major ecdysteroid reported from plants. These compounds have a polyhydroxylated cyclopentanoperhydrophenanthrene ring system. Most of the ecdysteroids

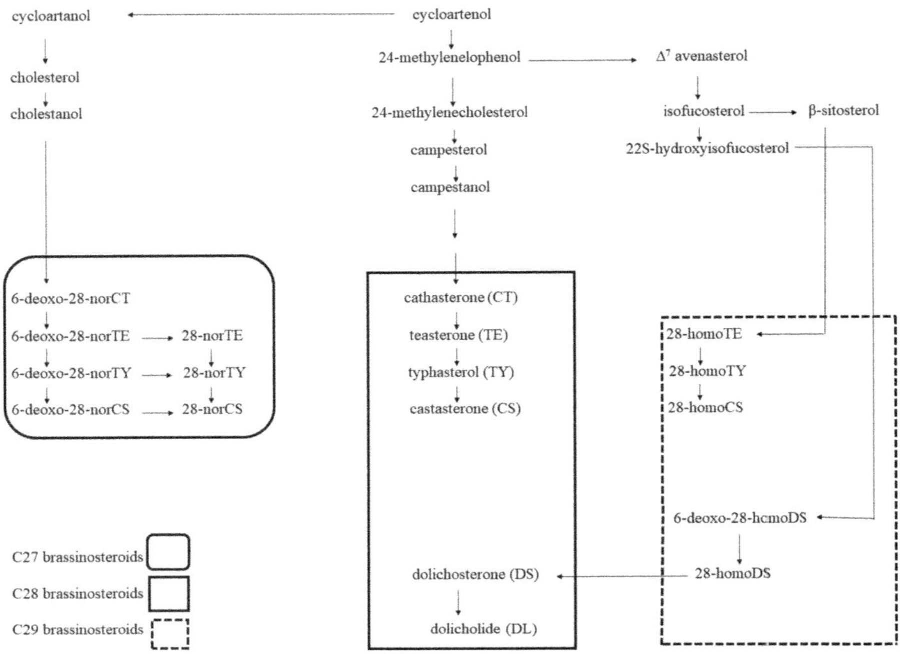

Figure 4.65 Biosynthesis of brassinosteroids.

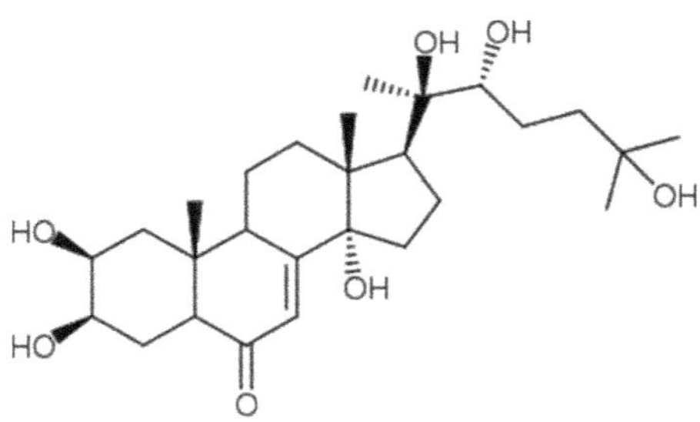

Figure 4.66 Phytoecdysteroid.

possesses a hydroxyl group at the 14α position. The phytoecdysteroids may exist in either free form or as conjugates or glucosides, phosphates, sulphates, acetates, or benzoates. These plant metabolites are structural analogues of the insect moulting hormone ecdysone. Their concentrations in plants are higher than those found in the arthropods. Phytoecdysteroids affect premature moulting, weight loss and cause metabolic damage and finally death of herbivore insects when they eat the plants containing these chemicals.[130–132]

4.6.6.9 Progesterone

Progesterone (Figure 4.67), a mammalian gonadal hormone, is naturally present in plants. It has been reported from *Pinus nigra* (Pinaceae), *Inula helenium* (Asteraceae), *Nicotiana tabacum* (Solanaceae), *Digitalis purpurea* (Plantaginaceae) and *Tribulus terrestris* (Zygophyllaceae). This steroid serves as the precursor in the biosynthetic processes of androstanes and oestranes.[133,134]

4.6.6.10 Testosterone and Structurally Related Androstanes

Testosterone (Figure 4.67), the steroid hormone, occurs in higher plants. Testosterone, epitestosterone, and androstenedione were first reported from pollen of *Pinus sylvestris* (Pinaceae) and then from the pollen of *Pinus nigra*. Dihydrotestosterone have been reported from about twenty plants including maize, barley and rhubarb. Testosterone was also reported from wheat, potatoes, soybeans and in oils such as olive oil, corn oil and safflower oil. Boldenone (1-dehydrotestosterone), boldione (androstadienedione) and androstenedione also occur naturally in plants. Androstadienedione and androstenedione (Figure 4.67) were detected in pollen of *Pinus sylvestris*

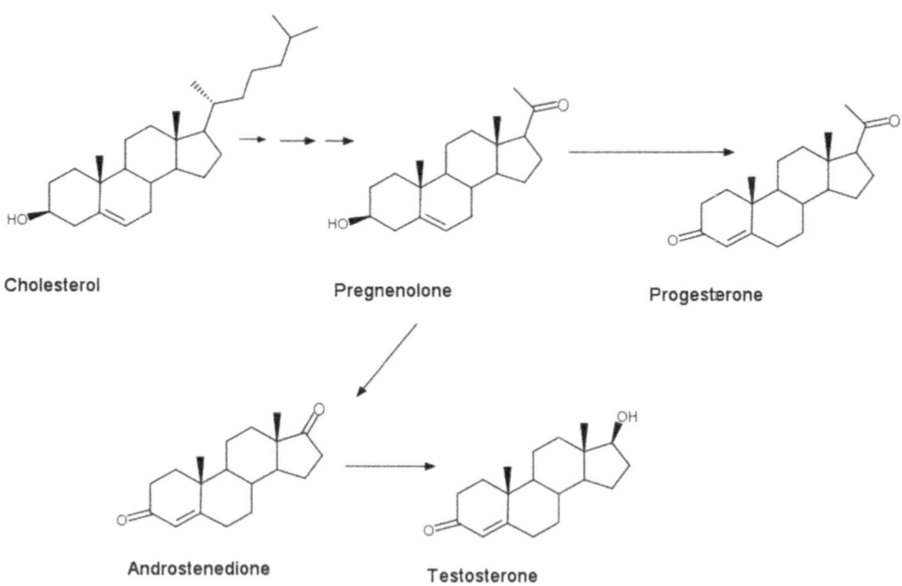

Figure 4.67 Biosynthesis of progesterone and testosterone.

and *Pinus nigra*. Androstenedione was also reported from wheat, potato *Nicotiana tabacum*, *Inula helenium*, and *Tribulus terrestris*.[133,134]

4.6.6.11 Oestrogens

Steroidal oestrogens, commonly produced by ovaries in adult females of higher vertebrates, were detected in many plant species. Oestrone was detected from *Punica granatum* var. *nana* (seeds, root, and flowers) (Lythraceae), *Hyphaene thebaica* (Arecaceae), *Zea mays* (Poaceae) oil, *Olea europaea* (Oleaceae) (oil and seed) and *Phoenix dactylifera* (Arecaceae) seeds. 17-β-Oestradiole was also reported from *Phaseolus vulgaris* (Fabaceae). Oestrogens are present in the reproductive parts of the plants (flowers, pollen grains, fruits, seeds) in higher quantity than in the vegetative parts (stems, leaves, roots) of the plants.[134]

4.6.6.12 Commercial Production of Steroidal Drugs from Natural Steroids

Steroids used as drugs may be of different types (Figure 4.68). The sex hormones, oral contraceptive steroids and diuretic steroids have no oxygen substitution in the ring C. Corticosteroids (adrenocortical hormones) are 11-oxosteroids. Corticosteroids have two groups *e.g.*, glucocorticoids and mineralocorticoids. Examples of glucocorticoids are hydrocortisone (cortisol) and corticosterone. Examples of mineralocorticoids are aldosterone and deoxycortisone (cortexone).[8]

Synthesis of steroidal drugs such as sex hormones and corticosteroids has been performed since around 1950 utilising natural steroids as the starting materials. Diosgenin, and solasodine were degraded by a series of chemical reactions known as Marker degradation. Cleavage of ring F and then ring E led to the formation of dehydropregnenolone acetate. Further chemical reactions produced progesterone. But it was difficult to introduce the 11β-hydroxyl group, necessary for glucocorticoid activity, by chemical means. Introduction of oxygen at the 11 position was possible by microbial conversion. *Rhizopus arrhizus* transformed progesterone to 11α-hydroxyprogesterone which can be converted to hydrocortisone by chemical means (Figure 4.69). Hecogenin (from *Agave sisalana* of family Asparagaceae) has a 12-keto group. Bromination converted hecogenin to 11α-bromo derivative. Further chemical reactions produced 12-hydroxy-11-ketone derivative and then to 11-keto sapogenin by side chain degradation using Marker degradation. Hydrocortisones are then produced chemically.[8]

However, utilization of sapogenins as the precursors of steroidal drugs have reduced significantly due to several factors such as high costs, multiple steps, low yields, and depletion of wild plant resources. Sapogenins are gradually being replaced by natural sterols such as cholesterol, campesterol, and β-sitosterol. Several phytosterols from soybean, pine and paper industry

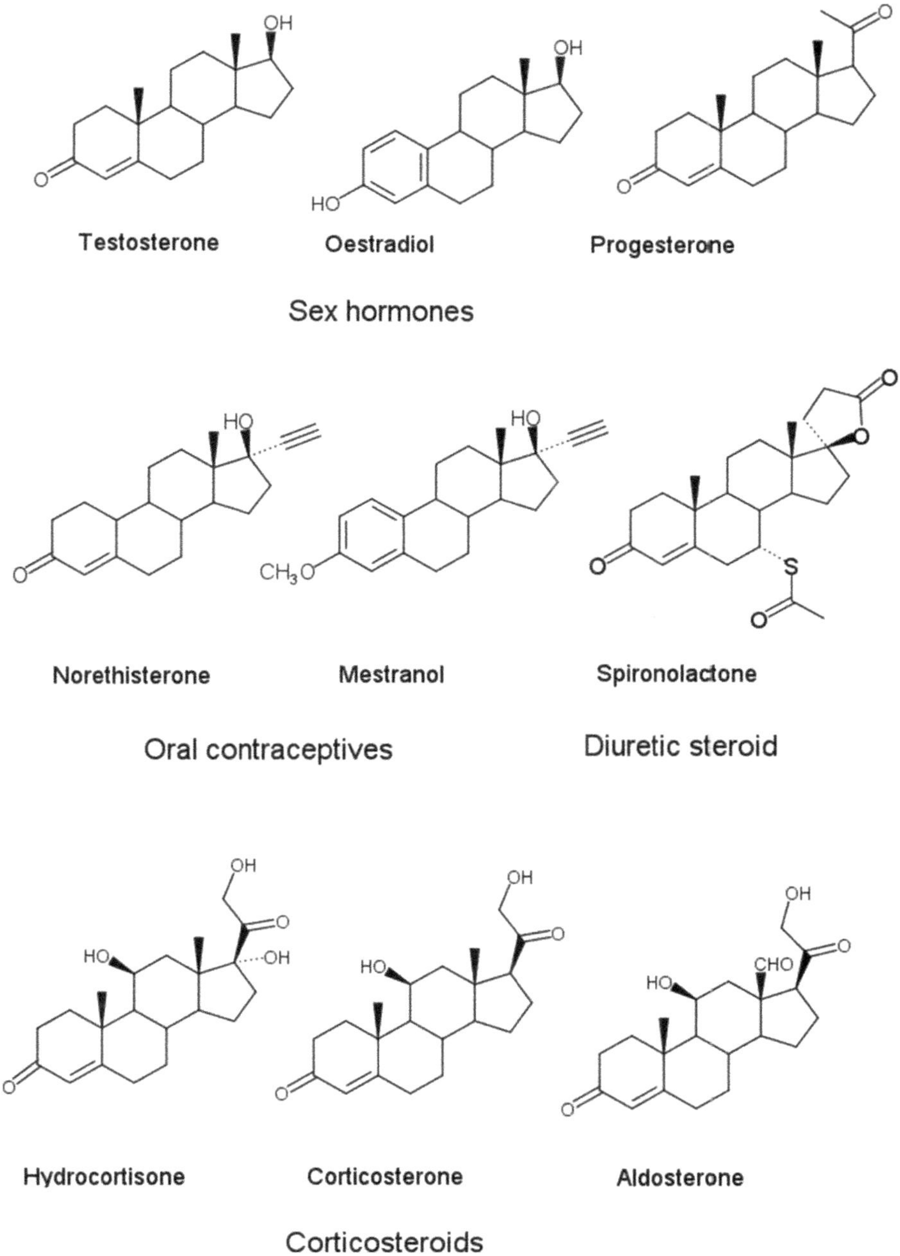

Figure 4.68 Different types of steroidal drugs.

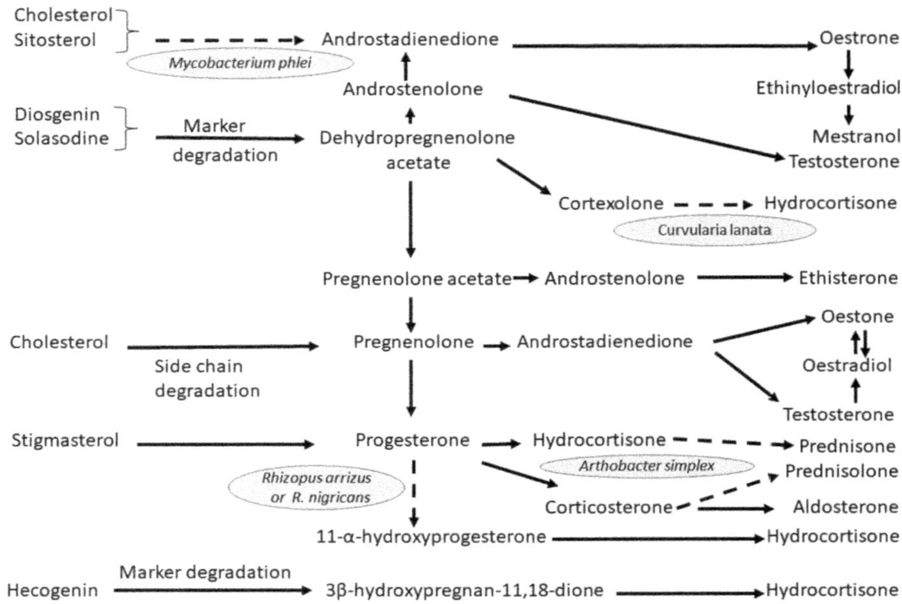

Figure 4.69 Commercial production of steroid drugs from plant steroids. ———: Chemical reactions; ------------: microbial reactions.

wastes are used for production of steroid-based drugs. Stigmasterol and the fungal sterol ergosterol can also be used. Several microbiological conversion steps are involved in the process of steroid manufacture. Many new steroid-producing strains have also been designed by metabolic engineering.[135] A diagrammatic presentation of the steps in the conversion of natural steroids to steroidal drugs is shown in Figure 4.69.

4.7 Tetraterpenes

Carotenoids are fat soluble tetraterpene (C40) pigments consisting of 8 isoprene units linked linearly. The characteristic long, conjugated, double-bond system produced characteristic orange, red, or yellow colour with UV absorption maxima between 450 and 570 nm. The carotenoids are distributed widely in some species of archaea and fungi, photosynthetic bacteria, algae, and different parts of plants (angiosperms), animals and humans. The animals are unable to synthesize carotenoids. However, they can modify plant carotenoids consumed from the diet. There are a few reports of C45 and C50 carotenoids called higher carotenoids. C45 carotenoids consisting of 9 isoprene units are produced by some bacteria. C50 carotenoids, produced by archaea and bacteria, comprise 10 isoprene units.[136,137]

Depending on the degree of substitution, two classes of carotenoids are found in nature: **carotenes** and **xanthophylls** (Table 4.8). Carotenes

Table 4.8 Classes of carotenoids.

Carotenoids	Examples
I. Carotenes	α-Carotene, β-carotene, γ-carotene, δ-carotene, lycopene
II. Xanthophylls	
With hydroxyl group	Cryptoxanthin, zeaxanthin, zeinoxanthin, lutein
With methoxyl group	Spirilloxanthin
With epoxide group	Luteochrome, violaxanthin, neoxanthin, antheraxanthin
With keto group	Astaxanthin, canthaxanthin, capsanthin, capsorubin

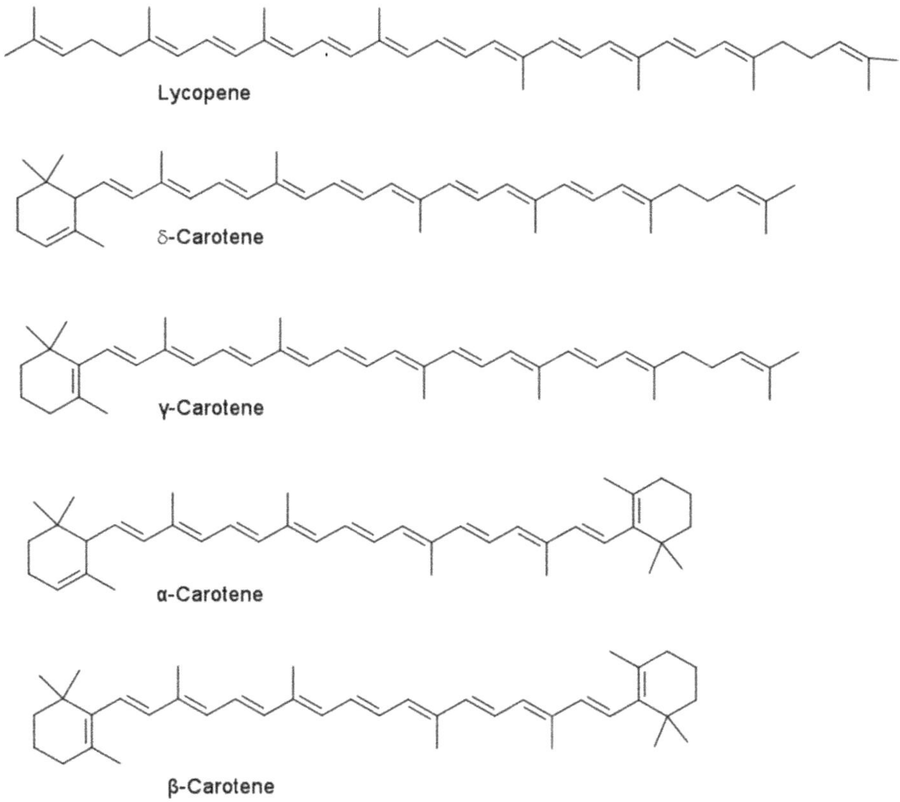

Figure 4.70 Carotenes.

(Figure 4.70) are highly unsaturated hydrocarbons with no oxygen and may be cyclized at one or both ends of the molecule. Xanthophylls (Figure 4.71) are oxygenated derivatives of carotenes containing one or more oxygen atoms as hydroxyl, methoxyl, epoxide, and keto groups on the terminal rings. About 50 kinds of carotenes and 800 kinds of xanthophylls have been reported from nature.[138]

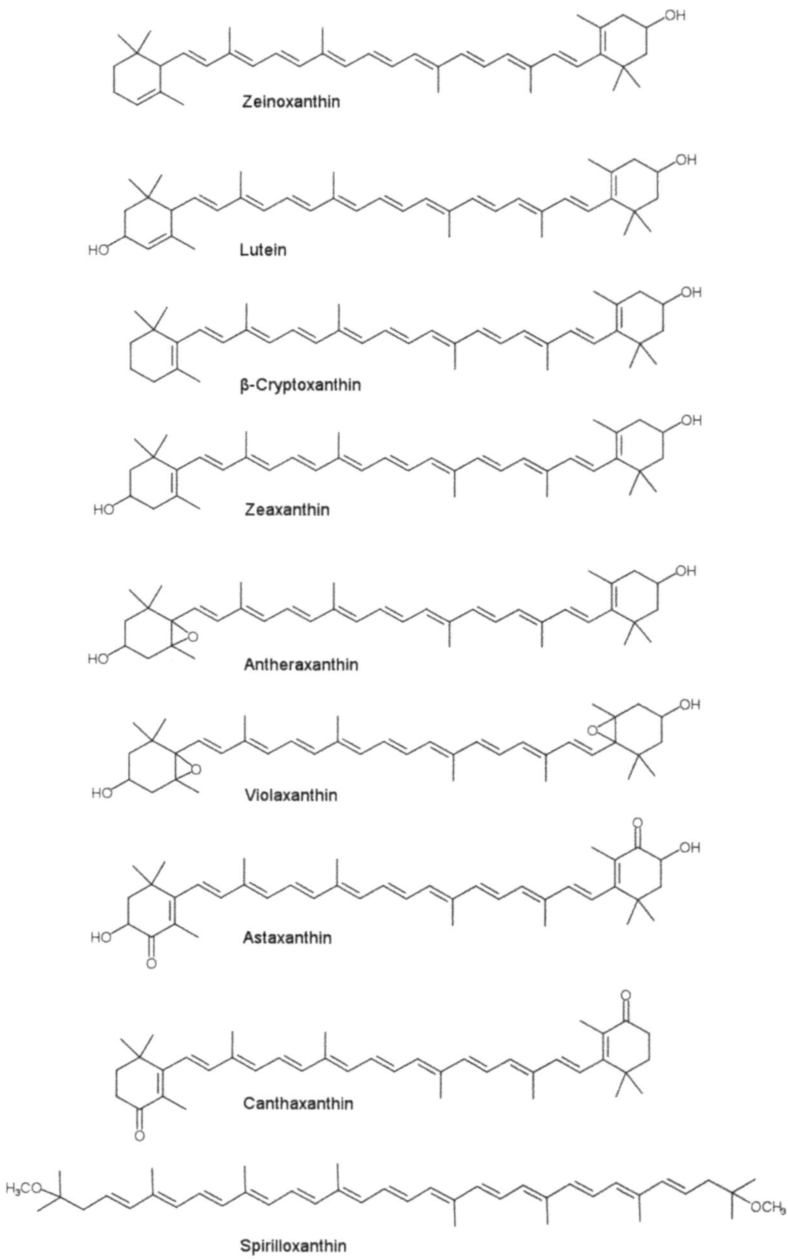

Figure 4.71 Xanthophylls.

Allene carotenoid and acetylene carotenoid structures are also found in nature by modification of the polyene chain (long conjugated double bond system) by the presence of one or two allene (C=C=C) (Figure 4.72) or acetylene (C≡C) (Figure 4.72) structures. Examples of allene carotenoids are peridinin in dinoflagellates, fucoxanthin in brown algae and neoxanthin in seed

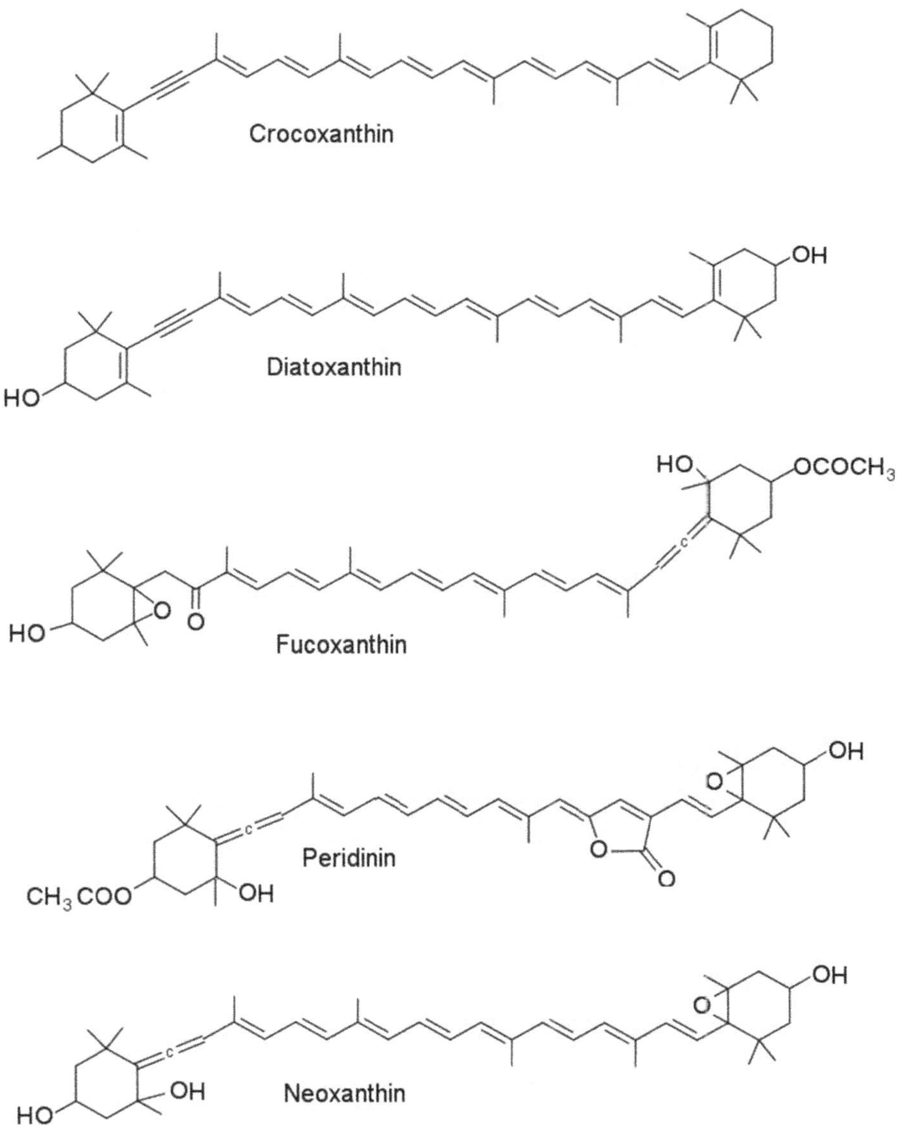

Figure 4.72 Acetylene and allene structures in xanthophylls.

plants and spore bearing plants. Acetylene carotenoids such as crocoxanthin and diatoxanthin are synthesized in microalgae.[139]

Carotenoids exist naturally either in free form or associated to other molecules *e.g.*, fatty acids (carotenoid esters), sugars (glycosylated carotenoids) or proteins (carotenoproteins). Xanthophylls in plants are found typically in an ester form by reaction of the carboxylic acid of fatty acids with an alcohol group of hydroxylated xanthophylls, with elimination of a water molecule. Crocins are found in glucosylated form. Carotenoprotein occurs in some Crustacea.[138,140]

4.7.1 Biosynthesis of Carotenoids

The tetraterpene structure of the carotenoids is formed by tail-to-tail coupling of two molecules of geranylgeranyl diphosphate (GGPP) produced by the MEP pathway. The reaction is catalyzed by phytoene synthase. The reactions synthesize phytoene through intermediate formation of prephytoene diphosphate. During phytoene biosynthesis, a double bond (*Z* or *cis* configuration in plants and fungi; *E* or *trans* configuration in bacteria) is generated in the centre of the molecule. Formation of the simplest carotene gives lycopene (all-*trans*-lycopene). Lycopene formation from phytoene requires four enzymes, phytoene desaturase, ζ-carotene isomerase, ζ-carotene desaturase, and carotene isomerase. Intermediates formed in the reactions are ζ-carotene (9,15,9′-tri-*cis*-ζ-carotene) and pro-lycopene (tetra-*cis*-lycopene). Lycopene undergoes additional cyclization of the chain ends by β-cyclase to produce α-carotene and β-carotene. Xanthophylls are produced initially by hydroxylation of α-, and β-carotenes. Subsequently epoxides are formed (Figure 4.73).[141–145]

Carotenoids are important constituents of plants. Carotenoids coordinate with plastidial isoprenoid pigments (chlorophylls) and regulate plant growth and development. Highly reactive intermediates, particularly reactive oxygen species, are generated due to photosynthesis and damage the photosynthetic apparatus decreasing the photosynthetic efficiency. Carotenoids as antioxidants along with other non-enzymatic (tocopherols, ascorbate, glutathione) and enzymatic antioxidants mitigate the generation of these reactive molecules. The biosynthesis of carotenoids in the dark grown etioplasts plays an important role in the assembly of membrane structures and formation of chloroplasts when seedlings emerge. Chlorophyll and carotenoid molecules coordinate formation of thylakoid membranes.[137]

Dietary carotenoids are claimed to have health beneficial properties. Carotenoids influence the signaling pathways as the mechanism of prevention of several diseases such as cancers and cardiovascular diseases, although there are some conflicting data. Carotenoids are antioxidants that quench singlet oxygen and scavenge peroxyl radicals thus preventing lipid peroxidation. β-Carotene supplements are considered as sun protectants. Lutein and zeaxanthin are available in the skin and eye, particularly in the macula lutea of the retina and protect the tissues from light exposure. The photoprotective activity of these carotenoids helps in protection of skin from sunburn.

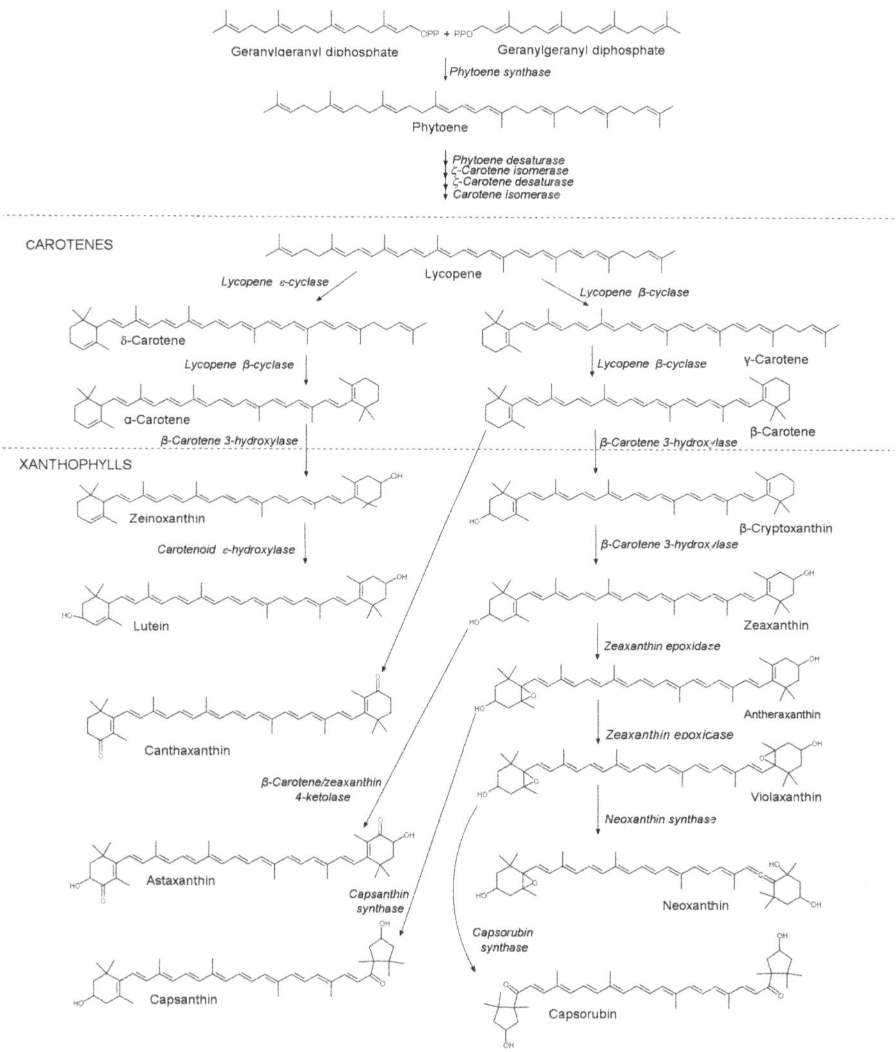

Figure 4.73 Biosynthesis of carotenoids.

β-Carotene, also known as a provitamin A substance, is the major carotene that contributes to the production of vitamin A. β-Cryptoxanthin also contributes to the supply of vitamin A.[137]

4.7.2 Apocarotenoids

Apocarotenoids are formed by enzymatic or non-enzymatic oxidative cleavage of double bonds in the carotenoid molecule. Carotenoid cleavage dioxygenases play a role in enzymatic conversion of C40 carotenoids into apocarotenoids. Other enzymatic oxidations by peroxidases or lipo-oxygenases also take place. Non-enzymatic photochemical oxidation in photosynthetic tissues is also involved in apocarotenoid production.[140,146–152] Examples and origin/importance of apocarotenoids are shown in Table 4.9. The precursors of main apocarotenoids are β-carotene (Figures 4.74 and 4.75) (for vitamin A and for strigolactones), violaxanthin (Figure 4.76) (for abscisic acid, xanthoxin), zeaxanthin (Figure 4.77) (for crocins), and lycopene (Figure 4.78) (for bixins).

Table 4.9 Apocarotenoids.

Apocarotenoids	
Vitamin A (retinal) and its derivatives	Oxidative cleavage of β-carotene results in the formation of two molecules of retinal
β-Ionone	Aroma constituent, derived from carotenoids, attract insects and animals for pollution and seed dispersal
Abscisic acid	A phytohormone derived from 9-*cis*-violaxanthin and 9-*cis*-neoxanthin
Strigolactones	Signaling molecules and shoot branching regulators, derived from carotenoids, released by plant roots, attract symbiotic arbuscular mycorrhizal fungi, trigger seed germination of root parasitic weeds such as *Striga* spp. (Orobanchaceae)
Mycorradicin	Derived from carotenoids and play an important role during arbuscular mycorrhizal fungi symbiosis and trigger hyphal branching in the rhizosphere
Bixin	Produced from lycopene, this dicarboxylic apocarotenoid is obtained from *Bixa orellana* (Bixaceae) seeds
Crocin, crocetin	Zeaxanthin derived crocetin is found in the stigmas of *Crocus sativus* (Iridaceae). The glycosidic ester of crocetin is crocin. It is responsible for the dark red colour of saffron. Picrocrocin and safranal are produced from the lateral cyclic moieties of zeaxanthin

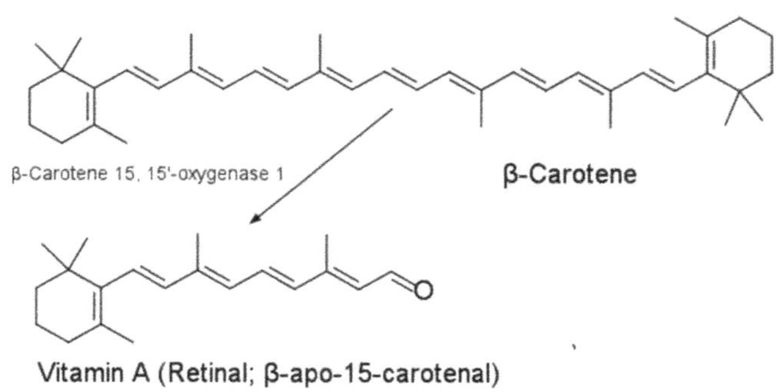

Figure 4.74 Formation of vitamin A from β-carotene.

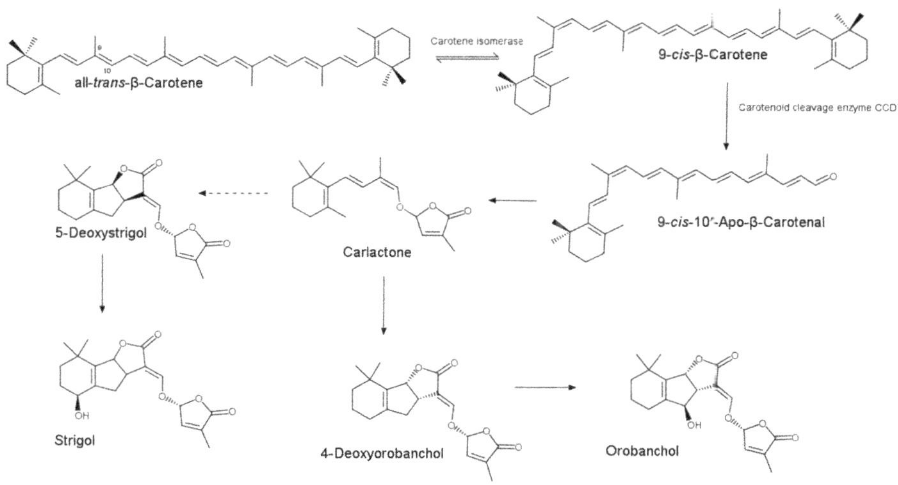

Figure 4.75 Formation of strigolactones from β-carotene.

4.8 Polyterpenes

Polyterpenes include **polyprenols** and **prenylated electron carriers** (ubiquinone and plastoquinone).

Polyprenols (polyisoprenoid alcohols) consist of long, linear isoprenes (greater than 45 carbons). They are typically found in bacteria and plants. Polyprenols are hydrophobic polymers consisting of from several up to 100 isoprene residues linked head-to-tail and have a hydroxyl group at one end (α-residue) and a hydrogen atom at the other end (ω-end). The general formula is H-$(C_5H_8)_n$-OH. Generally, polyprenols are classified into two main groups. These are *trans*-1,4-polyisoprene (poly-*trans* prenols) and *cis*-1,4-polyisoprene (poly-*cis* prenols). Hydrogenation of a double bond in the α-residue gives rise to dolichols, also known as dihydropolyprenols. Polyprenols with an unsaturated α-residue are also known as dehydrodolichols (Figure 4.79). Dolichols are found in animals and yeast and in some plants. Polyprenols are present mainly in the thylakoid membrane.[153–156]

Poly-*trans* prenols are solasenol (all-*trans*-Pren-9) belonging to Solanaceae and spadicol (all-*trans*-Pren-10) from the spadix of *Arum maculatum* (Araceae). Another example of all-*trans*-polyprenol is gutta percha (*trans*-1,4-polyisoprene). Gutta-percha is a purified and dry coagulated latex produced from different species of *Palaquium*. The Indian species that yield gutta-percha are *Palaquium obovatum*, *Palaquium polyanthum*, *Palaquium ellipticum*, and *Palaquium gutta* (Sapotaceae). *trans*-Polyisoprene is more rigid than *cis*-polyisoprene. It provides insulation, resistance to acid and alkali and has a low coefficient of thermal expansion/contraction. Gutta-percha is used for manufacture of corks, pipes, cements, thread, surgical instruments, window shades, carpets, gloves, mattresses, pillows, tents, umbrellas, golf balls *etc.*[155–157]

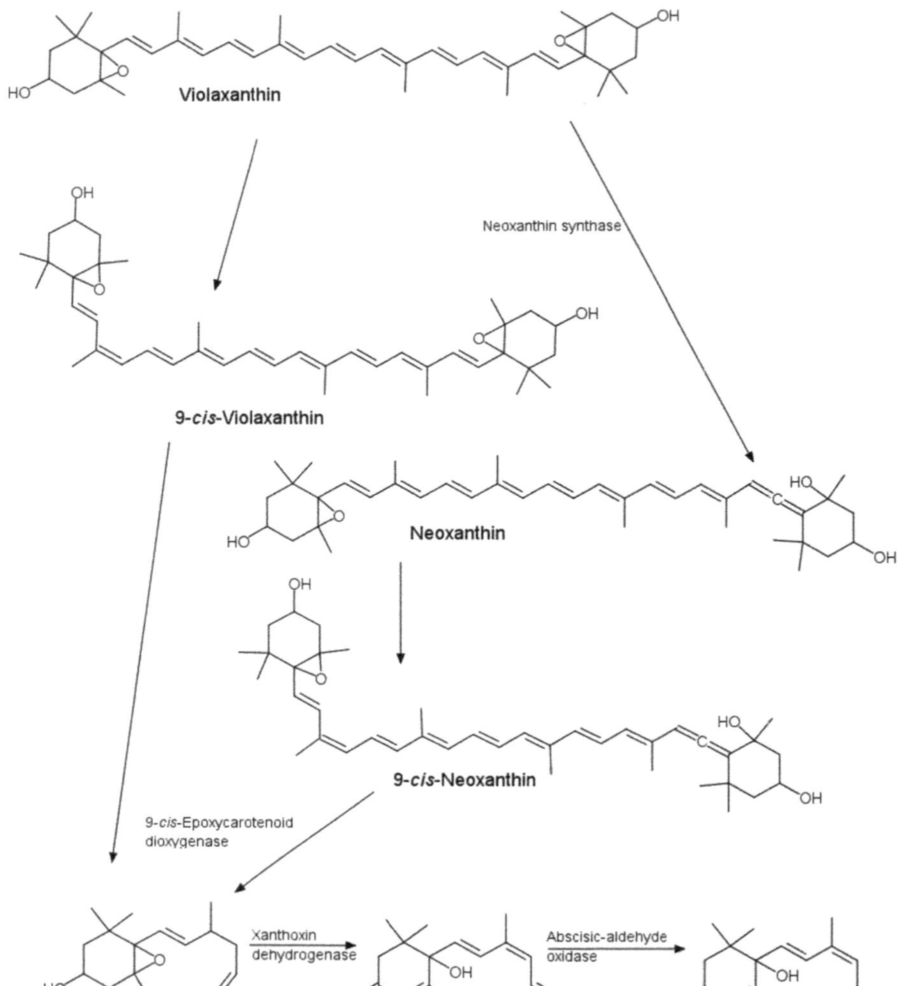

Figure 4.76 Formation of abscisic acid from violaxanthin and neoxanthin.

Different poly-*cis* prenols have been identified such as arachisprenol (C50–C60) from *Arachis hypogea* (Fabaceae). Rubber-like prenols (C400–C500) occur widely in plants. Natural rubber, obtained from the rubber tree *Hevea brasiliensis* (Euphorbiaceae) consists of a *cis*-1,4-polyisoprene polymer derived from isopentenyl monomers. Natural rubber has been reported to contain one epoxide group per 190–320 isoprene units. Rubber is a hydrophobic and biodegradable polymer. Rubber consists of 1500 to 60 000 isoprene units linked by *cis* double bonds.[158]

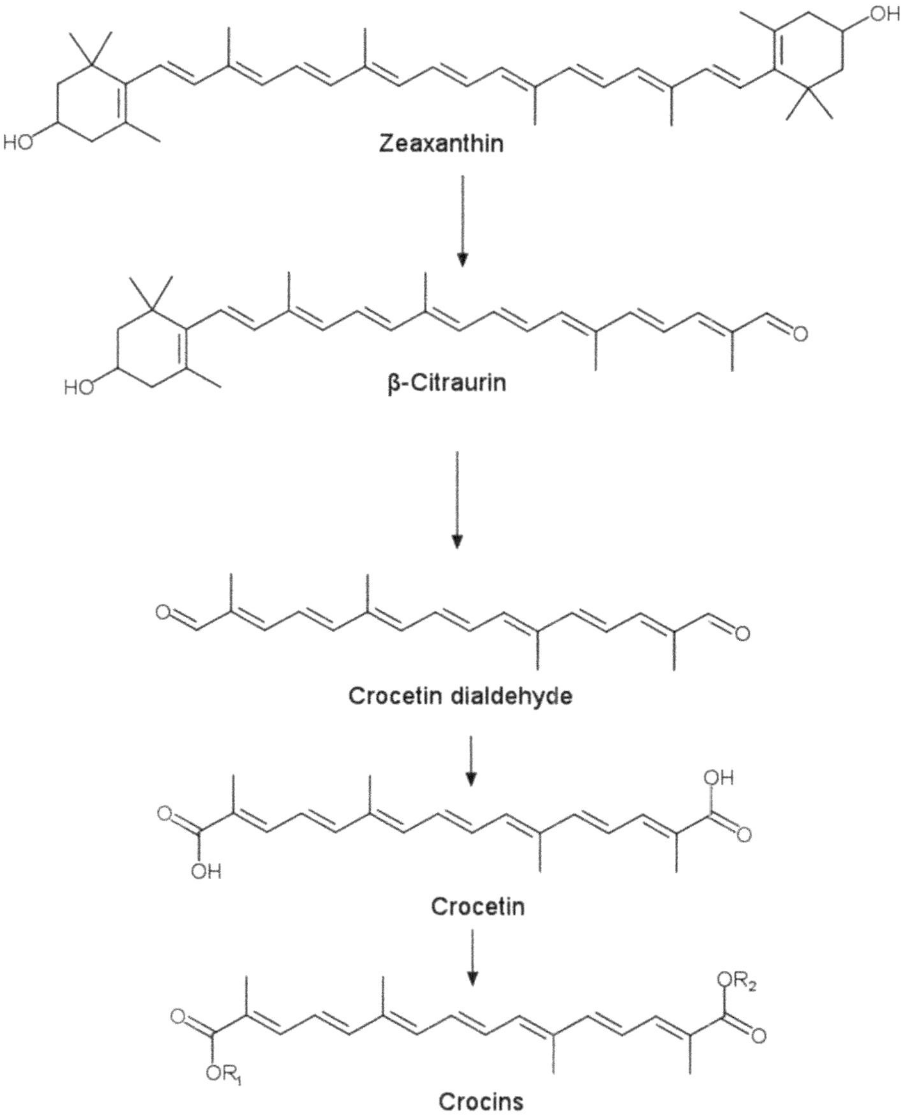

	R_1	R_2
Crocin 1	glucose	H
Crocin 2	gentiobiose	H
Crocin 3	gentiobiose	glucose
Crocin 4	gentiobiose	gentiobiose
Crocin 5	Three-glucose	gentiobiose

Figure 4.77 Formation of crocins from zeaxanthin.

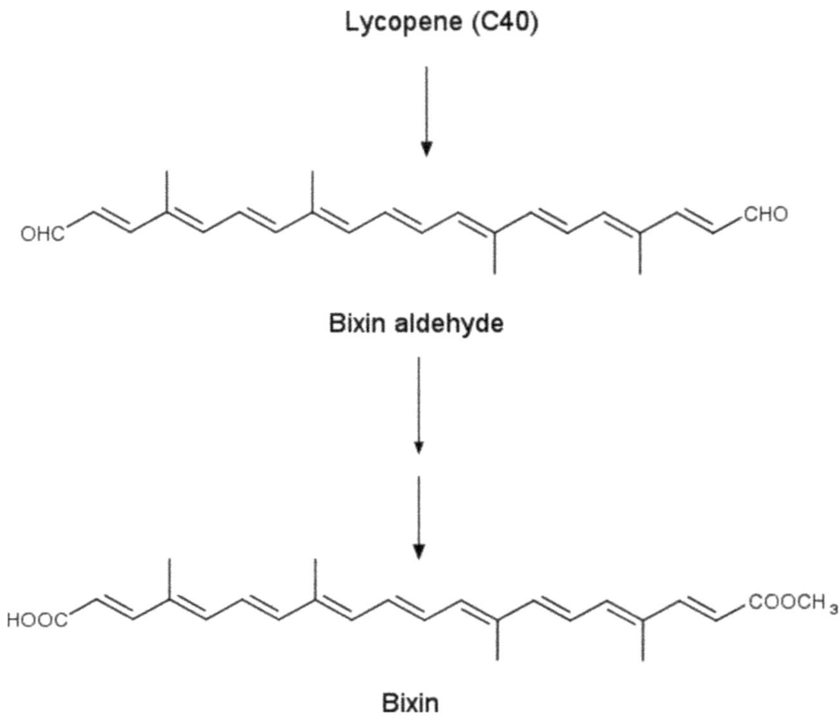

Figure 4.78 Formation of bixin from lycopene.

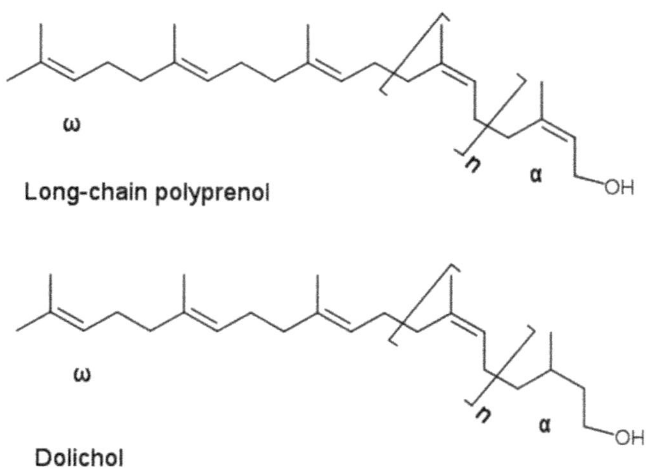

Figure 4.79 Polyprenols.

Dolichol is a poly-*cis*-prenol, hydrophobic in nature with C70 to C100 carbon atoms. It is an essential polyprenol within the endoplasmic reticulum of all eukaryotes. It is a membrane bound anchor onto which N-glycans are assembled prior to being transferred to nascent polypeptides.[156]

Prenylated Electron Carriers

Ubiquinone and plastoquinone (Figure 5.40) are two prenylquinones. They are made of a benzoquinone ring attached to a polyisoprenoid side chain. Plastoquinone and ubiquinone function as electron transporters in the electron transport chain in photosynthesis and the aerobic respiratory chain, respectively. They play an important role in plant growth and development, participate in biosynthesis and metabolism, are involved in plant stress response, act as antioxidants and regulate gene expression and cell signal transduction.[158,159]

References

1. Q. Wang, S. Quan and H. Xiao, Towards efficient terpenoid biosynthesis: manipulating IPP and DMAPP supply, *Bioresour. Bioprocess.*, 2019, **6**, 6.
2. A. O. Chatzivasoleiou, V. Ward, S. M. Edgar and G. Stephanopoulos, Two-step pathway for isoprenoid biosynthesis, *Proc. Natl. Acad. Sci. U. S. A.*, 2019, **116**, 506–511.
3. P. Pulido, C. Perello and M. Rodriguez-Concepcion, New insights into plant isoprenoid metabolism, *Mol. Plant*, 2012, **5**, 964–967.
4. M. A. Phillips, P. Leon, A. Boronat and M. Rodriguez-Concepcion, The plastidial MEP pathway: unified nomenclature and resources, *Trends Plant Sci.*, 2008, **13**, 619–623.
5. F. K. Bentley, A. Zurbriggen and A. Melis, Heterologous expression of the mevalonic acid pathway in cyanobacteria enhances endogenous carbon partitioning to isoprene, *Mol. Plant*, 2014, **7**, 71–86.
6. E. Corboda, M. Salmi and P. Leon, Unravelling the regulatory mechanisms that modulate the MEP pathway in higher plants, *J. Exp. Bot.*, 2009, **60**, 2933–2943.
7. P. M. Dewick, *Medicinal Natural Products: A Biosynthetic Approach*, John Wiley & Sons, Ltd, 3rd edn, 2009, pp. 187–310.
8. D. Yang, X. Du, X. Liang, R. Han, Z. Liang, Y. Liu, F. Liu and J. Zhao, Different roles of the mevalonate and methylerythritol phosphate pathways in cell growth and tanshinone production of *Salvia miltiorrhiza* hairy roots, *PLoS One*, 2012, **7**, e46797.
9. R. Chizzola, Regular monoterpenes and sesquiterpenes (essential oils), in *Natural Products*, ed. K. G. Ramawat and J. M. Merillon, Springer-Verlah Berlin Heidelberg, 2013, pp. 2973–3008.

10. M. n. Zielinska-Blajet and J. Feder-Kubis, Monoterpenes and their derivatives – recent development in biological and medical applications, *Int. J. Mol. Sci.*, 2020, **21**, 7078.
11. A. Koziol, A. Stryjewska, T. Librowski, K. Salat, M. Gawel, A. Moniczewski and S. Lochynski, An overview of the pharmacological properties and potential applications of natural monoterpenes, *Mini-Rev. Med. Chem.*, 2014, **14**, 1156–1168.
12. B. M. Lange and N. Srividya, Enzymology of monoterpene functionalization in glandular trichomes,*J. Exp. Bot.*, 2019, **70**, 1095–1108.
13. C. Wang, X. Gong, A. Bo, L. Zhang, M. Zhang, E. Zang, C. Zhang and M. Li, Iridoids: Research advances in their phytochemistry, biological activities, and pharmacokinetics, *Molecules*, 2020, **25**, 287.
14. K. Miettinen, L. Dong, N. Navrot, T. Schneider, V. Burlat, J. Pollier, L. Woittiez, S. van der Krol, R. Lugan, T. IlC, R. Verpoorte, K. Oksman-Caldentey, E. Martinoia, H. Bouwmeester, A. Goossens, J. Memelink and D. Werck-Reichhart, The seco-iridoid pathway from *Catharanthus roseus*, *Nat. Commun.*, 2014, **5**, 3606.
15. W. C. Evans, Miscellaneous isoprenoids, in *Trease and Evans Pharmacognosy*, Saunders, Elsevier, 16th edn, 2009.
16. Z. A. Demissie, L. A. E. Erland, M. R. Rheault and S. S. Mahamoud, The Biosynthetic origin of irregular monoterpenes in *Lavandula*, *J. Biol. Chem.*, 2013, **288**, 6333–6341.
17. W. Li, D. B. Lybrand, F. Zhoun, R. L. Last and E. Pichersky, Pyrethrin biosynthesis: The cytochrome P450 oxidoreductase CYP82Q3 converts Jasmolone to pyrethrolone, *Plant Physiol.*, 2019, **181**, 934–944.
18. E. Breitmaier, Sesquiterpenes, in *Terpenes: Flavours, Fragrances, Pharmaca, Pheromones*, Wiley-VCH Verlag GmbH & Co. KGaA, Weinheim, 2006, pp. 24–51.
19. X. Tian, X. Fang, J. Huang, L. Wang, Y. Mao and X. Chen, A gossypol biosynthetic intermediate disturbs plant defence response, *Philos. Trans. R. Soc., B*, 2019, **374**, 20180319.
20. M. Chadwick, H. Trewin, F. Gawthrop and C. Wagstaff, Sesquiterpenoid lactones: Benefits to plants and people, *Int. J. Mol. Sci.*, 2013, **14**, 12780–127805.
21. V. P. Sulsen and V. S. Martino, Overview, in *Sesquiterpene Lactones: Advances in Their Chemistry and Biological Aspects*, ed. V. P. Sulsen and V. S. Martino, Springer, 2018, ch. 1, pp. 3–18.
22. G. Eksi, S. Kurbanoglu and S. A. Erdem, Analysis of diterpenes and diterpenoids, in *Recent Advances in Natural Products Analysis*, Elsevier Inc., 2020, pp. 313–343.
23. I. P. de Sousa, M. V. S. Teixeira and J. C. Furtado, An overview of biotransformation and toxicity of diterpenes, *Molecules*, 2018, **23**, 1387.
24. D. S. Seigler, *Plant Secondary Metabolism*, SpringerLink, 1998, pp. 398–426.
25. V. Lanzotti, Diterpenes for therapeutic uses, in *Natural Products*, ed. K. G. Ramawat and J. M. Merillon, Springer-Verlag Berlin Heidelberg, 2013, pp. 3173–3191.

26. K. Kariya, N. Ube, M. Ueno, M. Teraishi, Y. Okumoto, N. Mori, K. Ueno and A. Ihihara, Natural variation of diterpenoid phytoalexins in cultivated and wild rice species, *Phytochemistry*, 2020, **190**, 112518.
27. X. Lu, J. Zhang, B. Brown, R. Li, J. Rodriguez-Romero, A. Berasategui, B. Liu, B. Xu, D. Luo, Z. Pan, S. R. Baerson, J. Gershenzon, Z. Li, A. Sesma, B. Yang and R. J. Peters, Inferring roles in defense from metabolic allocation of rice diterpenoids, *Plant Cell*, 2018, **30**, 1119–1131.
28. I. Pateraki, J. Anderson-Ranberg, B. Hamberger, A. M. Heskes, H. J. Martens, P. Zerbe, S. S. Bach, B. L. Moller, J. Bohlmann and B. Hamberger, Manoyl oxide (13R), the biosynthetic precursor of forskolin, is synthesized in specialized root cork cells in *Coleus forskohlii*, *Plant Physiol.*, 2014, **164**, 1222–1236.
29. A. D. Thingale, K. S. Shaikh, P. R. Channekar, U. C. Galgatte, P. D. Chaudhari and C. Bothiraja, Enhanced hepatoprotective activity of andrographolide complexed with a biomaterial, *Drug Delivery*, 2014, **22**, 117–124.
30. L. Gan, Y. Zheng, L. Deng, P. Sun, J. Ye, X. Wei, F. Liu, L. Yu, W. Ye, W. Fan, J. Liu and W. Zhang, Diterpenoid lactones with anti-inflammatory effects from the aerial parts of *Andrographis paniculata*, *Molecules*, 2019, **24**, 2726.
31. L. Silva, A. C. Gomes and J. M. L. Rodilla, Diterpene lactones with labdane, halimane and clerodane frameworks, *Nat. Prod. Commun.*, 2011, **6**, 497–504.
32. O. Grundmann, S. M. Phipps, I. Zadezensky and V. Butterweck, *Salvia divinorum* and salvinorin A: an update on pharmacology and analytical methodology, *Planta Med.*, 2007, **73**, 1039–1046.
33. B. Hamberger, T. Ohnishi, B. Hamberger, A. Seguin and J. Bohlmann, Evolution of diterpene metabolism: Sitka spruce CYP720B4 catalyzes multiple oxidations in resin acid biosynthesis of conifer defense against insects, *Plant Physiol.*, 2011, **157**, 1677–1695.
34. H. Saijo, H. Kofujita, K. Takahashi and T. Ashitani, Antioxidant activity and mechanism of the abietane-type diterpene ferruginol, *Nat. Prod. Res.*, 2015, **29**, 1739–1743.
35. I. Ara, B. S. Siddiqui, S. Faizi and S. Siddiqui, Tricyclic diterpenoids from the stem bark of *Azadirachta india*, *J. Nat. Prod.*, 1988, **51**, 1054–1061.
36. I. Ara, B. S. Siddiqui, S. Faizi and S. Siddiqui, Tricyclic diterpenoids from root bark of Azadirachta india, *Phytochemistry*, 1990, **29**, 911–914.
37. P. Hedden, Gibberellins, *Encyclopedia of Applied Plant Sciences*, Elsevier Ltd, 2nd edn, 2017, vol. 1, pp. 411–420, DOI: 10.1016/B978-0-12-394807-6.00219-7.
38. X. Gao, Y. Zhang, Z. He and X. Fu, Gibberellins, in *Hormone Metabolism and Plant Signaling in Plants*, ed. J. Li, C. Li and S. Smith, Elsevier Ltd, 2017, pp. 107–160, DOI: 10.1016/B978-0-12-811562-6.00004-9.
39. K. Hayashi, K. Horie, Y. Hiwatashi, H. Kawaide, S. Yamaguchi, A. Handa, T. Nakashima, M. Nakajima, L. N. Mander, H. Yamane, M. Hasebe and H. Nozaki, Endogenous diterpenes derived from ent-Kaurene, a common gibberellin precursor, regulate protonema differentiation of the moss *Physcomitrella patens*, *Plant Physiol.*, 2010, **153**, 1085–1097.

40. P. Hedden, The current status of research on gibberellin biosynthesis, *Plant Cell Physiol.*, 2020, **61**, 1832–1849.
41. K. Philippaert, A. Pironet, M. Mesuere, W. Sones, L. Vermeiren, S. Kerselaers, S. Pinto, A. Segal, N. Antoine, C. Gysemans, J. Laureys, K. Lemaire, P. Gilon, E. Cuypers, J. Tytgat, C. Mathieu, F. Schuit, P. Rorsman, K. Talavera, T. Voets and R. Vennekens, Steviol glycosides enhance pancreatic beta-cell function and taste sensation by potentiation of TRPM5 channel activity, *Nat. Commun.*, 2017, **8**, 14733.
42. S. G. Lee, E. Salomon, O. Yu and J. M. Jez, Molecular basis for branched steviol glucoside biosynthesis, *Proc. Natl. Acad. Sci. U. S. A.*, 2019, **116**, 13131–13136.
43. V. Lopez-Carbon, A. Sayago, R. Gonzalez-Dominguez and A. Fernandez-Recamales, Simple and efficient green extraction of steviol glycosides from *Stevia rebaudiana* leaves, *Foods*, 2019, **8**, 402.
44. S. Ceunen and J. M. C. Geuns, Steviol glycosides: Chemical diversity, metabolism, and function, *J. Nat. Prod.*, 2013, **76**, 1201–1228.
45. D. Matias, C. Bessa, M. F. Simoes, C. P. Reis, L. Saravia and P. Rijo, Natural products as lead protein kinase C modulators for cancer therapy, *Stud. Nat. Prod. Chem.*, 2016, **50**, 45–79.
46. G. Goel, H. P. S. Makkar, G. Francis and K. Becker, Phorbol esters: structure, biological activity, and toxicology, *Int. J. Toxicol.*, 2007, **26**, 279–288.
47. G. Appendino, Ingenane diterpenoids, in *Progress in the Chemistry of Organic Natural Product*, ed. A. D. Kinghorn, H. Falk, S. Gibbons and J. Kobayashi, Springer International Publishing Switzerland, 2016, vol. 102, pp. 1–86.
48. A. Vasas, D. Redei, D. Csupor, J. Molnar and J. Hohmann, Diterpenes from European *Euphorbia* species serving as prototypes for natural-product-based drug discovery, *Eur. J. Org. Chem.*, 2012, **27**, 5115–5130.
49. J. Li, I. Mutanda, K. Wang, L. Yang, J. Wang and Y. Wang, Chloroplast metabolic engineering coupled with isoprenoid pool enhancement for committed taxanes biosynthesis in *Nicotiana benthamiana*, *Nat. Commun.*, 2019, **10**, 4850.
50. U. Heinig and S. Jennewein, Taxol: a complex diterpenoid natural product with an evolutionarily obscure origin, *Afr. J. Biotechnol.*, 2009, **8**, 1370–1385.
51. C. Li, S. Liu, Q. Aerqin and X. Wu, Role of ginkgolides in the inflammatory immune response of neurological diseases: A review of current literatures, *Front. Syst. Neurosci.*, 2020, **14**, 45.
52. S. H. Omar, Ginkgolides and neuroprotective effects, in *Natural Products*, ed. K. G. Ramawat and J. M. Merillon, Springer-Verlag Berlin Heidelberg, 2013, pp. 3697–3740.
53. J. Cui, L. A. Hu, W. Shi, G. Cui, X. Zhang and Q. Zhang, Design, synthesis and Anti-platelet aggregation activity study of ginkgolide-1,2,3-triazole derivatives, *Molecules*, 2019, **24**, 2156.

54. Z. Feng, Q. Sun, W. Chen, Y. Bai, D. Hu and X. Xie, The neuroprotective mechanisms of ginkgolides and bilobalide in cerebral ischemic injury: a literature review, *Mol. Med.*, 2019, **25**, 57.
55. S. H. Omar, Ginkgolides and neuroprotective effects, in *Handbook of Natural Products*, ed. K. G. Ramawat and J. M. Merillon, Springer-Verlag Berlin Heidelberg, 2013, ch. 146, DOI: 10.1007/978-3-642-22144-6_146.
56. Y. Shen, W. Liang, Y. Shi, E. J. Kennelly and D. Zhao, Structural diversity, bioactivities, and biosynthesis of natural diterpenoid alkaloids, *Nat. Prod. Rep.*, 2020, **37**, 763–796.
57. K. Wada and H. Yamashita, Cytotoxic effects of diterpenoid alkaloids against human cancer cells, *Molecules*, 2019, **24**, 2317.
58. K. Li and K. R. Gustafson, Sesterterpenoids: chemistry, biology, and biosynthesis, *Nat. Prod. Rep.*, 2021, **38**, 1251–1281.
59. R. Thimmappa, K. Geisler, T. Louveau, P. O'Maille and A. Osbourn, Triterpene biosynthesis in plants, *Annu. Rev. Plant Biol.*, 2014, **65**, 225–257.
60. K. Miettinen, S. Inigo, L. Kreft, J. Pollier, C. De Bo, A. Botzki, F. Coppens, S. Bak and A. Goossens, The TriForC database: a comprehensive up-to-date resource of plant triterpene biosynthesis, *Nucleic Acids Res.*, 2017, **46**, D586–D594.
61. Y. Shang and S. Huang, Multi-omics data driven investigations of metabolic diversity of plant triterpenoids, *Plant J.*, 2019, **97**, 101–111.
62. P. D. Cardenas, A. Almeida and S. Bak, Evolution of structural diversity of triterpenoids, *Front. Plant Sci.*, 2019, **10**, 1523.
63. P. V. Welander, M. L. Coleman, A. L. Sessions, R. E. Summons and D. K. Newman, Identification of a methylase required for 2-methylpropanoid production and implications for the interpretation of sedimentary hopanes, *Proc. Natl. Acad. Sci. U. S. A.*, 2010, **107**, 8537–8542.
64. W. Chen, P. Balan and D. G. Popovich, Comparison of ginsenoside components of various tissues of New Zealand forest-grown Asian ginseng (*Panax ginseng*) and American ginseng (*Panax quinquefolium* L.), *Biomolecules*, 2020, **10**, 372.
65. L. Bai, J. Gao, F. Wei, J. Zhao, D. Wang and J. Wei, Therapeutic potential of ginsenosides as an adjuvant treatment for diabetes, *Front. Pharmacol.*, 2018, **9**, 423.
66. M. Zheng, Y. Xin, Y. Li, F. Xu, X. Xi, H. Guo, X. Cui, H. Cao, X. Zhang and C. Han, Ginsenosides: a potential neuroprotective agent, *BioMed Res. Int.*, 2018, 8174345.
67. S. Ramasamy, S. P. Chin, S. D. Sukumaran, M. J. C. Buckle, L. V. Kiew and L. Y. Chung, In silico and in vitro analysis of bacoside A aglycones and its derivatives as the constituents responsible for the cognitive effects of *Bacopa monnieri*, *PLoS One*, 2015, 0126565.
68. G. Gutierrez, L. M. Valencia, D. Giraldo-Davila, M. Y. Combariza, E. Galeano, N. Balcazar, A. J. Panay, A. M. Jerez and G. Montoya, Pentacyclic triterpene profile and its biosynthetic pathway in *Cecropia telenitida* as a prospective dietary supplement, *Molecules*, 2021, **26**, 1064.

69. S. Jager, H. Trojan, T. Kopp and M. N. Laszczyk, A. Scheffler, Pentacyclic triterpenoid distribution in various plants – rich sources for a new group of multi-potent plant extracts, *Molecules*, 2009, **14**, 2016–2031.
70. Sandeep, R. C. Misra, C. S. Chanotiya, P. Mukhopadhyay and S. Ghosh, Oxidosqualene cyclase and CYP716 enzymes contribute to triterpene structural diversity in the medicinal tree banaba, *New Phytol.*, 2019, **222**, 408–424.
71. M. H. Ghante and P. G. Jamkhande, Role of pentacyclic triterpenoids in chemoprevention and anticancer treatment: an overview on targets and underlying mechanisms, *J. Pharmacopuncture*, 2019, **22**, 55–67.
72. L. M. Bedoya, M. Beltran, J. Garcia-Perez, P. Obregon-Calderon, O. Callies, I. A. Jimenez, I. L. Bazzocchi and J. Alcami, Promiscuous, multi-target lupine-type triterpenoids inhibit wild type and drug resistant HIV-1 replication through the interference with several targets, *Front. Pharmacol.*, 2018, **9**, 358.
73. J. Zhao, R. Li, A. Pawlak, M. Henklewska, A. Sysak, L. Wen, J. Yi and B. Obminska-Mrukowicz, Antitumor activity of betulinic acid and betulin in canine cancer cell lines, *In Vivo*, 2018, **32**, 1081–1088.
74. Y. Takada, H. Sasama, T. Sayama, A. Kikuchi, S. Kato, M. Ishimoto and C. Tsukamoto, Genetic and chemical analysis of a key biosynthetic step for soyasapogenol A, an aglycone of group A saponins that influence soymilk flavor, *Theor. Appl. Genet.*, 2013, **126**, 721–731.
75. S. Saha, S. Walia, J. Kumar, S. Dhingra and B. S. Parmar, Screening for feeding deterrent and insect growth regulatory activity of triterpenic saponins from *Diploknema butyracea* and *sapindus mukorossi*, *J. Agric. Food Chem.*, 2010, **58**, 434–440.
76. G. S. Thakur, R. Sharma, B. S. Sanodiya, M. Pandey, G. B. K. S. Prasad and P. S. Bisen, *Gymnema sylvestre*: an alternative therapeutic agent for management of diabetes, *J. Appl. Pharm. Sci.*, 2012, **2**, 1–6.
77. K. Sanematsu, Y. Kusakabe, N. Shigemura, T. Hirokawa, S. Nakamura, T. Imoto and Y. Ninomiya, Molecular mechanisms for sweet-suppressing effect of gymnemic acids, *J. Biol. Chem.*, 2014, **289**, 25711–25720.
78. J. D. Fleck, A. H. Betti, F. P. da Silva, E. A. Troian, C. Olivaro, F. Ferreira and S. G. Verza, Saponins from *Quillaja Saponaria* and *Quillaja brasiliensis*: Particular chemical characteristics and biological activities, *Molecules*, 2019, **24**, 171.
79. W. E. Walkowicz, A. Fernandez-Tejada, C. George, F. Corzana, J. Jimenez-Barbero, G. Raghupathi, D. S. Tan and D. Y. Gin, *Quillaja* saponin variants with central glycosidic linkage modifications exhibit distinct conformations and adjuvant activities, *Chem. Sci.*, 2016, **7**, 2371.
80. C. L. Reichert, H. Salminen and J. Weiss, Quillaja saponin characteristics and functional properties, *Annu. Rev. Food Sci. Technol.*, 2019, **10**, 11.1–11.31.
81. K. Miettinen, J. Pollier, D. Byust, P. Arendt, R. Csuk, S. Sommerwerk, T. Moses, J. Mertens, P. D. Sonawane, L. Pauwels, A. Aharoni, J. Martins, D. R. Nelson and A. Goossens, CYP716 family is a major contributor to

the diversification of eudicot triterpenoid biosynthesis, *Nat. Commun.*, 2016, **81**, 4153.

82. A. Al-Harrasi, N. Ur-Rehman, A. L. Khan, M. Al-Broumi, I. Al-Amri, J. Hussain, H. Hussain and R. Csuk, Chemical, molecular and structural studies of *Boswellia* species: β-boswellic aldehyde and 3-epi-11β-dihydroxy BA as precursors in biosynthesis of boswellic acids, *PLoS One*, 2018, **13**, e0198666.
83. M. Schmiech, J. Ulrich, S. J. Lag, B. Buchele, C. Paetz, A. St-Gelais, T. Syrovets and T. Simmet, 11-Keto-α-boswellic acid, a noveltriterpenoid from *Boswellia* spp. with chemotaxonomic potential and antitumor activity against triple-negative breast cancer cells, *Molecules*, 2021, **26**, 366.
84. J. Zhou, T. Hu, L. Gao, P. Su, Y. Zhang, Y. Zhao, S. Chen, L. Tu, Y. Song, X. Wang, L. Huang and W. Gao, Friedelane-type triterpene cyclase in celastrol biosynthesis from *Tripterygium wilfordii* and its application for triterpenes biosynthesis in yeast, *New Phytol.*, 2019, **223**, 722–735.
85. J. Shi, J. Li, Z. Xu, L. Chen, R. Luo, C. Zhang, F. Gao, J. Zhang and C. Fu, Celastrol: A review of useful strategies overcoming its limitation in anticancer application, *Front. Pharmacol.*, 2020, **11**, 558741.
86. A. Almeida, L. Dong, G. Appendino and S. Bak, Plant triterpenoids with bond-missing skeletons: biogenesis, distribution and bioactivity, *Nat. Prod. Rep.*, 2020, **37**, 1207–1228.
87. A. Roy and S. Saraf, Limonoids: overview of significant bioactive triterpenes distributed in plants kingdom, *Biol. Pharm. Bull.*, 2006, **29**, 191–201.
88. A. J. Mordue (Luntz), E. D. Morgan and A. J. Nisbet, Azadirachtin, a natural product in insect control, in *Comprehensive Molecular Insect Science*, Elsevier BV, 2005, vol. 6, pp. 117–135.
89. I. J. C. Vieira and R. Braz-Filho, Quassinoids: structural diversity, Biological activity and synthetic studies, in *Studies in Natural Products Chemistry*, ed. Atta-ur-Rahman, Elsevier B.V., 2006, vol. 33, pp. 433–492.
90. I. Muhammad and V. Samoylenko, Antimalarial quassinoids: past, present and future, *Expert Opin. Drug Discovery*, 2007, **2**, 1065–1084.
91. G. Fiaschetti, M. A. Grotzer, T. Shalaby, D. Castelletti and A. Arcaro, Quassinoids: From traditional drugs to new cancer therapeutics, *Curr. Med. Chem.*, 2011, **18**, 316–328.
92. V. E. Tyler, L. R. Brady and J. E. Roberts, Lipids, in *Pharmacognosy*, Lee & Berger, 1988, ch. 7, pp. 156–173.
93. V. Piironen, D. G. Lindsay, T. A. Miettinen, J. Toivo and A. Lampi, Plant sterols: biosynthesis, biological function and their importance to human nutrition, *J. Sci. Food Agric.*, 2000, **80**, 939–966.
94. R. A. Moreau, B. D. Whitaker and K. B. Hicks, Phytosterols, phytostanols, and their conjugates in foods: structural diversity, quantitative analysis, and health promoting uses, *Prog. Lipid Res.*, 2002, **41**, 457–500.
95. S. I. Aboobucker and W. P. Suza, Why do plants convert sitosterol to stigmasterol?, *Front. Plant Sci.*, 2019, **10**, 354.

96. D. D. Bikle, Vitamin D metabolism, mechanism of action, and clinical applications, *Chem. Biol.*, 2014, **21**, 319–329.
97. J. L. Simmons-Boyce, and W. F. Tinto, Steroidal saponins from the Agavaceae family, *Nat. Prod. Commun.*, 2007, **2**, 99–114.
98. M. E. Wall, M. M. Krider, E. S. Rothman and C. R. Eddy, Steroidal sapogenins, *J. Biol. Chem.*, 1952, **198**, 533–543.
99. J. Cheng, J. Chen, X. Liu, X. Li, W. Zhang, Z. Dai, L. Lu, X. Zhou, J. Cai, H. Zhang and Y. Ma, The origin and evolution of the diosgenin biosynthetic pathway in yam, *Plant Commun.*, 2021, **2**, 100079.
100. P. Heretsch and A. Giannis, The Veratrum and Solanum alkaloids, in *The Alkaloids: Chemistry and Biology*, 2015, vol. 74, pp. 201–232.
101. J. McKenna, Steroidal alkaloids, *Q. Rev., Chem. Soc.*, 1953, **7**, 231–254.
102. E. L. Ghisalberti, Steroidal glycoalkaloids: Isolation, structure, analysis, and biosynthesis, *Nat. Prod. Commun.*, 2006, **1**, 859–884.
103. P. D. Cardenas, P. D. Sonawane, U. Heinig, S. E. Bocobza, S. Burdman and A. Aharoni, The bitter side of the nightshades: Genomics drives discovery in Solanacea steroidal alkaloid metabolism, *Phytochemistry*, 2015, **113**, 24–32.
104. P. Heretsch, S. Rabe and A. Giannis, A biomimetic approach to C-nor-D-homo-steroids, *J. Am. Chem. Soc.*, 2010, **132**, 9968–9969.
105. M. M. Augustin, D. R. Ruzika, A. K. Shukla, J. M. Augustin, C. M. Starks, M. O'neil-Johnson, M. R. McKain, B. S. Evans, M. D. Barrett, A. Smithson, G. K. Wong, M. K. Deyholos, P. P. Edger, J. C. Pires, J. M. Leebens-Mack, D. A. Mann and T. M. Kutchan, Elucidating steroid alkaloid biosynthesis in *Veratrum californicum*: production of verazine in Sf9 cells, *Plant J.*, 2015, **82**, 991–1003.
106. P. M. Dewick, Alkaloids, in *Medicinal Natural Products: A Biosynthetic Approach*, John Wiley & Sons, Ltd, 3rd edn, 2009, pp. 311–420.
107. S. Patel, Plant derived cardiac glycosides: Role in heart ailments and cancer management, *Biomed. Pharmacother.*, 2016, **84**, 1036–1041.
108. N. Morsy, Cardiac glycosides in medicinal plants, 2017, www.intechopen.com, DOI: 10.5772/65963
109. H. R. El-Seedi, S. A. M. Khalifa, E. A. Tather, M. A. Farag, A. Saeed, M. Gamal, M. F. Hegazy, D. Youssef, S. G. Musharaf, M. M. Alajlani, J. Xiao and T. Efferth, The cardenolides: Insights from chemical structure and pharmacological utility, *Pharmacol. Res.*, 2019, **141**, 123–175.
110. W. Kreis, The foxgloves (Digitalis) revisited, *Planta Med.*, 2017, **83**, 962–976.
111. L. H. A. Cavalcante-Silva, E. A. Lima, D. C. M. Carvalho, J. M. Sales-Neto, A. K. A. Alves, J. G. F. M. Galvao, J. S. F. da Silva and S. Rodrigues-Mascarenhas, Much more than a cardiotonic steroid: modulation of inflammation by ouabain, *Front. Physiol.*, 2017, **8**, 895.
112. N. F. Z. S. Schneider, C. Cerella, C. M. O. Simoes, and M. Diederich, Anticancer and immunogenic properties of cardiac glycosides, *Molecules*, 2017, **22**, 1932.

113. D. Reddy, R. Kumavath, D. Barh, V. Azevedo and P. Ghosh, Anticancer and antiviral properties of cardiac glycosides: A review to explore the mechanism of actions, *Molecules*, 2020, **25**, 3596.
114. W. Kreis and F. Muller-uri, Cardenolide aglycone formation in Digitalis, in *Isoprenoid Synthesis in Plants and Microorganisms: New Concepts and Experimental Approaches*, ed. T. J. Bach and M. Rohmer, Springer Science + Business Media New Work, 2013, pp. 425–438.
115. I. F. Makerevich and S. V. Kovalev, Cardiac glycosides from *Strophanthus kombe*, *Chem. Nat. Compd.*, 2006, **42**, 189–193.
116. P. S. Steyn and F. R. van Heerden, Bufadienolides of plant and animal origin, *Nat. Prod. Rep.*, 1998, **15**, 397–413.
117. M. Huang, J. He, H. Hu, K. Zhang, X. Wang, B. Zhao, H. Lou, D. Ren and T. Shen, Withanolides from the genus Physalis: a review on their phytochemical and pharmacological aspects, *J. Pharm. Pharmacol.*, 2020, **72**, 649–669.
118. L. Chen, H. He and F. Qiu, Natural withanolides: an overview, *Nat. Prod. Rep.*, 2011, **28**, 705.
119. E. Knoch, S. Sugawara, T. Mori, C. Poulsen, A. Fukushima, J. Harholt, Y. Fujimoto, N. Umemoto and K. Saito, Third DWF1 paralog in Solanaceae, sterol Δ^{24}-isomerase, branches withanolide biosynthesis from the general phytosterol pathway, *Proc. Natl. Acad. Sci. U. S. A.*, 2018, **115**, E8096–E8103.
120. S. Pal, S. Rastogi, D. A. Nagegowda, M. M. Gupta, A. K. Shasany and C. S. Chanotiya, RNAi of sterol methyl transferase 1 reveals its direct role in diverting intermediates towards withanolide/phytosterol biosynthesis in *Withania somnifera*, *Plant Cell Physiol.*, 2019, **60**, 672–686.
121. N. Dhar, S. Razdan, S. Rana, W. W. Bhat, R. Viswakarma and S. K. Lattoo, A decade of molecular understanding of withanolide biosynthesis and in vitro studies in Withania somnifera (L.) Dunal: Prospects and perspectives for pathway engineering, *Front. Plant Sci.*, 2015, **6**, 1031.
122. P. T. White, C. Subramanium, H. F. Motiwala and M. S. Cohen, Natural withanolides in the treatment of diseases, in *Anti-inflammatory Nutraceuticals and Chronic Diseases*, ed. S. C. Gupta, *et al.*, Advances in Experimental Medicine and Biology, Springer International Publishing, Switzerland, 2016, vol. 928, pp. 329–373.
123. M. H. Mirjalili, E. Moyano, M. Bonfill, R. M. Cusido and J. Palazon, Steroidal lactones from *Withania somnifera*, an ancient plant of novel medicine, *Molecules*, 2009, **14**, 2373–2393.
124. A. Singh, S. Duggal, H. Singh, J. Singh and S. Kakekhaye, Withanolides: Phytoconstituents with significant pharmacological activities, *Int. J. Green Pharm.*, 2010, **4**, 229–237.
125. S. Garg, S. C. Kaul and R. Wadhwa, Cucurbitacins B and cancer intervention: chemistry, biology and Mechanisms (review), *Int. J. Oncol.*, 2018, **52**, 19–37.

126. Y. Kim, D. Choi, A. Cha, Y. Lee, N. Baek, S. Rimal, J. Sang, Y. Lee and S. Lee, Critical enzymes for biosynthesis of cucurbitacin derivatives in watermelon and their biological significance, *Commun. Biol.*, 2020, **3**, 444.
127. J. C. Chen, M. H. Chiu, R. L. Nie, G. A. Cordell and S. X. Qiu, Cucurbitacins and cucurbitane glycosides: structures and biological activities, *Nat. Prod. Rep.*, 2005, **22**, 386–399.
128. A. Bajguz, M. Chmur and D. Gruszka, Comprehensive overview of the brassinosteroid biosynthesis pathways, substrates, products, inhibitors, and connections, *Front. Plant Sci.*, 2020, **11**, 1034.
129. J. Tang, Z. Han and J. Chai, Q&A: what are brassinosteroids and how do they act in plants?, *BMC Biol.*, 2016, **14**, 113.
130. L. Taha-Salaime, R. Davidovich-Rikanati, A. Sadeh, J. Abu-Nassar, S. Marzouk-Kheredin, Y. Yahyaa, M. Ibdah, M. Ghanim, E. Lewinsohn, M. Inbar and R. Aly, Phytoecdysteroids and clerodane content in three wild *Ajuga* species in Israel, *ACS Omega*, 2019, **4**, 2369–2376.
131. M. Bathori, N. Toth, A. Hunyadi, A. Markim and E. Zador, Phytoecdysteroids and anabolic-androgenic steroids – structure and effects on humans, *Curr. Med. Chem.*, 2008, **15**, 75–91.
132. K. H. Boo, D. Lee, G. L. Jeon, S. H. Ko, S. K. Cho, J. H. Kim, S. P. Park, Q. Hong, S. Lee, D. Lee and K. Z. Riu, Distribution and biosynthesis of 20-htdroxyecdysone in plants *Achyranthes japonica* Nakai, *Biosci., Biotechnol., Biochem.*, 2010, **74**, 2226–2231.
133. A. Janeczko and A. Skoczowski, Mammalian sex hormones in plants, *Folia Histochem. Cytobiol.*, 2005, **43**, 71–79.
134. D. Tarkowska, Plants are capable of synthesizing animal steroid hormones, *Molecules*, 2019, **24**, 2585.
135. L. Fernandez-Cabezon, B. Galan and J. L. Garcia, New insights on steroid biotechnology, *Front. Microbiol.*, 2018, **9**, 958.
136. A. Eroglu and E. H. Harrison, Carotenoid metabolism in mammals, including man: formation, occurrence, and function of apocarotenoids, *J. Lipid Res.*, 2013, **54**, 1719–1730.
137. T. Maoka, Carotenoids as natural functional pigments, *J. Nat. Med.*, 2020, **74**, 1–16.
138. F. Shahidi and J. A. Brown, Carotenoid pigments in seafoods and aquaculture, *Crit. Rev. Food Sci.*, 1998, **38**, 1–67,
139. M. F. De Jesus Raposo, A. M. M. B. de Morais and R. M. S. C. de Morais, Carotenoids from marine microalgae: A valuable natural source for the prevention of chronic diseases, *Mar. Drugs*, 2015, **13**, 5128–5155.
140. O. C. Demurtas, R. B. Francisco, G. Diretto, P. Ferrante, S. Frusciante, M. Pietralla, G. Aprea, L. Borghi, M. Feeney, L. Frigerio, A. Coricello, G. Costa, S. Alcaro, E. Martinoia and G. Giuliano, ABCC transporters mediate the vascular accumulation of crocins in saffron stigmas, *Plant Cell*, 2019, **31**, 2789–2804.
141. A. Hannoufa and Z. Hossain, Carotenoid biosynthesis and regulation in plants, *Agriculture and Agri-Food Canada*, 1391 Sandford Street, London,

Ontario, Canada, 2020, https://lipidlibrary.aocs.org/chemistry/physics/plant-lipid/carotenoid-biosynthesis-and-regulation-in-plants#:~:text=Carotenoids%20are%20lipophilic%20secondary%20metabolites,in%20attracting%20pollinators%20to%20flowers, (last seen on 4 June 2022).
142. N. Nisar, L. Li, S. Lu, N. C. Khin and B. J. Pogson, Carotenoid metabolism in plants, *Mol. Plant*, 2015, **8**, 68–82.
143. L. Stanley and Y. Yuan, Transcriptional regulation of carotenoid biosynthesis in plants: so many regulators, so little consensus, *Front. Plant Sci.*, 2019, **10**, 1017.
144. J. Hirschberg, Carotenoid biosynthesis in flowering plants, *Curr. Opin. Plant Biol.*, 2001, **4**, 2010–2018.
145. K. Jia, L. Baz and S. Al-Babili, From carotenoids to strigolactones, *J. Exp. Bot.*, 2018, **69**, 2189–2204.
146. T. H. Yeats and R. Nagel, Subcellular spice trade routes: Crocin biosynthesis in the saffron Crocus (*Crocus sativus*), *Plant Physiol.*, 2018, **177**, 869–870.
147. O. C. Demurtas, S. Frusciante, P. Ferrante, G. Diretto, N. H. Azad, M. Pietrella, G. Aprea, A. R. Taddei, E. Romano, J. Mi, S. Al-Babili, L. Frigerio and G. Giuliano, Candidate enzymes for saffron crocin biosynthesis are localized in multiple cellular compartments, *Plant Physiol.*, 2018, **177**, 990–1006.
148. F. Bouvier, O. Dogbo and B. Camara, Biosynthesis of food and cosmetic plant pigment bixin (Annatto), *Science*, 2003, **300**, 2089–2091.
149. Y. Li and D. C. Walton, Violaxanthin is an abscisic acid precursor in water stressed dark grown bean leaves, *Plant Physiol.*, 1990, **92**, 551–559.
150. S. H. Schwartz, B. C. Tan, D. A. Gage, J. A. D. Zeevaart, and D. R. McCarty, Specific oxidative cleavage of carotenoids by VP14 of maize, *Science*, 1997, **276**, 1872–1874.
151. B. V. Milborrow, The pathway of biosynthesis of abscisic acid in vascular plants: a review of the present state of knowledge of ABA Biosynthesis, *J. Exp. Bot.*, 2001, **52**, 1145–1164.
152. V. Fiorilli, J. Y. Wang, P. Bonfante and L. Lanfranco, Apocarotenoids: Old and new mediators of arbuscular mycorrhizal symbiosis, *Font. Plant Sci.*, 2019, **10**, 1186.
153. N. R. Hofmann, The who, what, and where of plant polyprenol biosynthesis point to thylakoid membranes and photosynthetic performance, *Plant Cell*, 2017, **29**, 1552–1553.
154. M. Enoki, Y. Doi and T. Iwata, Oxidative degradation of cis- and trans-1,4-polyisoprenes and vulcanized natural rubber with enzyme-mediator systems, *Biomacromolecules*, 2003, **4**, 314–320.
155. H. Sagami, E. Swiezewska and Y. Shidoji, The history and recent advances in research of polyprenol and its derivatives, *Biosci., Biotechnol., Biochem.*, 2018, **82**, 947–955.
156. E. Swiezewska and W. Danikiewicz, Polyisoprenoids: Structure, biosynthesis and function, *Prog. Lipid Res.*, 2005, **44**, 235–258.

157. V. Viswanath and H. M. Rao, Gutta-percha in endodontics – A comprehensive review of material science, *J. Conservative Dent.*, 2019, **22**, 2016–2222.
158. K. Cornish and W. Xie, Natural rubber biosynthesis in plants: Rubber transferase, *Methods Enzymol.*, 2012, **515**, 63–82.
159. M. Liu and S. Lu, Plastoquinone and ubiquinone in plants: Biosynthesis, physiological function and metabolic engineering, *Front. Plant Sci.*, 2016, 7, 1898.

CHAPTER 5

Shikimic Acid Pathway: Phenols

A **phenol** is a compound that has one or more hydroxyl groups attached to a benzene ring. Polyphenols are compounds that have at least two phenyl rings and more than one hydroxyl groups attached to one or more of the rings. Phenolic compounds comprise a large and diverse group of metabolites detected in all plants. Most of the phenolic compounds are biosynthesized *via* the **shikimic acid pathway**. In contrast, some of the polyphenol structures [coumarins (C6–C3), quinoids, flavonoids (C6–C3–C6), xanthones, stilbenes, curcuminoids, and phlobaphenes] are biosynthesized by a combination of the shikimic acid pathway originated phenylpropanoid pathway and the polyketide or MVA/MEP pathway (Figure 5.1).

The **shikimic acid pathway** begins with the condensation of phosphoenolpyruvate derived from glycolysis and erythrose 4-phosphate derived from the pentose phosphate pathway and ends with the synthesis of chorismic acid. The pathway consists of seven enzymatic steps. Plastid is the exclusive site of chorismic acid biosynthesis.[1] Animals do not possess the shikimic acid pathway. The intermediate steps and the enzymes involved in the pathway are shown in Figure 5.2.

Chorismic acid is the precursor of the three aromatic amino acids phenylalanine, tyrosine, and tryptophan in microorganisms and plants. Biosynthesis of phenylalanine and tyrosine requires conversion of chorismic acid to prephenic acid by chorismate mutase. Recent findings suggest that phenylalanine and tyrosine are produced either through the arogenate pathway or through the phenylpyruvate pathway (Figure 5.3). Experimental studies revealed that the arogenate pathway is the predominant pathway for the synthesis of these amino acids. In the arogenate pathway, the transamination reaction takes place when prephenic acid is converted to arogenic acid. The

Chemical Diversity of Plant Specialized Metabolites: A Biosynthetic Approach
By Bratati De

Published by the Royal Society of Chemistry, www.rsc.org

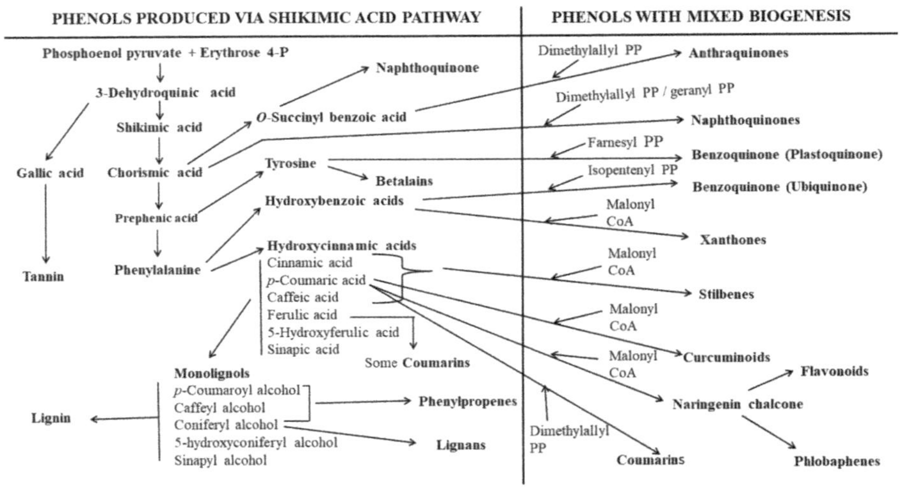

Figure 5.1 Schematic representation of phenol biosynthesis.

reaction is catalyzed by prephenate aminotransferase (PPA-AT). Arogenic acid then undergoes dehydration/decarboxylation to phenylalanine and tyrosine. Phenylalanine and tyrosine are exported to the cytosol. Chorismic acid is also exported to the cytosol. In case of the alternate phenylpyruvate pathway, dehydration/dehydrogenation/decarboxylation of prephenic acid leads to the formation of phenylpyruvate or 4-hydroxyphenylpyruvate. Then a transamination reaction occurs. The plastidial tyrosine acts as the amino donor utilizing cytosolic amino transferases *e.g.*, phenylpyruvate aminotransferase (PPY-AT) and tyrosine aminotransferase (TAT). Phenylpyruvate aminotransferase, converting phenylpyruvate to phenylalanine, is localized in the cytosol.[2–4]

The aromatic amino acids phenylalanine and tyrosine serve as the precursors of proteins and more than 8000 plant phenolic compounds. Lignin (the structural component of plant cell walls), quinones (ubiquinone), signaling molecules (salicylic acid, gentisic acid), pigments (anthocyanins), phytoalexins, feeding deterrents (tannins), defence chemicals and UV protectants (flavonoids) play crucial roles in the survival of the plants. Animals lack the ability to synthesize phenylalanine and depend on the dietary supply of this amino acid. The plant products derived from these amino acids also protect humans against many diseases.

Based on their structures, the phenolic compounds may be classified into following groups: simple phenols (C6), phenolic acids including hydroxybenzoic acids (C6–C1) and hydroxycinnamic acids (C6–C3), lignans, lignin, tannins, phenylpropenes, betalains, coumarins (C6–C3), quinoids (benzoquinones, naphthoquinones, and anthraquinones), flavonoids (C6–C3–C6), xanthones, stilbenes, curcuminoids and phlobaphenes.

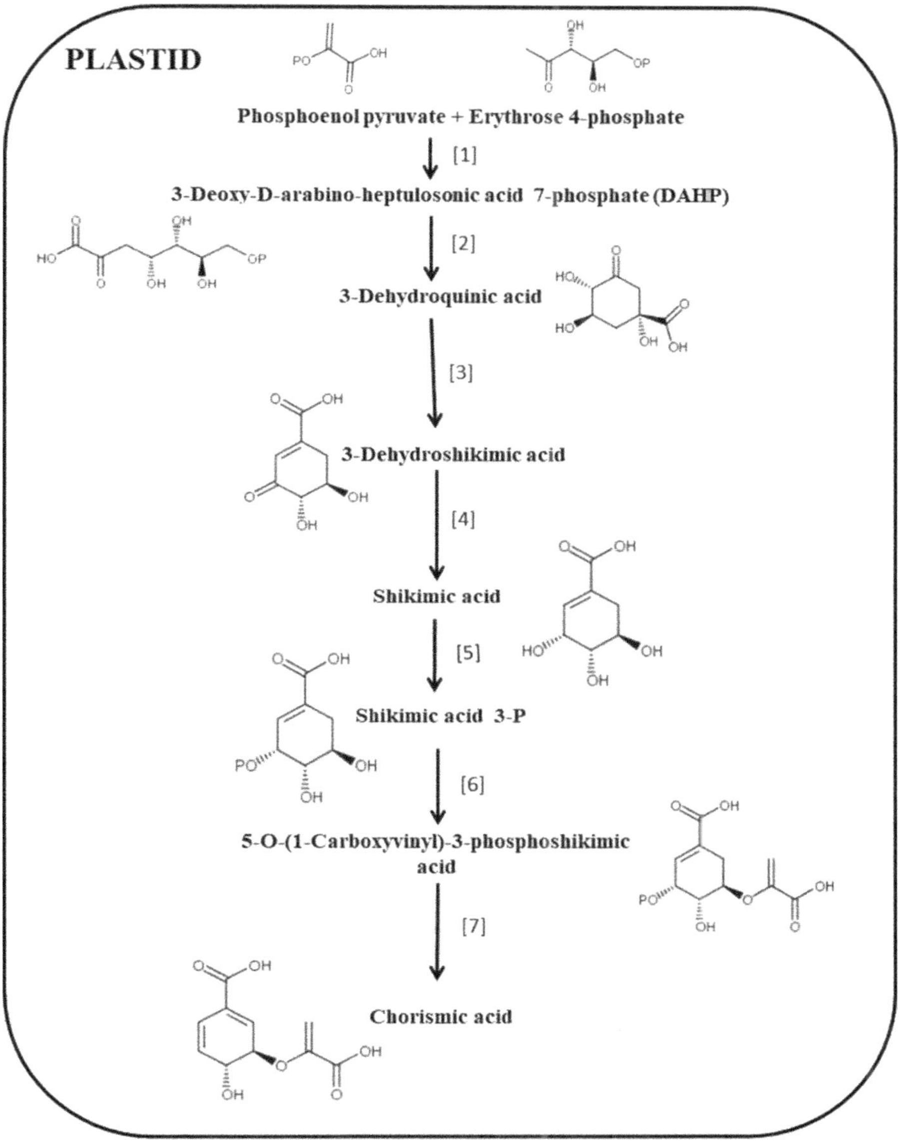

Figure 5.2 Shikimic acid pathway. [1]: 3-Deoxy-7-phosphoheptulonate synthase; [2]: 3-dehydroquinate synthase; [3]: 3-dehydroquinate dehydratase; [4]: shikimate dehydrogenase; [5]: shikimate kinase; [6]: 3-phosphoshikimate 1-carboxyvinyltransferase; [7]: chorismate synthase.

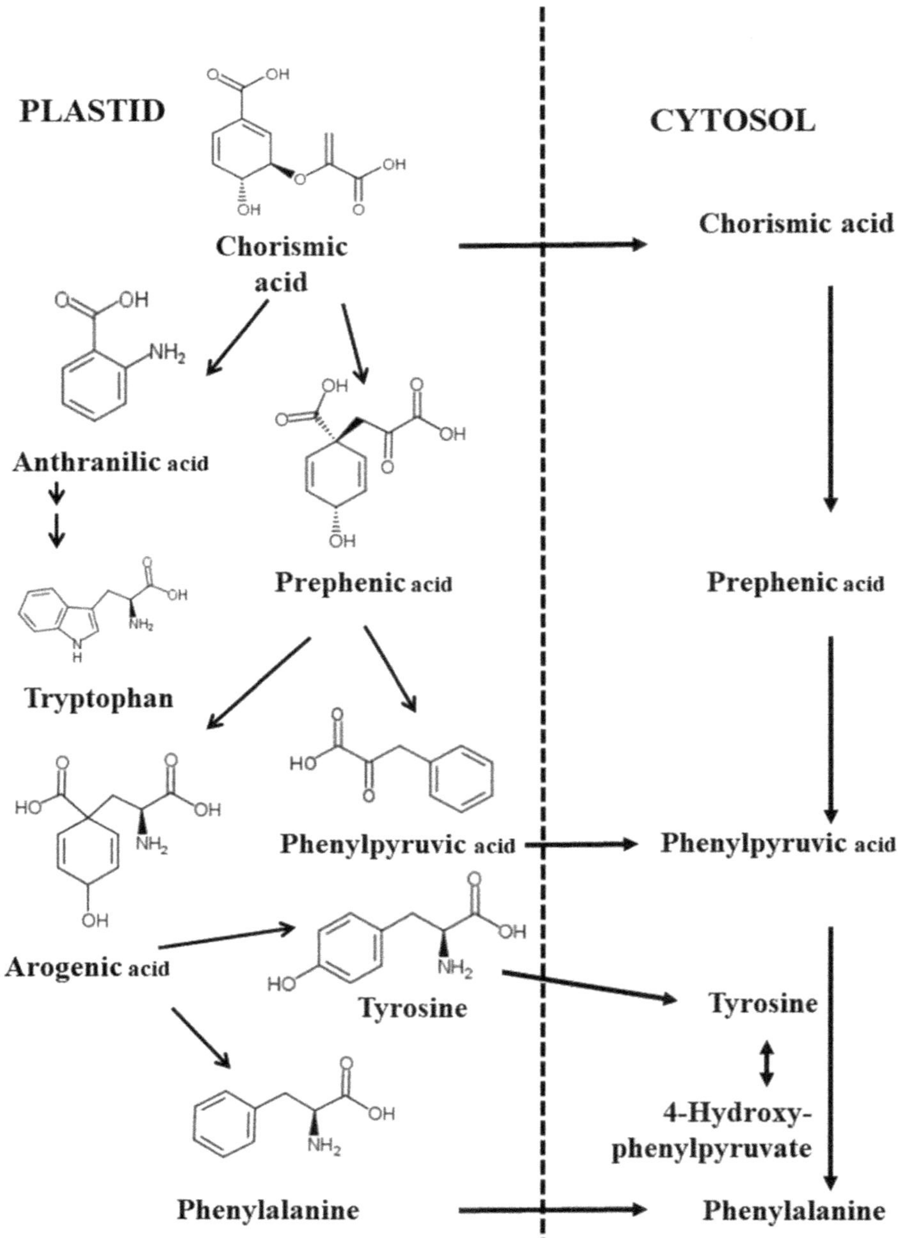

Figure 5.3 Biosynthesis of tryptophan, phenylalanine, and tyrosine.

5.1 Simple Phenols

Simple phenolics are substituted C6 phenols where at least one of the substituents is a hydroxyl group. Examples are **resorcinol**, **phloroglucinol** (Figure 5.4).

5.2 Phenolic Acids

Phenolic acids may belong to two groups *e.g.*, **hydroxybenzoic acids** and **hydroxycinnamic acids** (Figure 5.5). Hydroxybenzoic acids (C6–C1 skeleton) are phenols substituted by a carboxyl group. Examples are *p*-hydroxybenzoic acid, gallic acid, protocatechuic acid, salicylic acid, gentisic acid (2,5-dihydroxybenzoic acid) and vanillic acid. Vanillin is the aldehyde form of vanillic acid. Hydroxycinnamic acids (C6–C3 skeleton) are present in all plants.

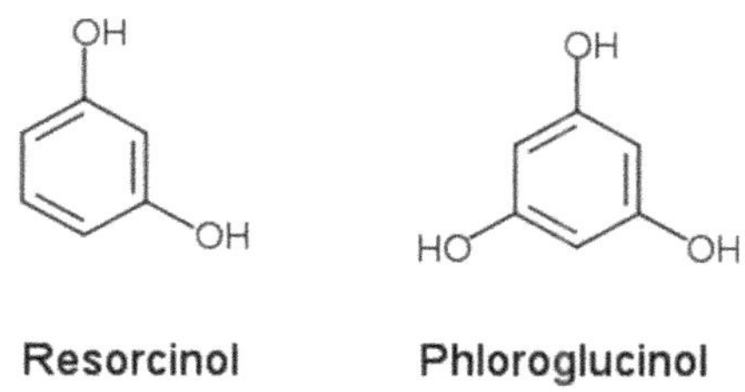

Figure 5.4 Simple phenols.

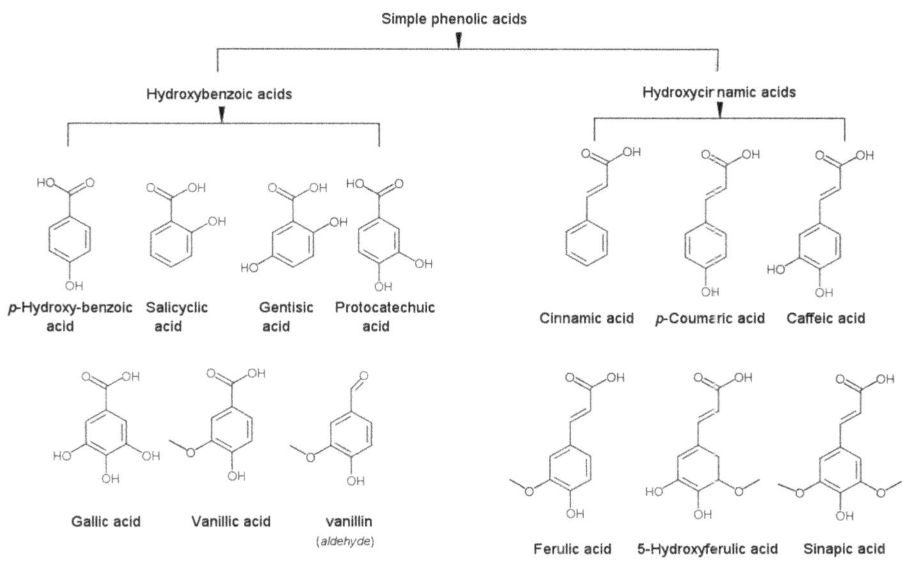

Figure 5.5 Phenolic acids and an aldehyde of vanillic acid.

Examples are cinnamic acid, *p*-coumaric acid, caffeic acid, ferulic acid, 5-hydroxyferulic acid, and sinapic acid *etc.*[1]

Phenylalanine in most plants or tyrosine in partial monocots leads to the biosynthesis of phenylpropanoids consisting of a 6-carbon aromatic phenyl ring and a 3-carbon propane side chain (C6–C3 or phenylpropane skeleton). Hydroxycinnamic acids are phenylpropanoids. The central phenylpropanoid pathway has three enzymatic activities:[5]

- Phenylalanine deamination by phenylalanine ammonia lyase (PAL) to *trans*-cinnamic acid
- Hydroxylation of *trans*-cinnamic acid to *p*-coumaric acid (4-coumaric acid) by the action of cinnamic acid 4-hydroxylase (CA4H)
- Conversion of *p*-coumaric acid to *p*-coumaroyl-CoA by 4-hydroxycinnamate-CoA ligase (4CL)

Biosynthesis of hydroxycinnamic acids is shown in Figure 5.6. The reactions take place in the cytosol. The P450 enzymes CA4H, C3′H, and F5H are anchored to the cytosolic face of the endoplasmic reticulum.[6]

Phenylpropanoid derivatives are a large class of phenolic metabolites which can be divided into several groups *e.g.*, phenolic acids, flavonoids, stilbenes, coumarins, monolignols, and others.

Hydroxybenzoic acids may be produced either in the plastid directly from shikimic acid pathway metabolites or in the cytosol from the hydroxycinnamic

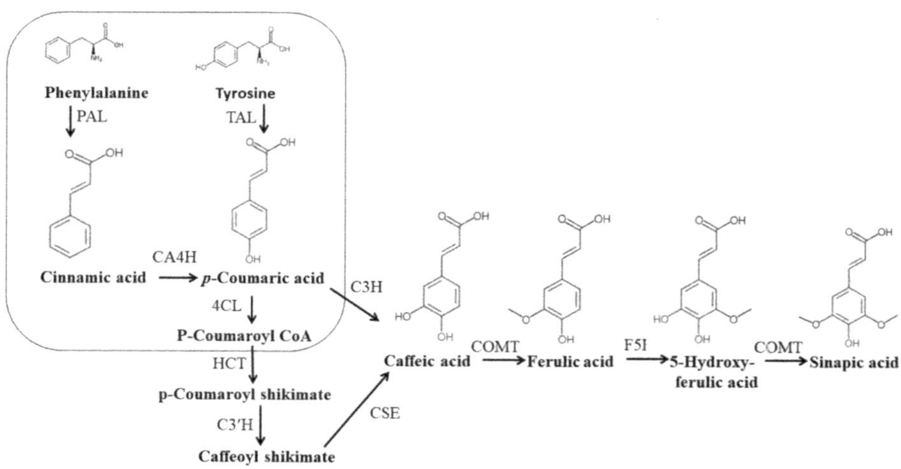

Figure 5.6 Phenylpropanoid biosynthesis leading to different hydroxycinnamic acids. PAL: phenylalanine ammonia lyase; TAL: tyrosine ammonia lyase; CA4H: cinnamic acid 4-hydroxylase; 4CL: 4-hydroxycinnamate: CoA-ligase; C3′H: 4-coumaroyl shikimate 3′-hydroxylase; C3H: 4-coumarate 3-hydroxylase; HCT: shikimate *O*-hydroxycinnamoyltransferase; F5H: ferulate 5-hydroxylase; COMT: caffeic acid 3-*O*-methyltransferase; CSE: caffeoyl shikimate esterase.

acids produced following the phenylpropanoid pathway. Gallic acid and protocatechuic acid are produced in the plastid from 3-dehydroshikimic acid, a shikimic acid pathway metabolite. Salicylic acid may be produced either in the plastid from chorismic acid *via* isochorismic acid (also known as the isochorismic acid pathway) or from phenylalanine derived metabolites such as benzoic acid or *ortho*-coumaric acid in the cytosol (known as the PAL pathway).[1,2,7–9] The metabolic pathways of the hydroxybenzoic acids are shown in Figure 5.7.

Phenolic acids are widely found in plants and plant foods. They have good water solubility and high bioavailability. The hydroxyl, methoxyl, and carboxylic acid groups are important to their antioxidant properties.[10] Salicylic acid is a hormone that regulates plant growth and development directly or indirectly. It also regulates responses to different abiotic and biotic stresses.[7] The role of the phenolic acids as defense signaling hormones is documented. Phenolic acids play an important role in lignin formation. Phenolic acids also regulate plant-microbe symbiosis, allelopathic activities and resistance to pathogen attacks in plants.[9] Major food flavouring agents in vanilla

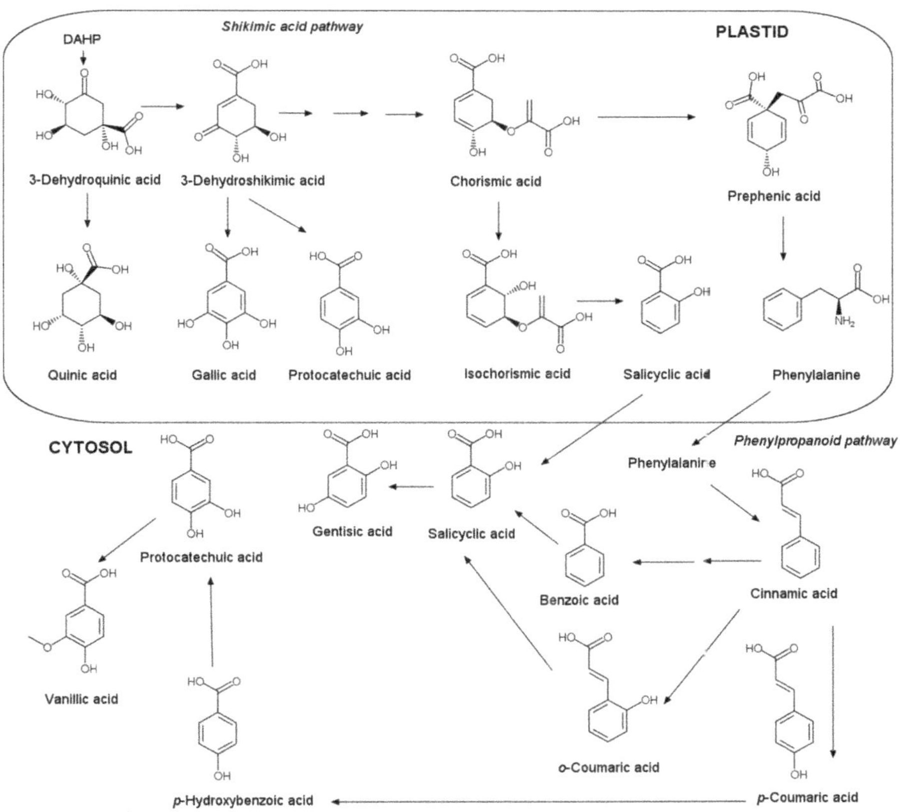

Figure 5.7 Biosynthesis of hydroxybenzoic acid.

(obtained from the pods of different species of *Vanilla* mostly from *Vanilla planifolia* of Orchidaceae) are vanillic acid and its aldehyde vanillin.[11]

5.3 Lignin

Lignin is a high molecular weight complex racemic heteropolymer derived mainly from the oxidative combinatorial coupling of three **hydroxycinnamyl alcohol monomers** (or **monolignols**). The monolignols differ in their degree of methoxylation, namely **coniferyl alcohol, sinapyl alcohol** and ***p*-coumaryl alcohol**. Beside cellulose, lignin is the second most abundant terrestrial biopolymer on the earth. Lignin accounts for 30% of organic carbon in the biosphere.[12]

The three monolignols coniferyl, sinapyl, and *p*-coumaryl alcohols, when incorporated in the lignin polymer, produce **guaiacyl (G), syringyl (S), and *p*-hydroxyphenyl (H) units**, respectively. The amount and composition of lignin vary among individual cell wall layers, cell types and taxa. Lignin from softwood of gymnosperms is composed of mostly G units with low levels of S units. The lignin of the hardwood of dicotyledonous angiosperms consists principally of G and S units and traces of H units. Lignin in grasses (monocotyledonous angiosperms) is composed of G and S units at comparable levels and more H units than dicotyledonous angiosperms.[12,13] Biosynthesis of monolignols is shown in Figure 5.8.

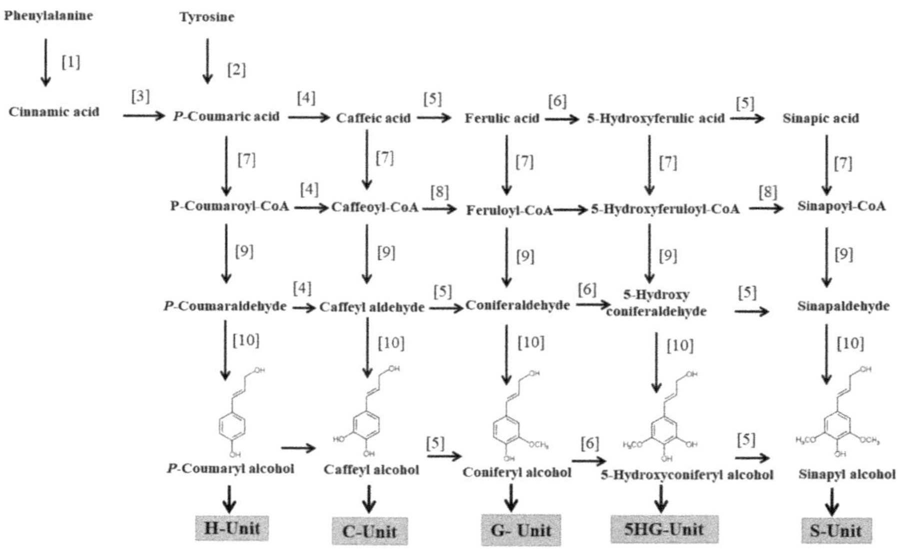

Figure 5.8 Biosynthesis of monolignols. [1]: phenylalanine ammonia lyase; [2]: tyrosine ammonia lyase; [3]: cinnamic acid 4-hydroxylase; [4]: *p*-coumarate 3-hydroxylase; [5]: caffeic acid 3-*O*-methyltransferase; [6]: ferulate-5-hydroxylase; [7]: 4-coumarate:CoA-ligase; [8] caffeoyl-CoA *O*-methyltransferase; [9]: cinnamoyl-CoA reductase; [10]: cinnamyl-alcohol dehydrogenase.

Apart from these three monomers, other units are also reported from lignin from different sources.[12–16]

- High levels of **5-hydroxyguaiacyl (5HG)** units occur in the seed coats of cacti.
- **Acylated monolignols** produce acylated lignin units **Y**. Examples are *p*-coumarates **Y3** units in grass lignin; *p*-hydroxybenzoates **Y2** units in poplars, palms, willows; acetates acylated monolignols **Y1** in palms, kenaf (*Hibiscus cannabinus*, Malvaceae). In kenaf bast fiber, more than 50% of lignin consists of acylated units.
- **Cinnamyl aldehyde (X2)** and **benzyl aldehyde (X3)** units are always detected in lignin. **Diconiferyl alcohol (X5G)** and **guaiacylpropane-1,3-diol (X6G)** units are always detected in gymnosperm lignin.
- **Ferulated (M4G)** units are reported in grass lignin.
- **C-lignin** is a homopolymer of caffeoyl alcohol linked by benzodioxane units. This type of lignin is observed in the seed coats of vanilla orchid, some cacti, and *Jatropha* species (Euphorbiaceae).
- **Tricin**, a flavonoid, is found in all grass lignin.
- **Hydroxystilbenes** and/or their glycosides, including resveratrol, are found naturally in lignin of palm (*e.g.*, coconut) fruit endocarp.

The monolignols and lignin precursors are formed in the cytosol, transported to the apoplast, and polymerized into lignin in the apoplast.[14,15] It has been suggested that the key enzymes for the oxidative polymerization of monolignol units are peroxidases and laccases.[16] Laccase and peroxidase, the oxidative enzymes, catalyze formation of phenoxy radicals from monolignols (Figure 5.9). Hydrogen peroxide is produced in the apoplast from apoplastic oxygen. In addition, dirigent-domain-containing proteins are implicated in directing the correct lignin deposition in the casparian strips of *Arabidopsis*.[17] An oxidized monolignol radical cross-couples with a radical formed on the free phenolic end of a growing lignin polymer. A radical coupling step and variations in the mode of the post-coupling quinone methide rearomatization step allow numerous possible inter-molecular linkages in the final lignin polymer.[17] Extension of the lignin polymer chain has been explained with Figure 5.10.

Lignin is indispensable for plant growth. Lignin provides structural support to terrestrial plants to grow upright. During deposition on a carbohydrate matrix, lignin and hemicellulose components may form bonds eliminating water and forming a hydrophobic environment. Lignin confers mechanical properties to the cells for structural support as well as for water transport. Lignin is a well-known defense polymer with antimicrobial activity. Lignin forms a barrier to block movement and multiplication of pathogens and prevents diffusion of toxins from pathogens. The phenolic compounds in the lignin counter pathogenic diseases. It has been observed that lignin deposition and lignin biosynthetic genes are induced during attack by pathogens.[13,16,17] Lignin residues may be used to produce bioplastics and some value-added dyes, adhesive, and fertilizers.[18,19]

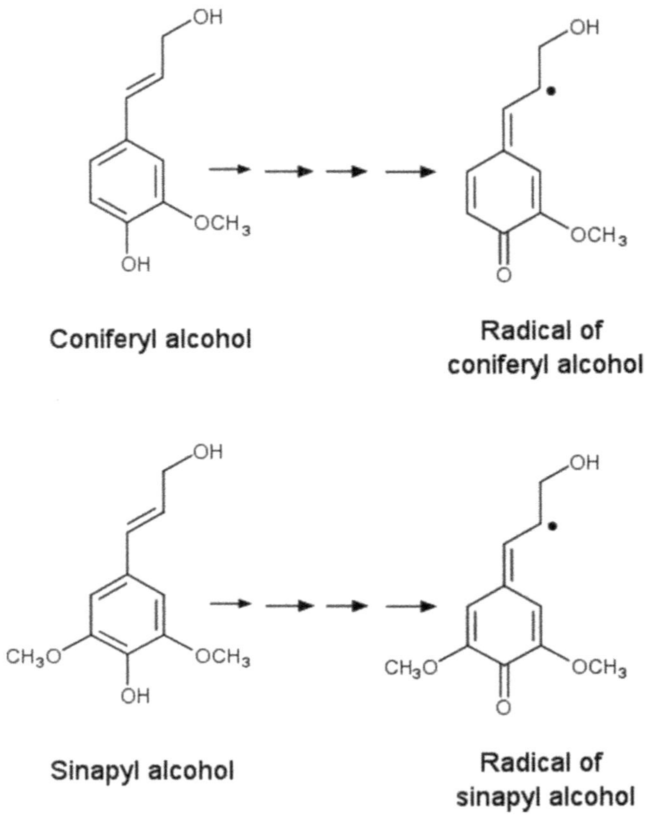

Figure 5.9 Radical formation in monolignols.

5.4 Lignans

Lignans are polyphenols which consist of two phenylpropanoid derived structures (dimeric structure of two C6–C3 units) linked together by β-β′ bonds (positions 8 and 8′). Lignans are formed by coupling of two coniferyl alcohol residues that are present in the plant cell wall. In the absence of a C-8 to C-8′ bond, the dimer formed from two C6–C3 units is known as **neolignan.** When an ether oxygen atom links two C6–C3 units in the case of neolignan, the lignan is known as **oxyneolignan**. The lignan oligimers are known as **oligolignans.** Oligolignans consist of neolignan or two or three guaiacylglycerol units linked to each other. **Flavonolignans** are formed by oxidative coupling between a flavonoid (taxifolin) and a phenylpropanoid derivative monolignol (usually coniferyl alcohol).[20–24]

Important plant **lignans** are **pinoresinol, lariciresinol, secoisolariciresinol, matairesinol, syringaresinol, medioresinol, arctigenin** and **sesamin** (Figure 5.11). These lignans occur mainly as glycosides in whole grain cereals (rye, buckwheat, oat, wheat, barley, millet), seeds (of flax, sesame, pumpkin,

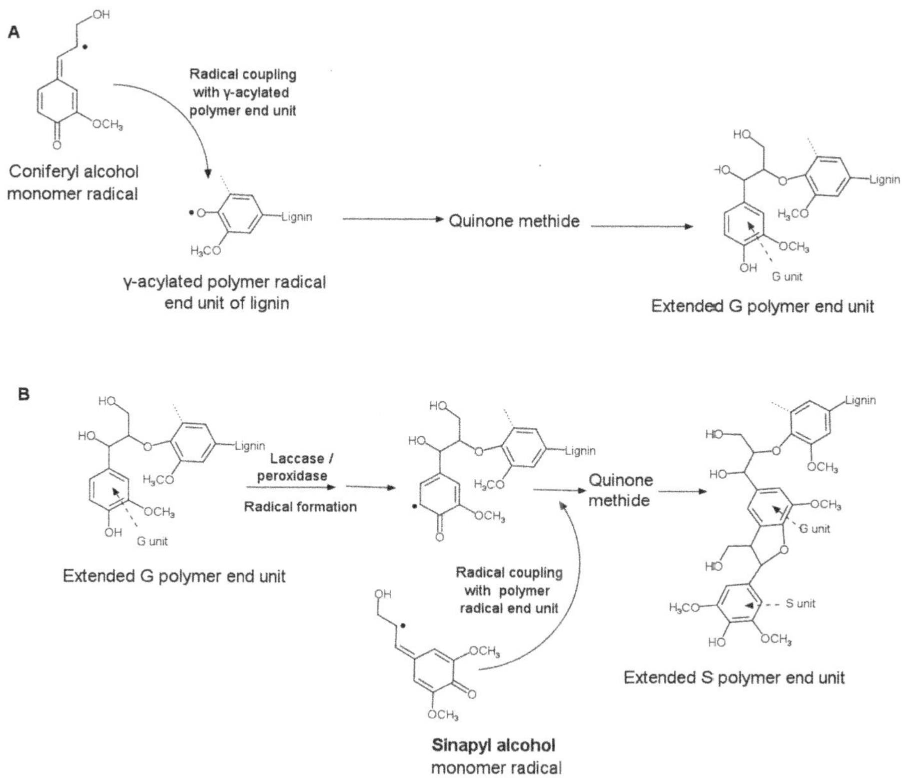

Figure 5.10 Polymerization for extension of lignin polymer. A: Addition of coniferyl alcohol radical; B: addition of sinapyl alcohol radical.

sunflower), nuts (almond, cashew, chestnut, hazelnut, peanut, pistachio, walnut), beans, vegetables, berries, and some fruits.[20,23,25]

Biosynthesis of lignans (Figure 5.11) starts with 2 molecules of coniferyl alcohol that are formed by the reduction of coumaric acid. Dimerization of two coniferyl alcohols takes place with the help of dirigent protein to give rise to pinoresinol. Pinoresinol is reduced to secoresinol *via* lariciresinol. Podophyllotoxin is produced from matairesinol.[20,23,26]

Dietary plant lignans are metabolized to **enterodiol** and **enterolactone**, known as enterolignans or mammalian lignans, by bacteria residing in the intestine. Enterodiol and enterolactone are structurally similar to that of the sex hormone 17β-estradiol. So, the two enterolignans bind to the estrogen receptor and exert a weak estrogenic or anti-estrogenic effect and are often classified as **phytoestrogens**. Experimental evidence indicates that the enterolignans may affect the risk of several chronic diseases through antiproliferative, anti-inflammatory, and apoptotic properties.[27]

Podophylotoxin (Figure 5.11) is a lignan obtained from the resin of several species of *Podophyllum* (Berberidaceae) such as *Podophyllum peltatum* and

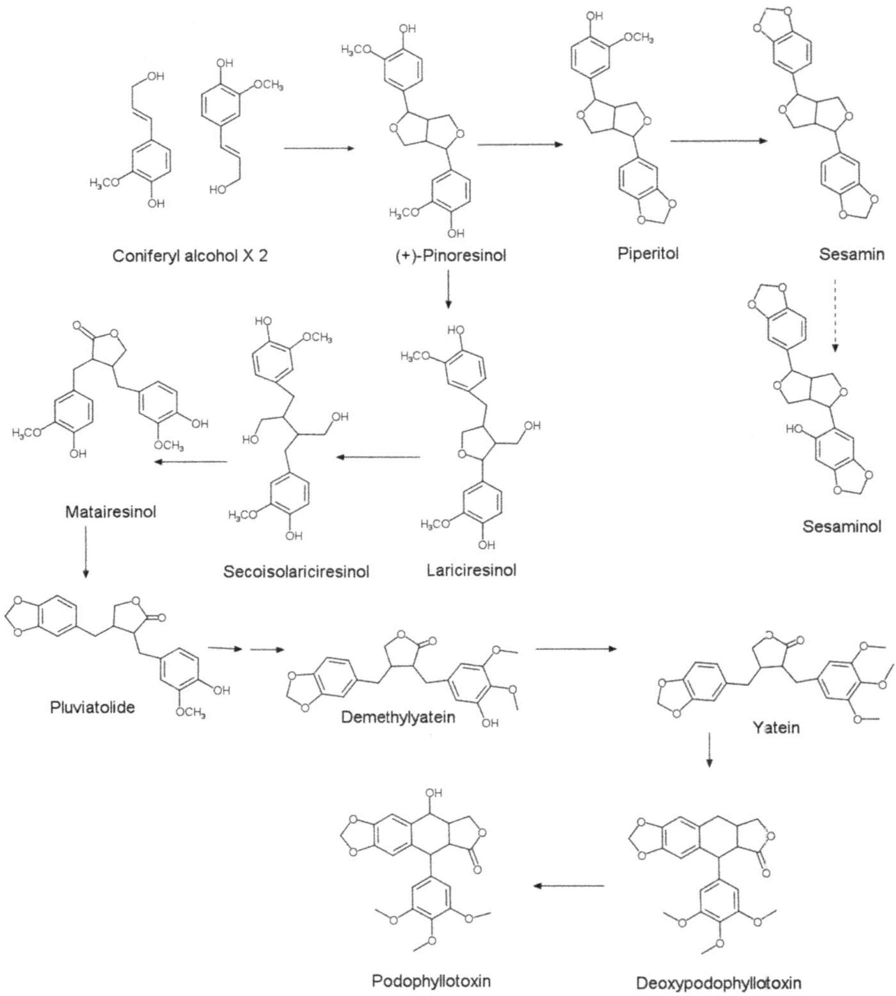

Figure 5.11 Biosynthesis of lignans.

Podophyllum emodi (Syn. *Podophyllum hexandrum*). Etoposide is a semisynthetic podophyllotoxin. Etoposide is used in patients with lung cancer, lymphoma, testicular cancer, and other cancers. It causes single strand breaks in DNA and is a topoisomerase II inhibitor. Teniposide is a second semisynthetic derivative of podophyllotoxin and an analogue of etoposide. The mechanism of action of teniposide is like that of etoposide.[28–30]

Neolignans (Figure 5.12) and their biosynthesis are not well studied. Two examples are **magnolol** and **honokiol**. These neolignans were isolated and identified from the cortex of *Magnolia officinalis* (Magnoliaceae), which has been used in traditional Chinese and Japanese Medicine. Magnolol has been reported to possess anti-inflammatory, cardiovascular protection,

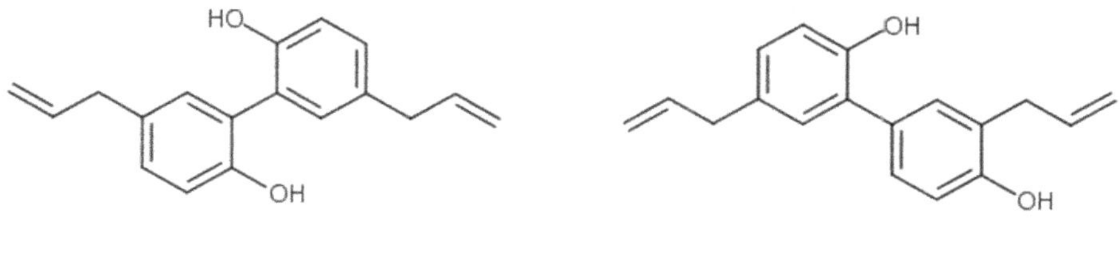

Magnolol Honokiol

Figure 5.12 Neolignans.

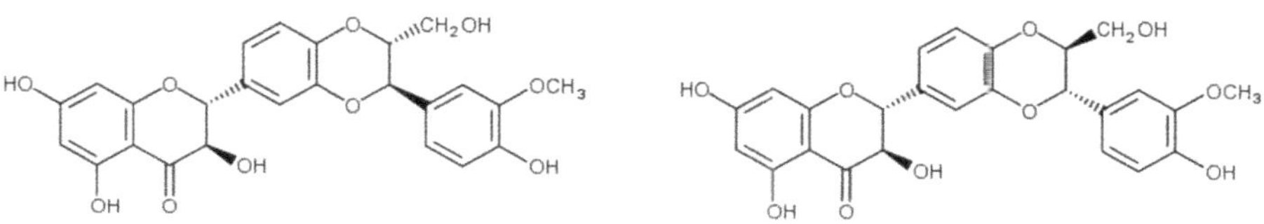

Silybin A
(Silibinin A)

Silybin B
(Silibinin B)

Figure 5.13 Flavonolignans.

antiangiogenesis, antidiabetic and antioxidant properties. Honokiol has therapeutic potential for anxiety, pain, cerebrovascular damage, epilepsy, and cognitive disorders.[31]

Flavonolignans (Figure 5.13) are formed by oxidative coupling between the flavonoid taxifolin and coniferyl alcohol. **Silymarin**, a mixture of flavonolignans, has long been used as a hepatoprotective remedy. It is made from the fruit extract of the edible Mediterranean herb *Silybum marianum* (Asteraceae) or milk thistle. Silymarin is a mixture of seven closely related flavonolignans **silybin A (silibinin A), silybin B (silibinin B), isosilibinin A (isosilybin A), isosilibinin B (isosilybin B), silicristin, isosilicristin**, and **silidianin.** In addition, silymarin also contains the flavonoid taxifolin. Silibinin (silybin), a mixture of silibinin A and silibinin B in an almost equimolar ratio, is the most biologically active component. Silymarin exerts its hepatoprotective activity through antiviral, anti-inflammatory, antioxidant, and immunomodulatory actions in liver and immune cells. Silymarin acts as a free radical scavenger and modulator of enzymes associated with cellular damage, fibrosis and cirrhosis in patients with alcoholic and non-alcoholic fatty liver diseases in clinical studies.[32–34]

5.5 Tannins

Tannins are structurally complex polyphenols. Many tannins are glycosides. Usually, their molecular weights are >500 Dalton. Tannins can aggregate with alkaloids, gelatin, and other proteins to form precipitates. They are astringent in taste.[35,36]

Tannins may be classified into four groups. These are **hydrolysable tannins, condensed tannins, complex tannins** and **phlorotannins.**[37]

5.5.1 Hydrolysable Tannins

Hydrolysable tannins may be of two types: gallotannins and ellagitannins. Hydrolysable tannins containing a glucose molecule, esterified with gallic acid are called **gallotannins**. Pentagalloylglucose is the basic unit. **Ellagitannins** contain hexahydroxydiphenic acid or its dilactone form known as ellagic acid around a central glucose (Figure 5.14). Intermolecular oxidative couplings resulting in C–C or C–O–C linkages between galloyl residues of glucogalloyls may generate diverse structures with molecular weight between 2000 and 5000 Dalton.[35–38]

5.5.2 Condensed Tannins

Condensed tannins are oligomers or polymers of flavon-3-ol compounds (Figure 5.15). The most frequent flavon-3-ols found in condensed tannins are (+)-catechin, (−)-epicatechin and (+)-gallocatechin. Condensed tannin

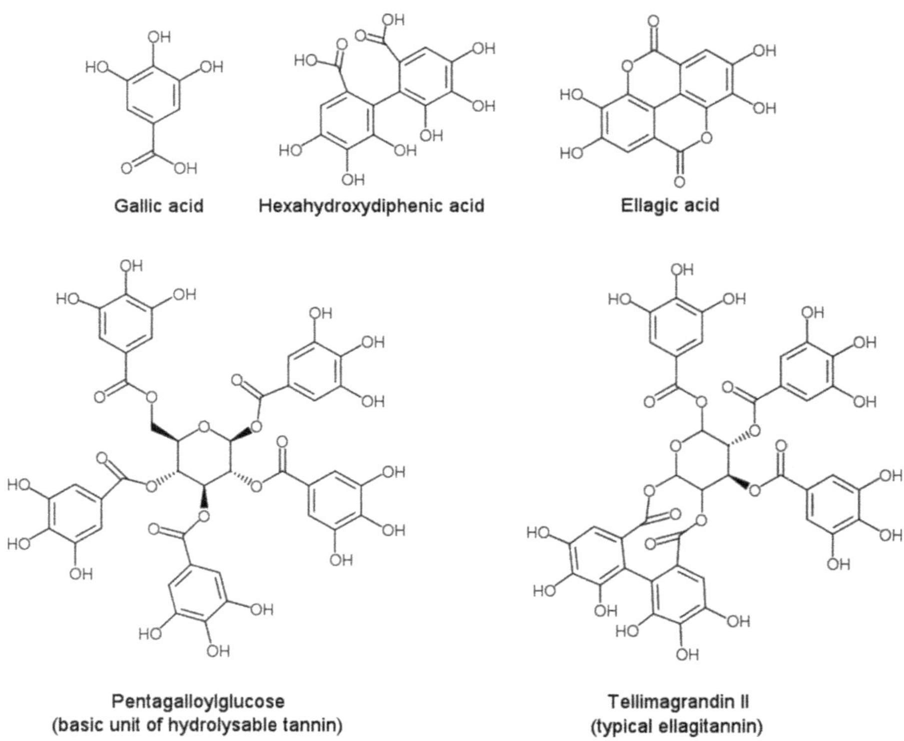

Figure 5.14 Structures associated with hydrolysable tannins.

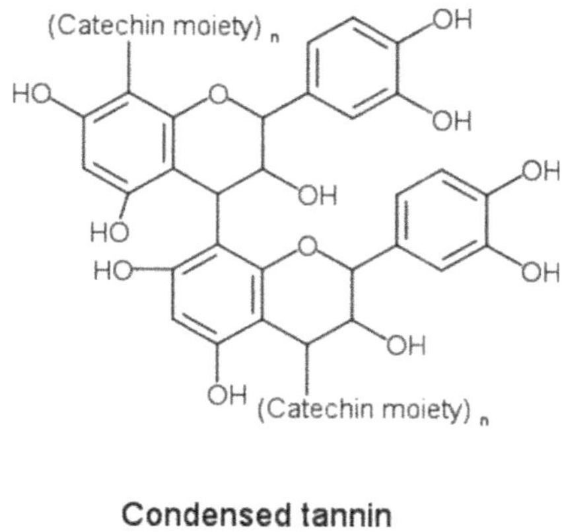

Figure 5.15 Condensed tannin.

monomers are joined by C-4→C-8 bonds or C-4→C-6 linkages. Two other flavon-3-ols, mainly found in green tea, are (−)-epigallocatechin and (−)-epigallocatechin gallate. In addition, *O*-methylation, *O*-galloylation or *C*- and *O*-glycosylation contributes to the diversity in the structures. Condensed tannins are also known as **proanthocyanidins** as they are decomposed to anthocyanidins in heated ethanol solutions under acidic conditions. Proanthocyanidins thus may be further classified into procyanidins, prodelphinidins, propelargonidins *etc.*[35–38]

5.5.3 Complex Tannins

Complex tannins are formed by conjugation of proanthocyanidins and hydrolysable tannins *via* C–C bonds.

5.5.4 Phlorotannins

Phlorotannins are based on differently bound phloroglucinol units (Figure 5.16), with molecular weight ranging between 126 and 650 kDa. These types of tannins are found in marine brown algae (Phaeophyceae).[37,38]

Tannins are present in different parts of plants such as seeds, grains, roots, wood, bark, leaves, and fruits. Tannins act as defensive molecules to protect plants from herbivorous attacks and against insects and infection.[39]

Tannin based adsorbents displayed proficiency at removing heavy metals and other chemical pollutants from polluted water. Tannins are used as textile dyes, and in the production of inks (iron gallate ink). Tannins are tanning agents for the leather industry.[39,40]

Tannin containing food materials have various biological activities such as antioxidant (free radical scavenging and metal chelating properties), antitumour, hepatoprotective, hypolipidemic, hypoglycaemic, immunomodulating,

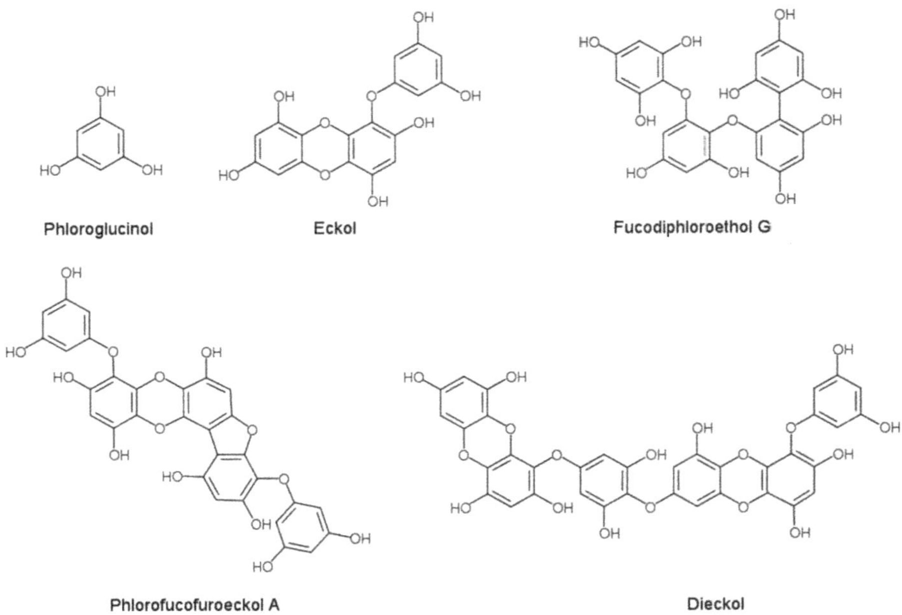

Figure 5.16 Structures associated with phlorotannins.

anti-inflammatory, cardioprotective, vasodilating, antithrombotic, antiprotozoal, antiviral, and antibacterial activities.[35,38,41] Some tannins are considered antinutritional and are removed by food processing techniques. Tannins are used as food additives and food preservatives.[42]

5.6 Phenylpropenes

Phenylpropenes are plant volatiles derived from the amino acid phenylalanine.[43] These volatile (essential) oil constituents are produced by plants in addition to the volatile monoterpenes and sesquiterpenes. These phenolic constituents may remain in the volatile oil mixture along with the terpene derived essential oil constituents. The phenylpropenes are structurally diverse due to different substitution patterns on the benzene ring and the position of double bond in the propenyl side chain. The names of the plants where these phenylpropanoid derived essential oil constituents (Figure 5.17) are present as one of the important constituents are mentioned in Table 5.1.

Deamination of phenylalanine leads to the formation of cinnamic acid. The enzyme catalyzing this step is phenylalanine ammonia lyase. Next *p*-coumaric acid and ferulic acids are formed from cinnamic acid. Monolignols are formed in the next steps. Finally, many phenylpropenes are formed from such monolignols[43] (Figure 5.18).

The widespread distribution of the phenylpropenes among angiosperms is probably due to the defensive properties of the compounds. Humans have further used this property to protect plants and food stocks.

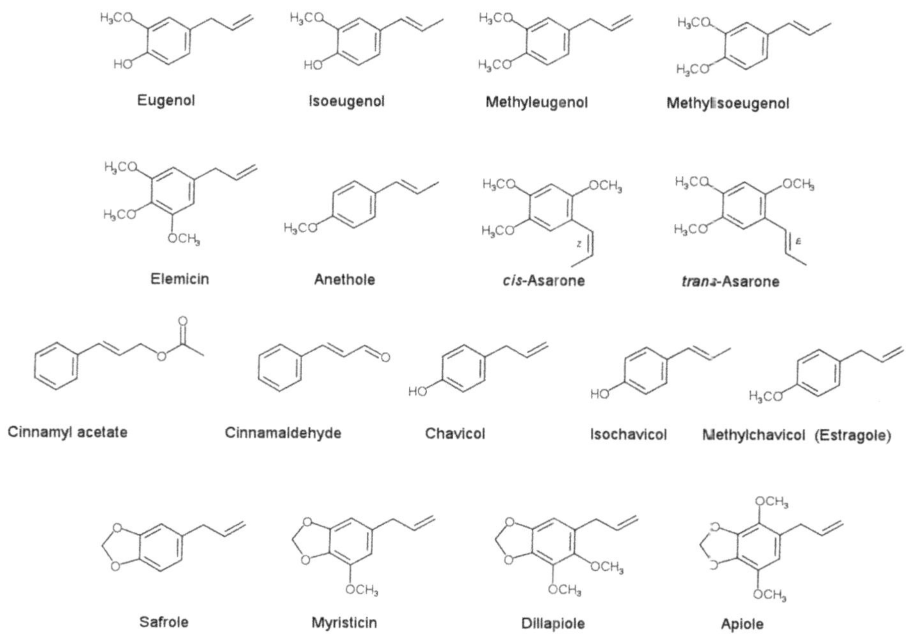

Figure 5.17 Structures of different phenylpropenes.

Table 5.1 Phenylpropanoid essential oil constituents.

Essential oil (phenolic)	Source
Eugenol (4-allylguaiacol)	Flowers of *Syzygium aromaticum* (Myrtaceae) or clove, *Piper betle* (Piperaceae) leaves, leaves of cinnamon (several species of *Cinnamomum* of family Lauraceae), *Ocimum basilicum* (Lamiaceae)
Isoeugenol	*Petunia hybrida* (Solanaceae) flower, *Clarkia breweri* (Onagraceae)
Methyleugenol	*Ocimum basilicum*, *Piper betle* leaves
Isomethyleugenol	*Clarkia breweri* (Onagraceae)
Safrole	*Sassafras* species (Lauraceae), a few varieties of *Piper betle* leaves
Myristicin	*Myristica fragrance* (Myristicaceae)
Elemicin	*Myristica fragrance* (nutmeg)
Anethole	*Pimpinella anisum* (Apiaceae) (aniseed) fruits, *Illicium verum* (Schisandraceae) (star anise), *Foeniculum vulgare* (Apiaceae) (fennel)
(*Z*)-Asarone, (*E*)-asarone	*Acorus calamus* (Acoraceae), *Asarum europaeum* (Aristolochiaceae)
Cinnamyl acetate	*Cinnamomum zeylanicum* (Lauraceae) (cinnamon) bark
Cinnamaldehyde	*Cinnamomum zeylanicum* (cinnamon) bark
Chavicol (4-allylphenol)	*Ocimum basilicum*, some varieties of *Piper betle* leaves
Isochavicol	Some varieties of apple (*Malus domestica*, Rosaceae)
Methylchavicol (estragole)	*Foeniculum vulgare* (fennel), *Ocimum basilicum*, some varieties of *Piper betle* leaf
Dillapiole	*Piper aduncum* (Piperaceae) leaves,
Apiole	*Anethum graveolens* (Apiaceae) (dill)

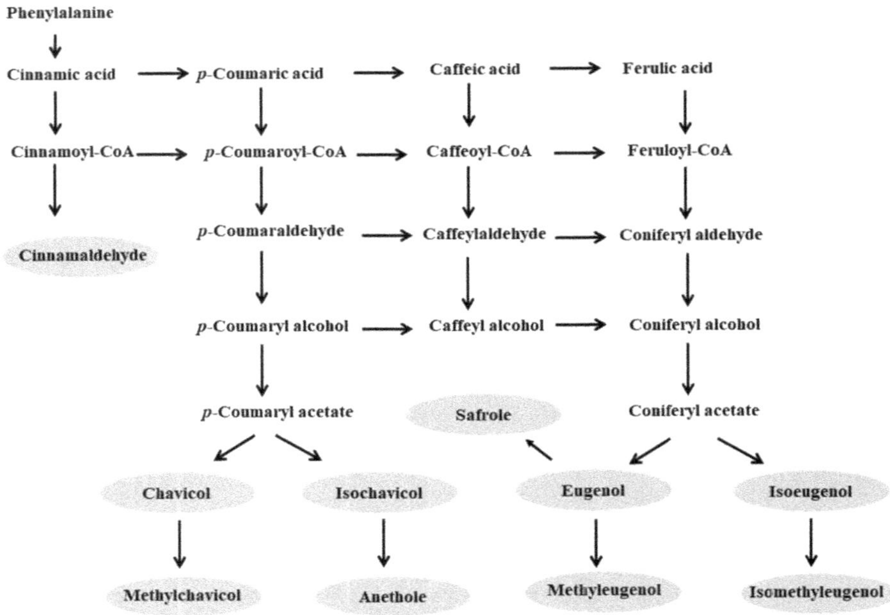

Figure 5.18 Biosynthesis of some phenylpropenes.

Eugenol inhibits many food-borne pathogenic bacteria significantly. Eugenol also has fungicidal and nematocidal properties and has insecticidal properties due to its pungent odour. Methyleugenol is responsible for the pleasant and fresh scents of many flowers, thus attracting pollinators.[43]

Eugenol possesses antibacterial, antiviral, antioxidant, analgesic, and anti-inflammatory properties. This phenylpropene has been used to treat toothache and pulpitis in dentistry. Eugenol, in combination with zinc oxide, is frequently used as a root canal sealer in endodontics and has a restorative effect on dental erosion and demineralization. Eugenol is an ingredient in different oral care products.[44]

Safrole was used as a flavouring agent in beer and soft drinks. However, later it was found to be a hepatocarcinogenic material. It is also a precursor to an illicit drug so different government authorities have restricted procurement and use of this chemical except for synthesis of some fine chemicals. Safrole is the raw material used to produce heliotropin, and piperonyl butoxide in the fragrance and pesticide industries, respectively.[45]

5.7 Betalains

Betalains are nitrogen containing water soluble pigments dissolved in the vacuolar sap of plants. Betalains are classified into two groups: red violet **betacyanins** and yellow to orange **betaxanthins**. This group of pigments are found in the families of Caryophyllales, except Caryophyllaceae and Molluginaceae which accumulate anthocyanins instead. The pigments are also detected in some genera of fungi such as *Amanita*, *Hygrocybe* and *Hygrophorus*.[46,47]

Betalains are immonium derivatives of **betalamic acid**. Condensation of betalamic acid with *cyclo*-dopa [*cyclo*-L-(3,4-dihydroxyphenylalanine)] or its glucosyl derivatives generates betacyanins. Betacyanins also may form glucosyl derivatives. Betaxanthins are formed by condensation of **betalamic acid** with amines and/or their derivatives[46,48] (Figure 5.19 and Table 5.2).

5.7.1 Betacyanins

The aglycone betanidin is the backbone of all betacyanins. Betacyanins can be divided into four structural types, betanin, gomphrenin, amaranthin, and bougainvillein[46–48] (Figure 5.20).

- **Betanin type** of betacyanin has a hydroxyl group attached to the carbon 6 and a glucosyl or derivative attached to the carbon 5. Examples are betanin (betanidin 5-*O*-β-glucoside) and isobetanin from red beet root, phyllocactin in flowers and fruits of the Cactaceae and hylocerenin in some edible fruits of cacti.

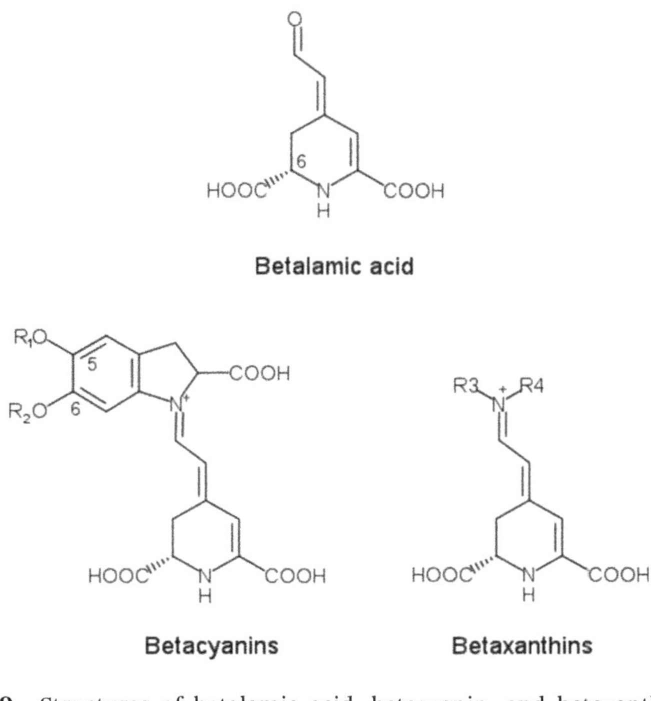

Figure 5.19 Structures of betalamic acid, betacyanin, and betaxanthin. R_1 and R_2 = H and glucosyl derivatives; R_3 and R_4 = H and amines or their derivatives.

Table 5.2 Different groups of betalains.

Betalains			
Betacyanin	Source	Betaxanthin	Source
Betain group		**Amino acid derived group**	
Betanidin	*Beta vulgaris* (Amaranthaceae)	Indicaxanthin	*Opuntia ficus-indica* (Cactaceae)
Betanin	*Beta vulgaris*	Portulaccaxanthin I	*Portulaca grandiflora*
Phyllocactin	*Phyllocactus hybridus* (Cactaceae)	Portulaccaxanthin II	*Portulaca grandiflora* (Portulacaceae)
Hylocerenin	*Hylocereus polyrhizus* (Cactaceae)	Vulgaxanthin I	*Beta vulgaris*
Gomphrenin group		Miraxanthin I	*Mirabilis jalapa* (Nyctaginaceae)
Gomphrenin I	*Gomphrena globosa* (Amaranthaceae)	Dopaxanthin	*Glottiphyllum longum* (Aizoaceae)
Gomphrenin II	*Gomphrena globosa*	**Amino derived group**	
Amaranthin group		3-Methoxytyramine-betaxanthin	*Celosia argentea* (Amaranthaceae)
Amaranthin	*Celosia cristata* (Amaranthaceae)		
Celosianin I	*Celosia cristata*		
Celosianin II	*Celosia cristata*		
Bougainvillein group			
Bougainvillein v	*Bougainvillea glabra* (Nyctaginaceae)		
Gomphrenin group			
Gomphrenin I	*Gomphrena globosa* (Amaranthaceae)		

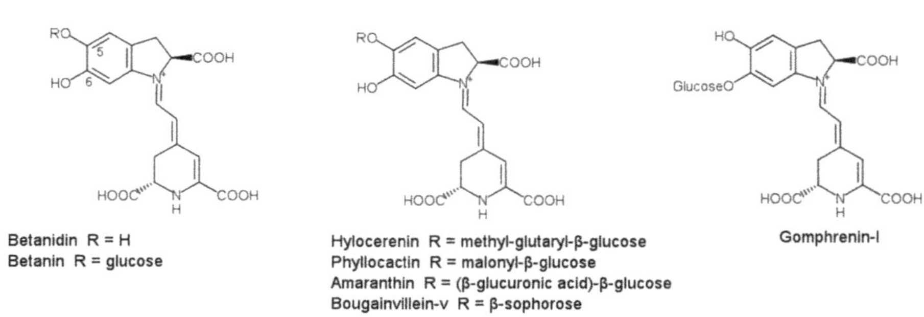

Figure 5.20 Different betacyanins.

- **Gomphrenin type** of betacyanin contains a hydroxyl group attached to carbon 5 and glucosyl or derivative attached to the carbon 6.Examples are Gomphrenin I, II, and III (from the inflorescence of *Gomphrena globosa*, Amaranthaceae).
- **Amaranthin type** of betacyanin possesses a glucuronylglucosyl moiety or derivative attached to carbon 5. An example is amaranthin (the characteristic pigment of Amaranthaceae, inflorescence of *Celosia argentea*)
- **Bougainvillein type** of betacyanin consists of diglucosyl moiety or derivative attached to carbon 5 or to carbon 6 of a carboxylated or decarboxylated structure. One example is bougainvillein-v from *Bougainvillea glabra* (Nyctaginaceae).

5.7.2 Betaxanthins

Betaxanthins (Figure 5.21) are divided into two groups.[46,47]

- Amino-acid-derived conjugate group of betalamic acid *e.g.*, portulacaxanthin II and III from *Portulaca grandiflora* (Portulacaceae) and tryptophan-betaxanthin from yellow and orange inflorescences of *Celosia argentea.*
- Amine derived conjugate group of betalamic acid *e.g.*, 3-methoxytyraminebetaxanthin from yellow and orange inflorescences of *Celosia argentea.*

Betalains are **tyrosine** derived pigments. Shikimic acid pathway derived tyrosine is converted to **L-dopa** (an aromatic amino acid) by 3-hydroxylation

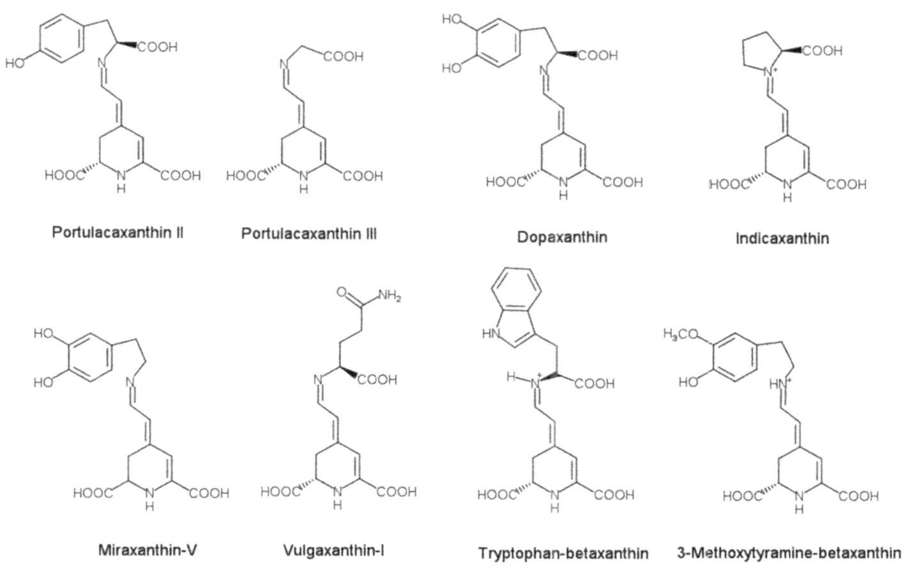

Figure 5.21 Different betaxanthins.

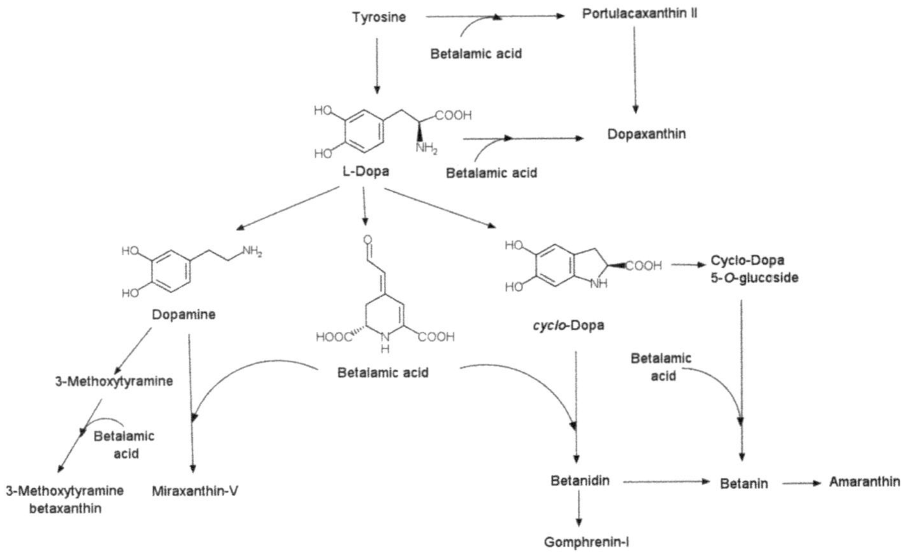

Figure 5.22 Biosynthetic steps of betalains.

and subsequently to ***cyclo*-dopa**. Dopa dioxygenase catalyzes formation of betalamic acid from L-dopa. Betalamic acid spontaneously reacts between the amine group of the amine and the aldehyde group of betalamic acid to form betaxanthins. On the other hand, *cyclo*-dopa condenses with betalamic acid forming betanidin, the betacyanin precursor[49,50] (Figure 5.22).

Betalain containing plant parts like flowers, fruits, stems, seeds and leaves attract animals and thus help plant propagation through pollen transfer, and seed dispersal. The pigments also protect the plant tissues against UV radiation induced damage, induce pathogen resistance and provide higher tolerance to heat and salt stress, most likely through their strong antioxidant capacity. Betalains have potent antioxidant and health promoting potential. Because of these properties, betalains are used as colourants and food additives in the food industry and as dietary supplements.[50]

5.8 Coumarins

Coumarins are organic heterocycles represented by benzo-α-pyrone (Figure 5.23). They are ubiquitously found throughout the plant kingdom. About 1300 coumarins have been reported from higher plants, fungi, and microorganisms.[51]

Coumarins are classified into three groups: simple coumarins, furanocoumarins, and pyranocoumarins.

Simple coumarins: These are hydroxylated, alkoxylated, and alkylated derivatives of coumarin and their glycosides. Examples are umbelliferone (7-hydroxycoumarin), skimmin (a glucoside of umbelliferone), limettin,

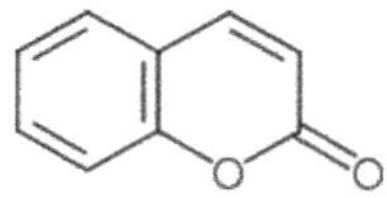

Coumarin

Figure 5.23 Structure of coumarin nucleus.

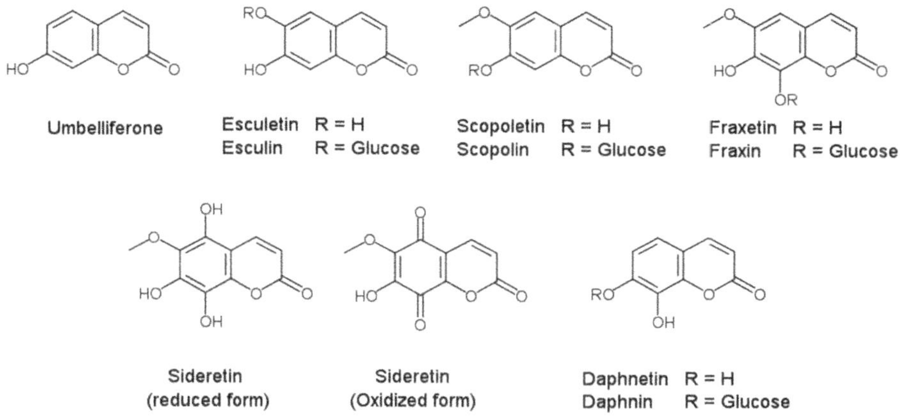

Figure 5.24 Simple coumarins.

esculetin, esculin, fraxetin, fraxin (fraxetin-8-*O*-glucoside), scopoletin, scopolin, sideretin, daphnetin, and daphnin (Figure 5.24).

Furanocoumarins: Here a furan ring is fused with a coumarin. Furanocoumarins are of two types:[52,53] (i) C6/C7 (linear type) such as psoralen, bergapten (5-methoxypsoralen), and xanthotoxin (ii) C7/C8 (angular type) such as angelicin, isobergapten, and pimpinellin (Figure 5.25).

Pyranocoumarins: In pyranocoumarins, a six membered pyran ring is fused with the benzene ring and these are of two types: linear and angular (Figure 5.26).

It is well established that the phenylpropanoid pathway leads to the formation of simple coumarins (Figure 5.27).[54] It has also been suggested from experimental evidence in members of family Rutaceae that the MEP pathway provides hemiterpenes for the biosynthesis of pyranocoumarins and the furan ring of psoralen[55] (Figure 5.28).

Coumarins are suggested to possess phytoalexin and allelopathic properties. Scopoletin, a coumarin, is excreted in the rhizosphere and inhibits the growth of fungal pathogens. However, it does not affect the growth promoting bacteria that are tolerant to the antimicrobial effect of scopoletin. Iron is poorly available for plants despite being abundant in the soil except the iron depleted ones. Secretion of coumarins such as esculetin, scopoletin, fraxetin, and sideretin into the rhizosphere help in iron acquisition by non-grass plants, when iron availability is low.[56]

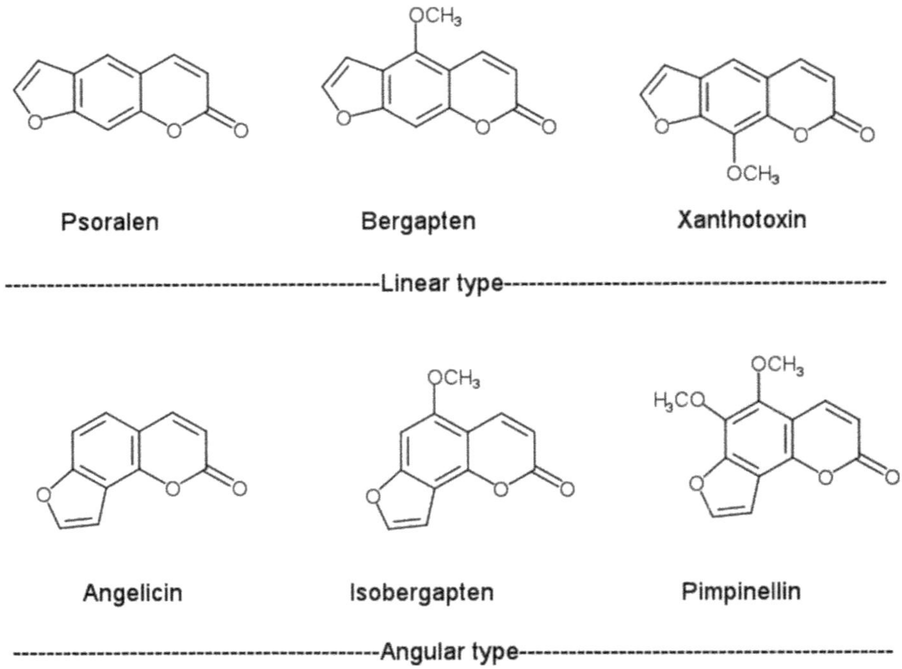

Figure 5.25 Furanocoumarins.

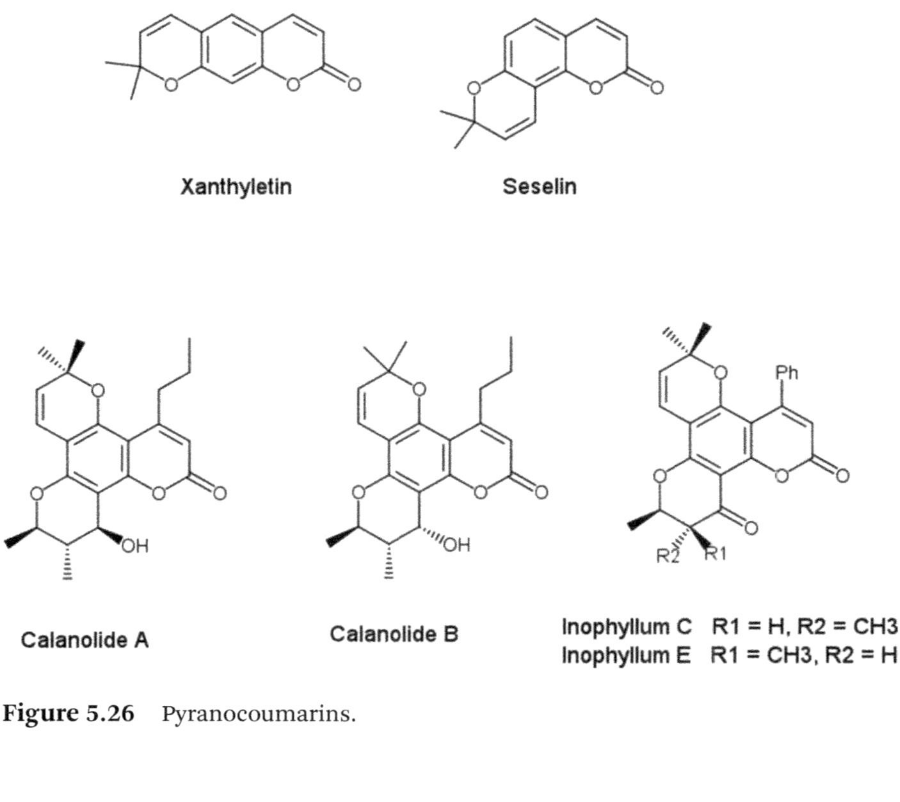

Figure 5.26 Pyranocoumarins.

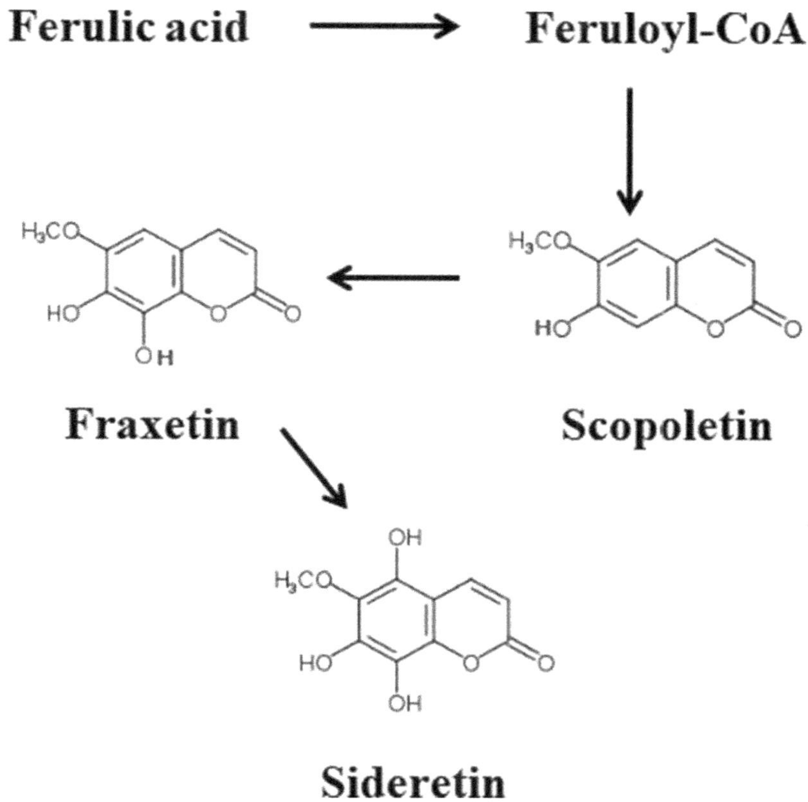

Figure 5.27 Biosynthesis of simple coumarins.

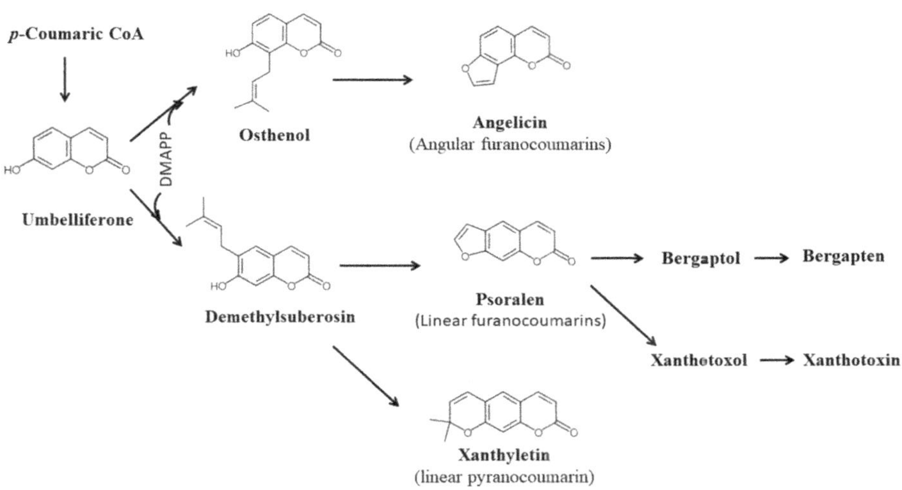

Figure 5.28 Biosynthesis of pyranocoumarins and furanocoumarins.

Coumarins are vitamin K antagonists. They have an anticoagulant effect as they interfere with cyclic conversion of vitamin K and vitamin K epoxide. Calanolide A and some other calanolides isolated from *Calophyllum lanigerum* var. *austrocoriaceum* (Calophyllaceae) have protective activity against HIV-1 replication. Calanolide A is a reverse transcriptase inhibitor and is now under development for the treatment of HIV infections. Calanolide A has also been reported from *Calophyllum inophyllum* seed oil. Inophyllum A, B, C, D and P from the oil of *Calophyllum inophyllum* also have anti-HIV properties. Coumarins also show antifungal and anticancer properties.[52]

5.9 Flavonoids

Flavonoids (C15 compounds) have C6–C3–C6 skeleton, two benzene rings linked by three carbons. The flavonoids are one of the major groups of specialized metabolites. More than 9000 such metabolites belonging to this group are so far reported. Such metabolites are classified into three classes based on the arrangement of the three carbons. These are chalcones, aurones and flavonoids.[57] Flavonoids have been found in liverworts and mosses, several algal genera, ferns, gymnosperms, and angiosperms. Flavonoids play important roles in plants *e.g.*, they are UV-protectants, signaling molecules, regulators of auxin transport and fertility, pollinators and phytoalexins.[57]

5.9.1 Chalcones

Chalcones have two phenyl rings connected through a three carbon α, β-unsaturated carbonyl chain (Figure 5.29). Chalcones are key intermediates in biosynthesis of all flavonoids.[58]

Plants containing chalcones such as *Angelica* (Apiaceae), *Glycyrrhiza* (Fabaceae), *Piper* (Piperaceae) and *Ruscus* (Asparagaceae) have long been utilized for their medicinal properties. Apples and sour fruits are nutritional sources of chalcones and dihydrochalcones. Chalcones, such as glucosides of **butein**, provide yellow colour in many flowers in plants. Chalcones have been reported to have important biological activities.[59,60] Some examples of chalcones are shown in Figure 5.30.

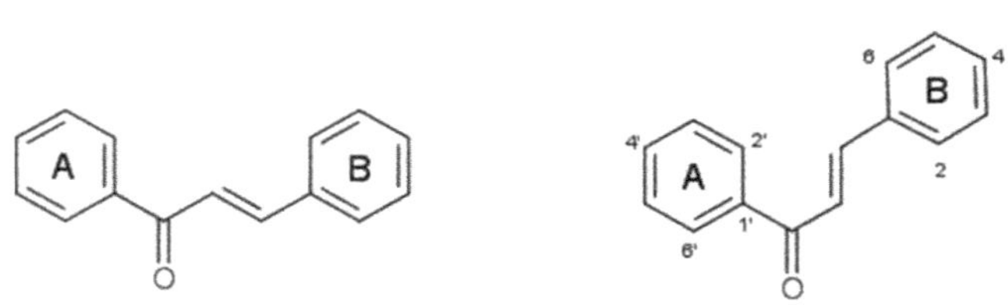

Figure 5.29 Structures of chalcones.

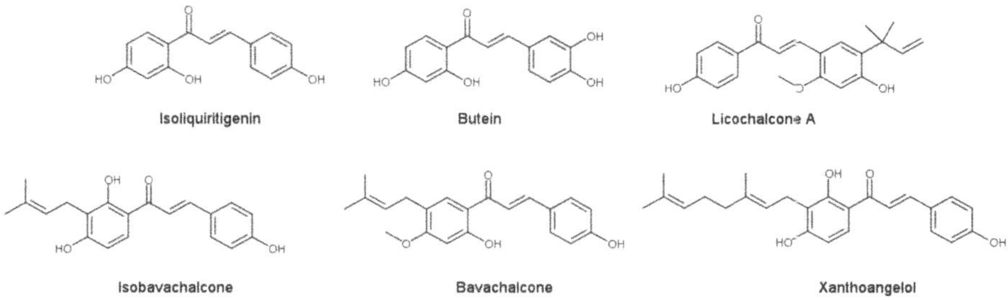

Figure 5.30 Examples of chalcones.

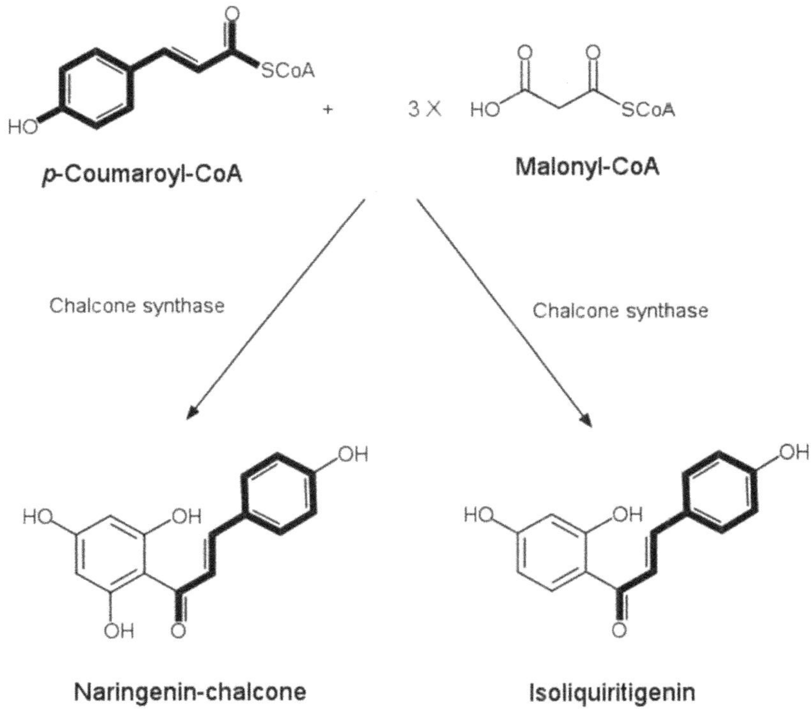

Figure 5.31 Biosynthesis of chalcones.

Biosynthesis of chalcones involves both acetate malonate and shikimate pathways (Figure 5.31). Chalcone synthase, a type III polyketide synthase, catalyzes condensation of *p*-coumaroyl-CoA derived from phenylalanine and three units of malonyl-CoA to produce the chalcones naringenin-chalcone and isoliquiritigenin. A series of condensation reactions of three acetate units derived from three malonyl CoA forms a thioester linked tetraketide. Intramolecular Claisen reaction forms the new ring system chalcone.[61–63]

5.9.2 Aurones

Aurones have the 2-benzylidene-3(2*H*)-benzofuranone system (Figure 5.32). Aurones are present in flowers and fruits as pigments. Bright yellow ornamental flower colours of *Antirrhinum majus* (Plantaginaceae), *Coreopsis, Cosmos,* and *Dahlia* (Asteraceae) are due to presence of aurones. Aurones are also detected in Oxalidaceae, Gesneriaceae, Rosaceae *etc.* In some plants aurones are also found in bark, heartwood (Fabaceae, Anacardiaceae, and Rhamnaceae), leaves, seedlings, and nectar. Aurones may exist as glucosides, diglucosides or acetylated glucosides. Generally, aurones are of three types. **4′-Monohydroxyaurones** (*e.g.*, hispidol and hispidol 4′-*O*-glucoside), **3′,4′-dihydroxyaurones** [*e.g.*, aureusidin and its glucosides aureusin (aureusidin 6-*O*-glucoside) cernuoside (a 4-*O*-glucoside of aureusidin); sulfuretin and sulfurein (sulfuretin 6-*O*-glucoside); maritimetin and maritimein (maritimetin 6-*O*-glucoside); leptosidine and leptosin (leptosidin 6-*O*-glucoside)], and **3′,4′,5′-trihydroxyaurones** (bracteatin and bracteatin 6-*O*-glucoside). Examples of 4-hydroxyaurones are bractein, and cernuoside. Aurones have also been grouped as **4-deoxyaurones** (sulfuretin, maritimetin, leptosidin, and their 6-*O*-glycosides) and **4-hydroxyaurones** (aureusidin, bracteatin and their glycosides).[64,65]

Aurones are biosynthesized from chalcones. Monohydroxy chalcones such as isoliquiretigenin and 4,2′,4′,6′-tetrahydroxychalcone (THC) (also known as naringenin chalcone) and their 4′-*O*-glucosides are used as substrates. 3,4-Dihydroxychalcones such as 3,4,2′,4′,6′-pentahydroxychalcone (PHC) (also known as eriodictyol chalcone) and butein and their glycosides are also used as substrates. The biosynthetic pathway of aurones is compartmentalized into cytoplasm, vacuole, and plastid (Figure 5.33).[64,65]

Aurones play a major role in pigmentation, particularly bright yellow colour in flowers. The colour helps in pollination. Aurones present in nectar could provide antibacterial activity against nectar infestation. Some aurones are also phytotoalexins and play a role in plant defense. The aurone based phytoalexin cephalocerone was reported from the cactus *Cephalocereus senilis* (Cactaceae). This compound inhibits the growth of a cactus pathogen responsible for rot. Hispidol was found to be a phytoalexin in *Medicago truncatulata* (Fabaceae) and is an active antifungal agent for the fungus responsible for spring black stem. Castelliene D and sulfuretin are natural antifungal aurones being active against *Lenzites trabea* which is responsible for vegetal rotting, and the plant pathogen bacteria *Ralstonia solanacearum*.[64,66,67]

5.9.3 Flavonoids

A flavonoid skeleton consists of two benzene rings (A and B) linked by a closed pyran ring C (Figure 5.34). Ring B is attached to C ring at C2 position (except isoflavone). Flavonoids can be divided into various subclasses based on the level of oxidation and hydroxylation pattern in ring C, presence or absence of a double bond between C2 and C3 and formation of a carbonyl group at C4.

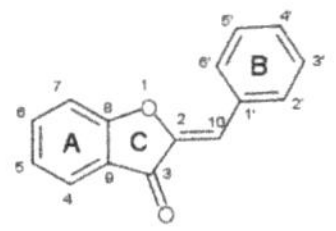

Aurone

4′-Monohydroxyaurones

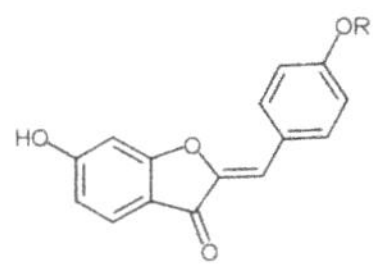

Hispidol R = H
Hispidol 4′-*O*-glucoside R = Glucose

3′,4′-Dihydroxyaurones

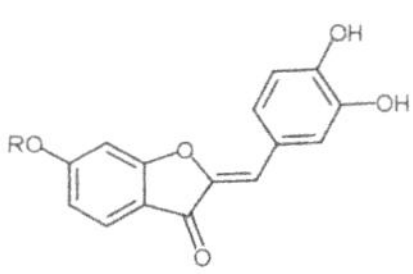

Sulfuretin R = H
Sulfuretin-6-*O*-glucoside R = Glucose
(Sulfurein)

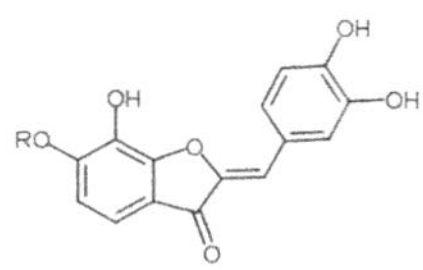

Maritimetin R = H
Maritimetin-6-*O*-glucoside R = Glucose
(Maritimein)

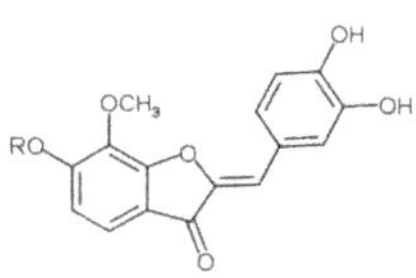

Leptosidin R = H
Leptosidine-6-*O*-glucoside R = Glucose
(Leptosin)

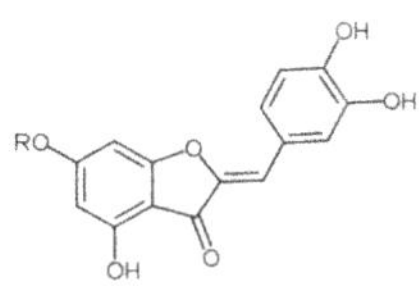

Aureusidin R = H
Aureusidine-6-*O*-glucoside R = Glucose
(Aureusin)

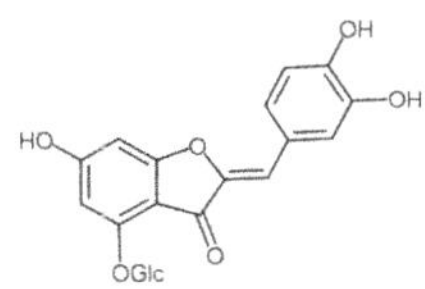

Cernuoside

3′,4′,5′-Trihydroxyaurones

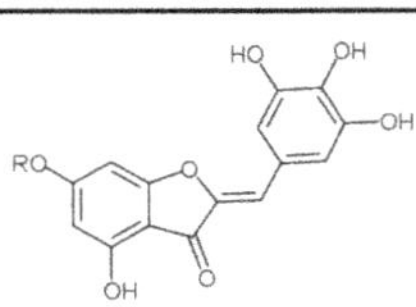

Bracteatin R = H
Bracteatin-6-*O*-glucoside R = Glucose

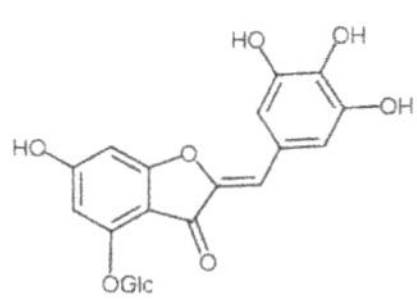

Bractein

Figure 5.32 Aurones.

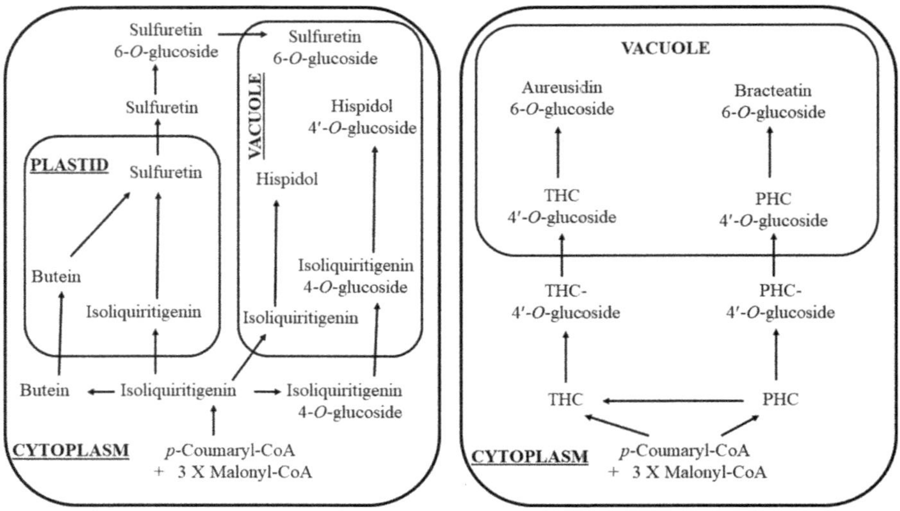

Figure 5.33 Biosynthesis of aurones. THC: 4,2′,4′,6′-tetrahydroxychalcone; PHC: 3,4,2′,4′,6′-pentahydroxychalcone.

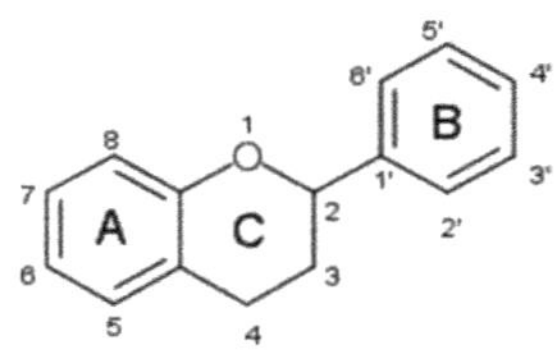

Figure 5.34 Basic flavonoid structure.

These subclasses are flavones, isoflavones, flavonols, flavanonol (dihydroflavanol), flavanones, flavan-3-ols, leucoanthocyanidins and anthocyanidins. Flavonoids may occur as aglycones, glycosides, and methylated derivatives. In the case of glycosides, the glycosidic linkages are usually located at 3 or 7 positions.[68–70]

(i) **Flavonoids without a double bond between C2 and C3 positions** (Figure 5.35)

Flavanones: Flavanones are without a double bond in ring C and without a hydroxyl group at the C3 position. Flavanones are usually found in citrus fruits. Examples are **eriodictyol, hesperetin** and **naringenin.**

Flavanonols (dihydroflavonols): These are 3-hydroxy derivatives of flavanones. Examples are **taxifolin** (dihydroquercetin), **aromadendrin** (dihydrokaempferol) and **ampelopsin** (dihydromyricetin).

Flavan-3-ols (flavanols): This group of flavonoids contain a saturated C ring (without double bond), and a hydroxyl group at C3 position without a carbonyl group at C4 position. Examples are **catechin** and **epicatechin.**

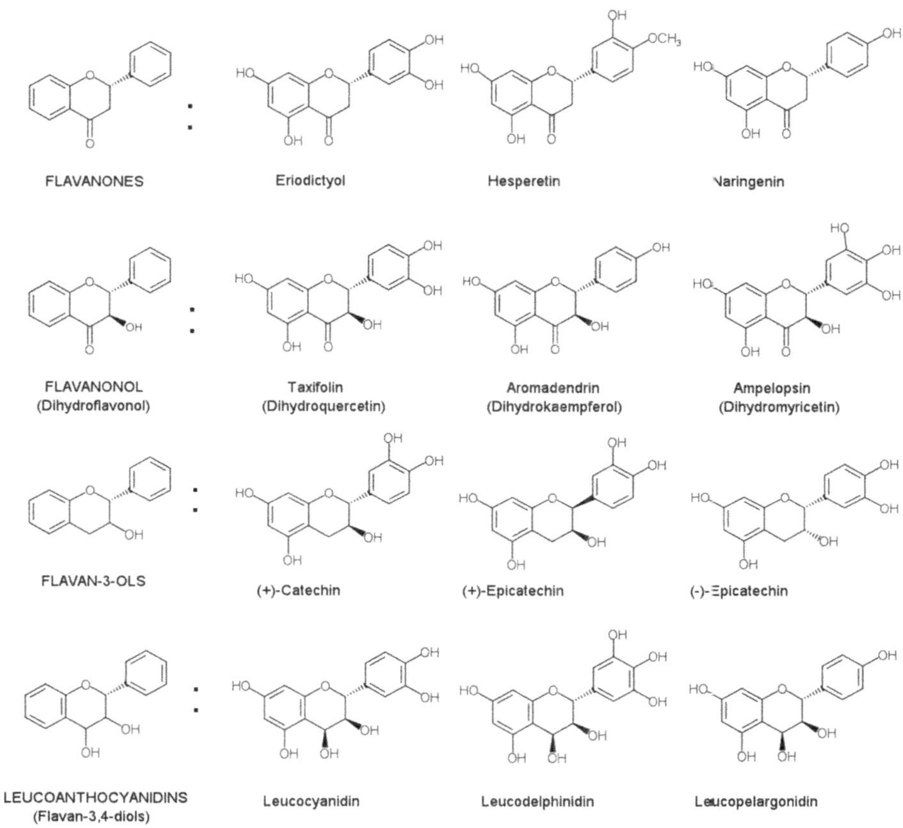

Figure 5.35 Flavonoids without a double bond between C2 and C3 positions.

These compounds may exist in two stereoisomeric forms each, giving (+)-catechin, (+)-epicatechin, (−)-epicatechin *etc.*

Leucoanthocyanidins (flavan-3,4-diols): Here the ketone group at C4 position is reduced forming a hydroxyl group. Examples are **leucocyanidin, leucodelphinidin** and **leocopelargonidin**. These compounds are found in wood, taking part in the formation of condensed tannins.

(ii) **Flavonoids with a double bond between C2 and C3 positions** (Figure 5.36)

Flavones: Flavones contain a ketone group at the C4 position and a double bond between C2 and C3.Examples are **luteolin** and **apigenin**.

Isoflavones: In isoflavones, ring B is attached at the C3 position of ring C in contrast to the usual C2 position in other flavonoids. Examples are **genistein, daidzein, glycitein, formononetin**, and **biochanin A**.

Flavonols: Flavonols are 3-hydroxy derivatives of flavones. Examples are **kaempferol, quercetin, myricetin, galangin**, and their glycosidic forms.

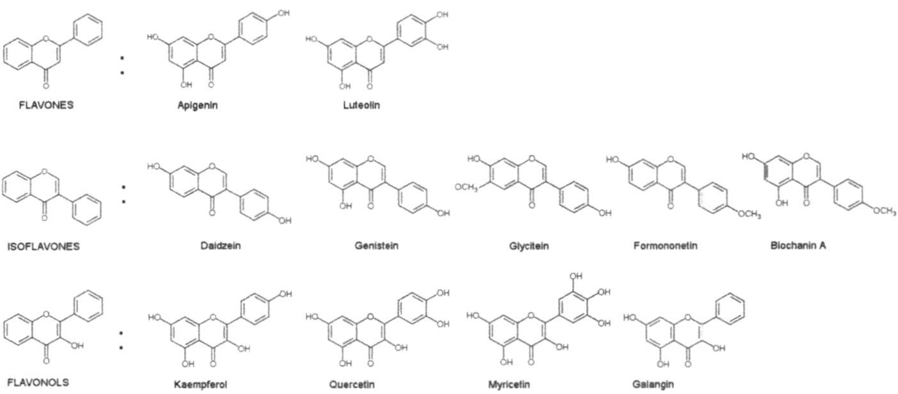

Figure 5.36 Flavonoids with a double bond between C2 and C3 positions.

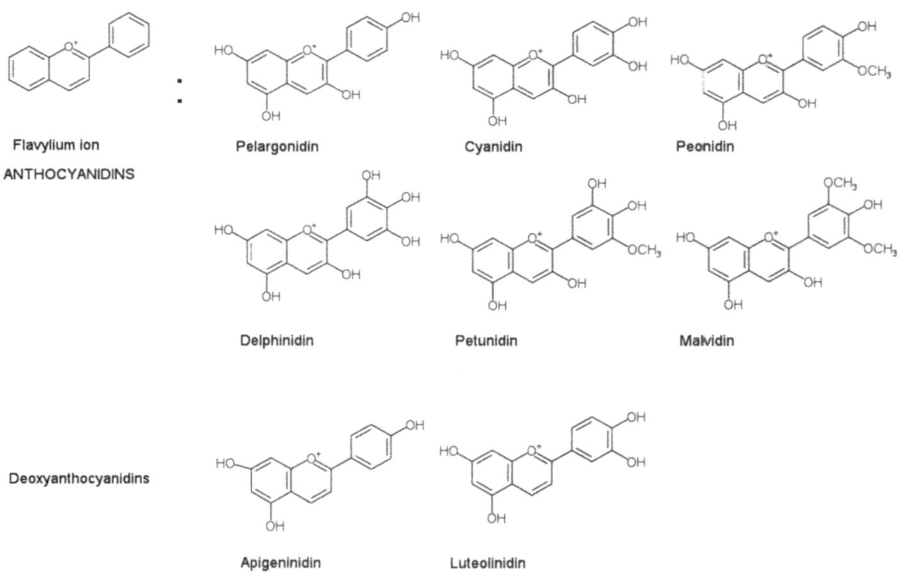

Figure 5.37 Flavonoids with the flavylium ion as the basic structure.

(iii) **Flavonoids with the flavylium ion as the basic structure** (Figure 5.37)

Anthocyanidins: Anthocyanidins have a positive charge at the oxygen atom of the C ring of the flavonoid structure. It is known as **flavylium (2-phenylchromenylium) cation**. Anthocyanidins lack a ketone oxygen at the C4 position. Glycosylated forms of anthocyanidins are known as **anthocyanins**. The conjugated bond of anthocyanins produces blue, red, and purple colours in plants particularly flowers and fruits. The colour may change with pH. There are over 700 anthocyanin derivatives of 27 aglycones (anthocyanidins). The most common anthocyanidins found in plants are **cyanidin, delphinidin, pelargonidin, petunidin, peonidin,** and **malvidin.** An increased number of hydroxyl group and/or methoxyl groups on the B ring of anthocyanidin gives blue colour to the anthocyanin.[71,72]

3-Deoxyanthocyanidins lack the hydroxyl group at the C3 position. Examples are **luteolinidin** and **apigeninidin.** These types of anthocyanins are rare in higher plants. However, pigmented sorghum bran contains high levels of 3-deoxyanthocyanidins.[73]

Anthocyanins are used as natural colourants and food additives to attract consumers and are alternatives to synthetic dyes. These coloured pigments have potential nutraceutical and pharmaceutical values.

As discussed before, flavonoids are derived from *p*-coumaryl-CoA (a phenylpropanoid produced following the shikimate pathway) and malonyl-CoA (from the acetate malonate derived polyketide pathway) (Figure 5.3). Chalcone synthase catalyzes condensation of *p*-coumaryl-CoA and three units of malonyl-CoA to produce naringenin-chalcone. Naringenin-chalcone is cyclized to the flavanone naringenin. The reaction is catalyzed by chalcone isomerase. On the other hand, the chalcone isoliquiritigenin is converted to liquiritigenin. Both **naringenin** and **liquiritigenin** are **flavanones.**[57] These flavanones are precursors of all the flavonoids (Figure 5.38)

1. Naringenin is converted to dihydrokaempferol (aromadendrin), a **flavanonol/dihydroflavonol,** by flavone 3β-hydroxylase. Further hydroxylations of dihydrokaempferol take place *via* the two enzymes flavone 3′β-hydroxylase and flavone 3′,5′–β-hydroxylase.
   - The dihydroflavonols are reduced and converted to **leucoanthocyanidins** by dihydroflavonol 4-reductase.
   - **Flavonols** are synthesized from the dihydroflavonols by the enzyme flavonol synthase.
   - Leucoanthocyanidins are converted to **anthocyanidins** by leucoanthocyanidin dioxygenase/anthocyanidin synthase.
   - Reduction of leucoanthocyanidins and anthocyanidins produce **flavan-3-ols** by leucoanthocyanidin reductase and anthocyanidin reductase.
2. **Flavones** such as apigenin and luteolin are synthesized from naringenin by flavone synthase.
3. **Isoflavones** are synthesized from both naringenin and liquiritigenin. Through formation of hydroxyisoflavones, naringenin is converted to genistein and biochanin A and liquiritigenin is converted to glycitein and daidzein.

In plants, flavonoids play important roles in combating oxidative stress and as growth regulators. Flavonoids also protect plants from UV radiation. They can create auxin gradients that translate into phenotypes with different morphoanatomical features. Flavonoids regulate the IAA-oxidase activity. They also play a significant role in plant-environment interactions. In plants,

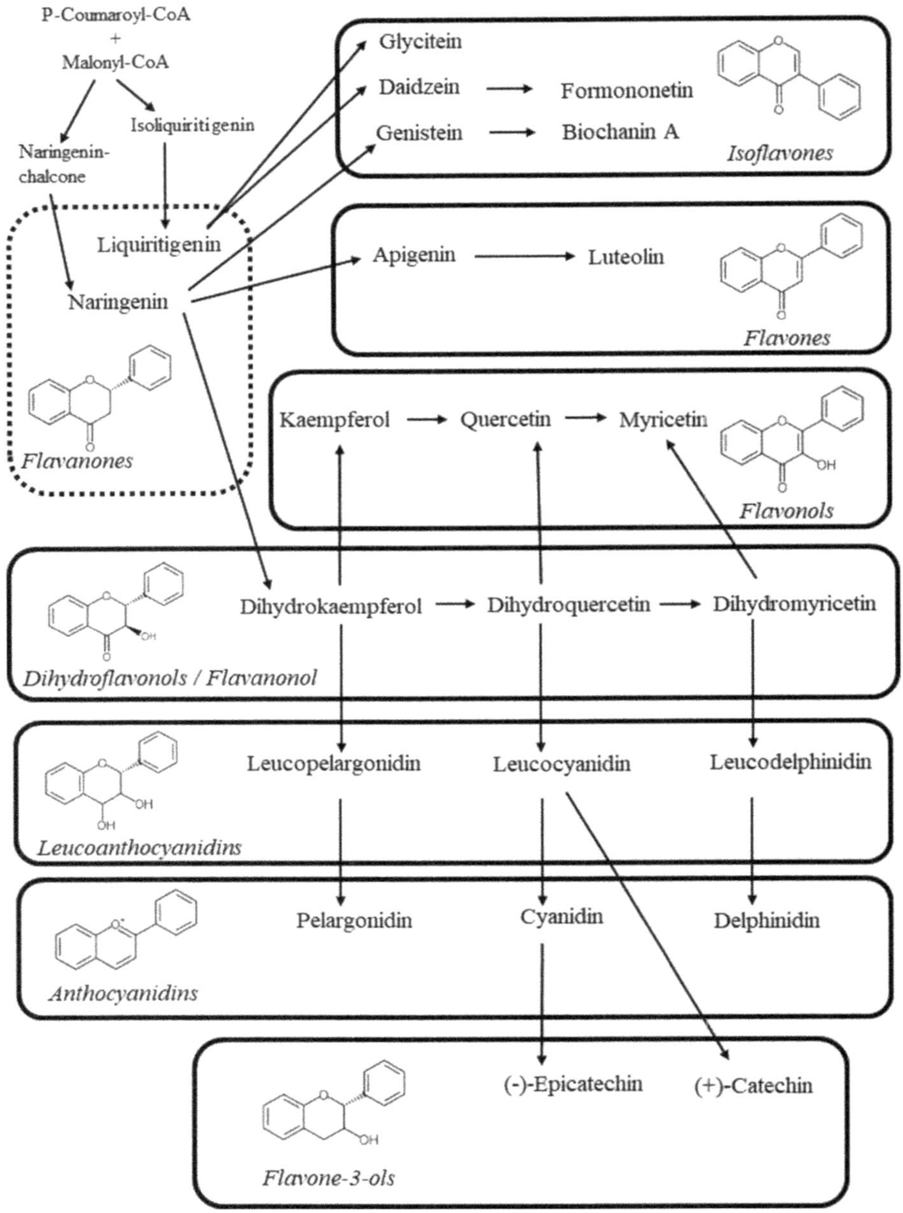

Figure 5.38 Biosynthesis of flavonoids.

isoflavones act as phytoalexins or act as precursors of phytoalexins. These defense compounds have fungistatic, antibacterial, antiviral and antioxidant properties. Isoflavones are essential for nodulation as they induce the nodulation genes.[69,74]

Food and medicinal plants are rich in flavonoids. Most of the flavonoids are highly antioxidant. Thus, dietary flavonoids help prevent stress related diseases.

The isoflavonoids, in addition to some non-flavonoids like lignans, are known as **phytoestogens** due to their molecular structure and size resembling estrogens, in particular estradiol. These phytoestrogens exert estrogenic and/or antiestrogenic effects. Phytoestrogens have a negative effect on female reproductive ability. Genistein, daidzein, glycitein, biochanin A, and formononetin are isoflavonoid phytoestrogens. The main source of isoflavonoids is legumes, including soybean (*Glycine max*). Isoflavones are considered chemopreventive. These metabolites are used as an alternative therapy for a wide range of hormonal disorders including several cancer types (breast cancer, prostate cancer), cardiovascular diseases, osteoporosis, or menopausal symptoms. The phenol group of isoflavones enable attachment to and activation of the estrogen receptors. Phytoestrogens show lower estrogenic potency than estradiol. However, phytoestrogens are consumed in large amounts. Many of them are also considered to be endocrine disruptors.[75]

5.10 Quinoids

Quinones have a hexacyclic saturated dione or diketone system. They are generally derived from oxidation of hydroquinones. *o*-Quinones are generated by tyrosinase that catalyzes phenols and catechols (*o*-diphenols). *o*-Quinones are reactive compounds responsible for oxidative browning of plant products. *p*-Quinones are reasonably stable compounds. Naturally occurring quinones may be benzoquinones, naphthoquinones, and anthraquinones.

5.10.1 Benzoquinones

Benzoquinones (simple quinones) are monocyclic quinones containing carbonyl groups usually at the *ortho* or *para* position. There are many quinones reported from plants. Some are of importance to plants. For example, plastoquinone and ubiquinone have a benzoquinone ring attached to a polyisoprenoid side chain (Figure 5.39). Plastoquinone and ubiquinone function as electron transporters in the electron transport chain being involved in photophosphorylation and oxidative phosphorylation, respectively.

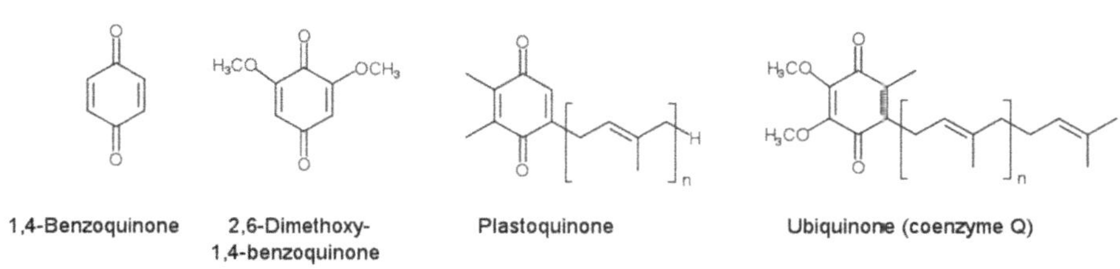

Figure 5.39 Structures of some benzoquinones.

Some quinones play the role of signaling agent. Orobanchaceae plants that parasitize roots, develop the feeding organs haustoria. Host derived 2,6-dimethoxy-1,4-benzoquinone is a potent haustorium-inducing factor.[76]

The benzoquinone ring of plastoquinone originates from homogentisic acid[77] that is produced from tyrosine by the catalytic actions of the enzymes tyrosine aminotransferase and 4-hydroxyprenylpyruvate reductase. The precursor of the benzoquinone ring of ubiquinone is 4-hydroxybenzoic acid synthesized from tyrosine and phenylalanine. The prenyl side chain of these two quinones is derived from the MEP/MVA pathway (Figure 5.40).

5.10.2 Naphthoquinones

Naphthoquinones are widely distributed in plants and animals. Each compound (C6–C4 skeleton) has a benzene moiety (ring A) with a conjugated cyclic diketone ring (ring B) having carbonyl groups. Carbonyl groups occur as the *para* isomer (1,4-naphthoquinone) or *ortho* isomer (1,2-napthoquinone).

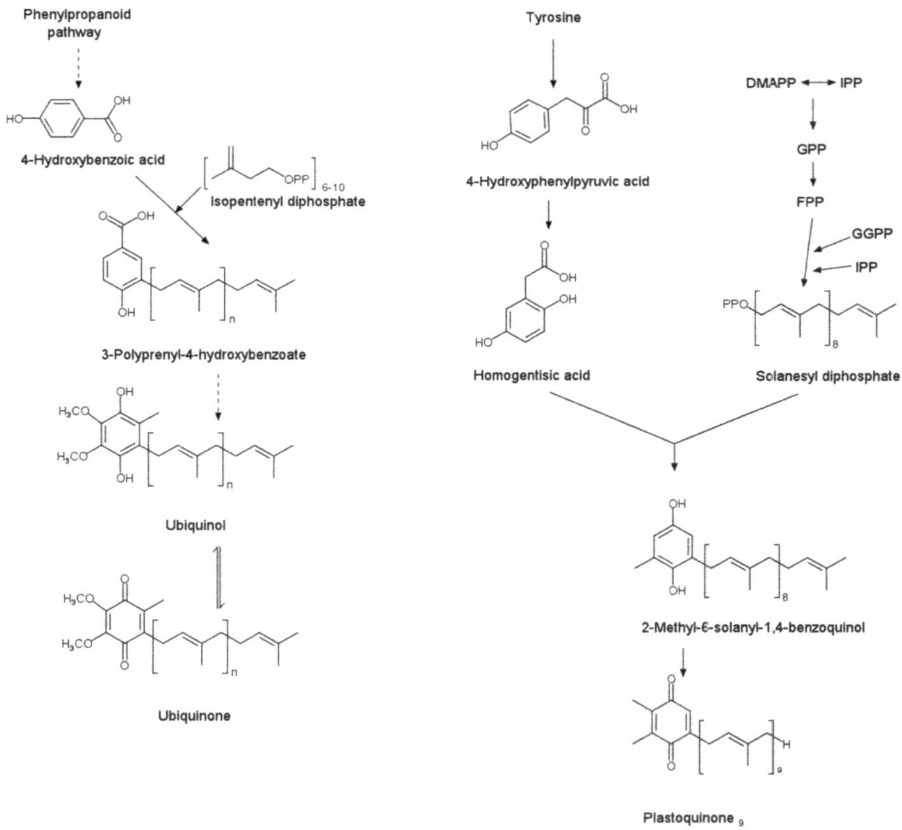

Figure 5.40 Biosynthesis of ubiquinone and plastoquinone.

They may exist as monomers, dimers, and tetramers. The compounds may also exist as furano and pyrano derivatives. Plants show great diversity of naphthoquinones. These compounds are also produced by bacteria, fungi, algae, and diatoms.[78–80] Different examples of naphthoquinones are shown in Table 5.3 and Figure 5.41.

Table 5.3 Examples of naphthoquinones and source plants.

Napthoquinones	Plants
Plumbagin	Insectivorous plants (*Nepenthes* spp., Nepenthaceae), some plants of Ebenaceae, Plumbaginaceae
Droserone	Insectivorous plants (*Nepenthes* spp.)
Diospyrin	*Diospyros chamaethamnus* (Ebenaceae)
Juglone	*Diospyros lycioides*, *Juglans nigra* (Juglandaceae) (black walnut), *Juglans regia* (English walnut), *Sesamum indicum* (Pedaliaceae) (sesame)
Lawsone	*Lawsonia inermis* (Lythraceae)
Alkannin	Species of *Alkanna* (Boraginaceae)
Shikonin	Species of *Alkanna*, *Lithospermum erythrorhizon*, and some other plants of Boraginaceae
Lapachol	*Tectona grandis* (Lamiaceae), some genera of Bignoniaceae
Phylloquinone (vitamin K_1)	Many plants

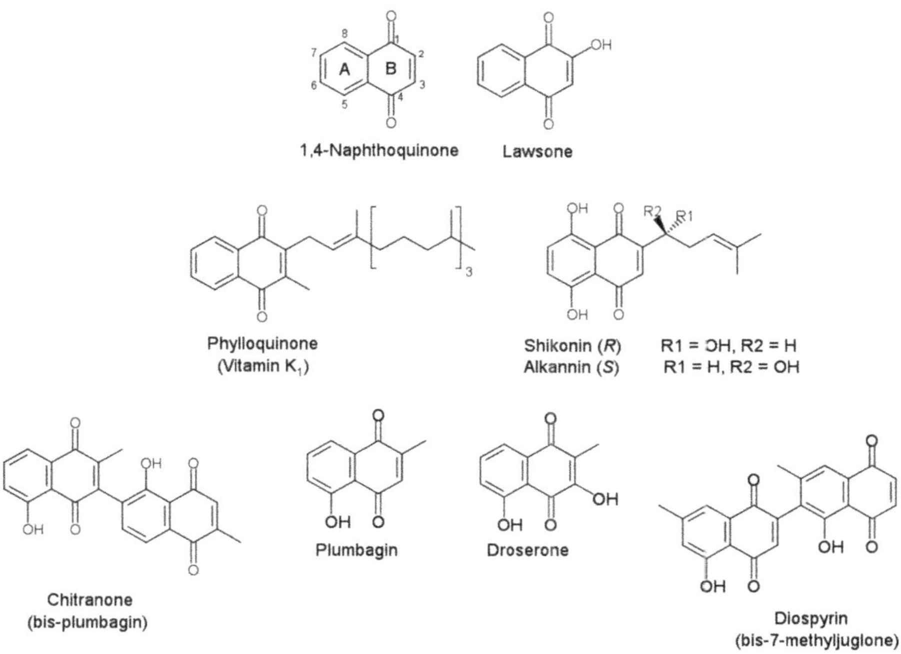

Figure 5.41 Structures of some naphthoquinones.

Naphthoquinones have properties which allow them to play ecological roles in plant–plant (allelopathy), plant-insect and plant-microbe interactions. One classic example of plant–plant interaction occurs in black walnut which releases the phytotoxic allelopathic chemicals 1,4-napthoquinone and juglone. Phylloquinone is vitamin K_1. *Lawsonia inermis* (Lythraceae) leaf (commonly known as henna) extracts are used as a dye[80] due to presence of lawsone. Shikonin exists as an enantiomeric pair with alkannin. These two metabolites are obtained from the roots of different species of *Lithospermum, Alkanna, Onosma, Echium, Cynoglossum*, and *Anchusa* of Boraginaceae. Shikonin has potential antimicrobial, anticancer and wound healing properties.[78] The carnivorous plants belonging to Droseraceae and Nepenthaceae contain important naphthoquinones like plumbagin and 7-methyljuglone with antimicrobial activity.[79] Plumbagin is also available in *Plumbago zeylanica* (Plumbaginaceae).

Naphthoquinones are derived from several metabolic pathways *e.g.*, *o*-succinyl benzoate pathway (OSB pathway), 4-hydroxybenzoic acid/geranyl diphosphate pathway (4HBA/MVA pathway), acetate polymalonate pathway and homogentisate/MVA (HGA/MVA) pathway in plants.[80] The biosynthetic pathways for different naphthoquinones are presented schematically in Figure 5.42.

5.10.3 Anthraquinones

Anthraquinones are tricyclic compounds having an anthracene nucleus with two carbonyl groups usually on ring B. A few common anthraquinones are chrysophanol, emodin, rhein, physcion, aloe-emodin and alizarin (Figure 5.43). The anthraquinones may exist as aglycones or glycosides.

There are several examples of medicinally important plants containing anthraquinones. Mostly anthraquinone containing plants are used as laxatives. Senna (*Cassia angustifolia*, synonym: *Senna alexandria*, Fabaceae) leaves are used worldwide as a natural laxative. The anthraquinone glycosides sennosides (sennoside A, B, C, and D) are found in large quantities in the leaves. Sennosides A and B, the homo-dianthrones of rhein, are mostly responsible for the bioactivity of senna. Sennoside C and D are hetero-dianthrones of rhein and aloe-emodin anthrones.[81] *Aloe vera* (syn. *Aloe barbadensis*), of family Asphodelaceae, extracts are used in herbal medicine as cosmetics to be used externally as emollients, and moisturizers. Aloe is also a laxative. The active constituents are aloin A (barbaloin), aloin B (isobarbaloin), and aloe-emodin. The real active constituent is aloe-emodin. Aloin is converted to aloe-emodin in the gut. Other constituents are aloinosides (A and B) and aloeresin A and B.[82,83] Other anthraquinone containing laxative plants are rhubarb roots (dried rhizomes and roots of *Rheum palmatum*, *Rheum tanguticum* and *Rheum officinale* from family Polygonaceae) and cascara bark derived from the bark of *Rhamnus purshiana*, synonym *Frangula purshiana* (Rhamnaceae).

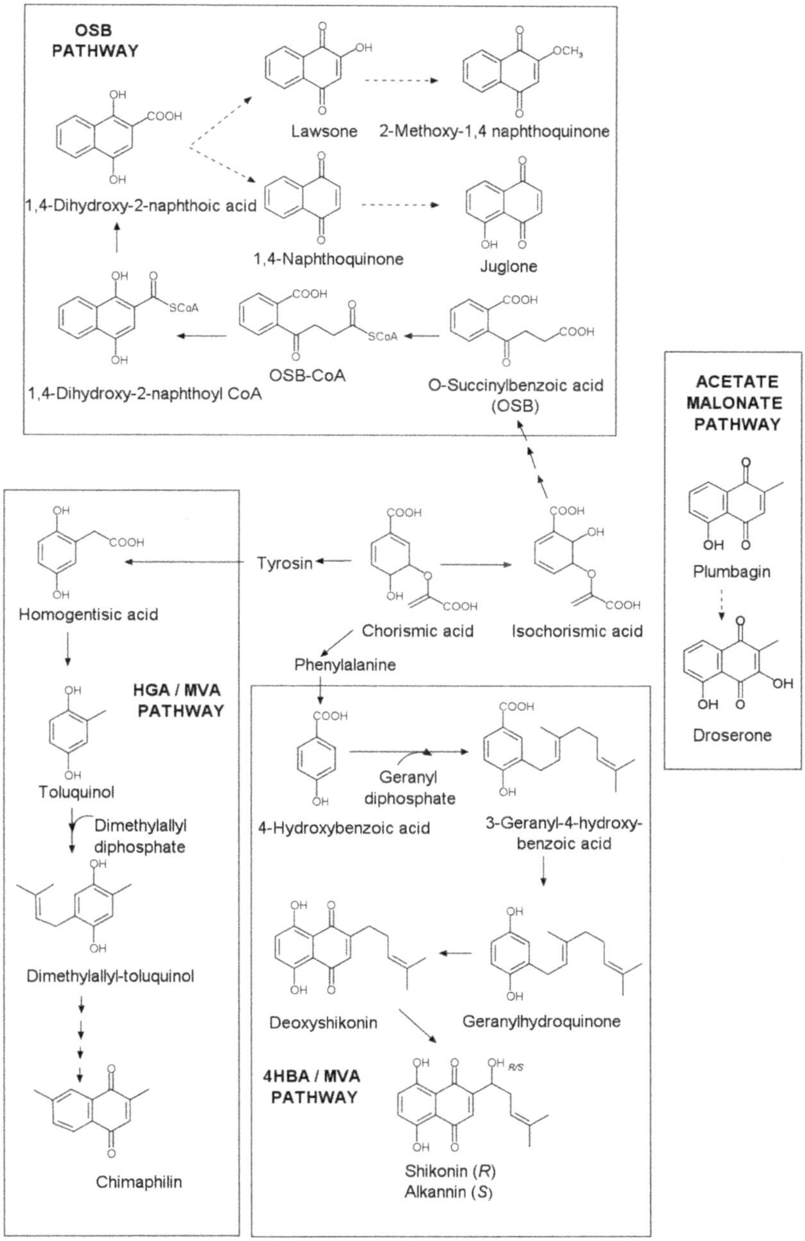

Figure 5.42 Biosynthesis of naphthoquinones.

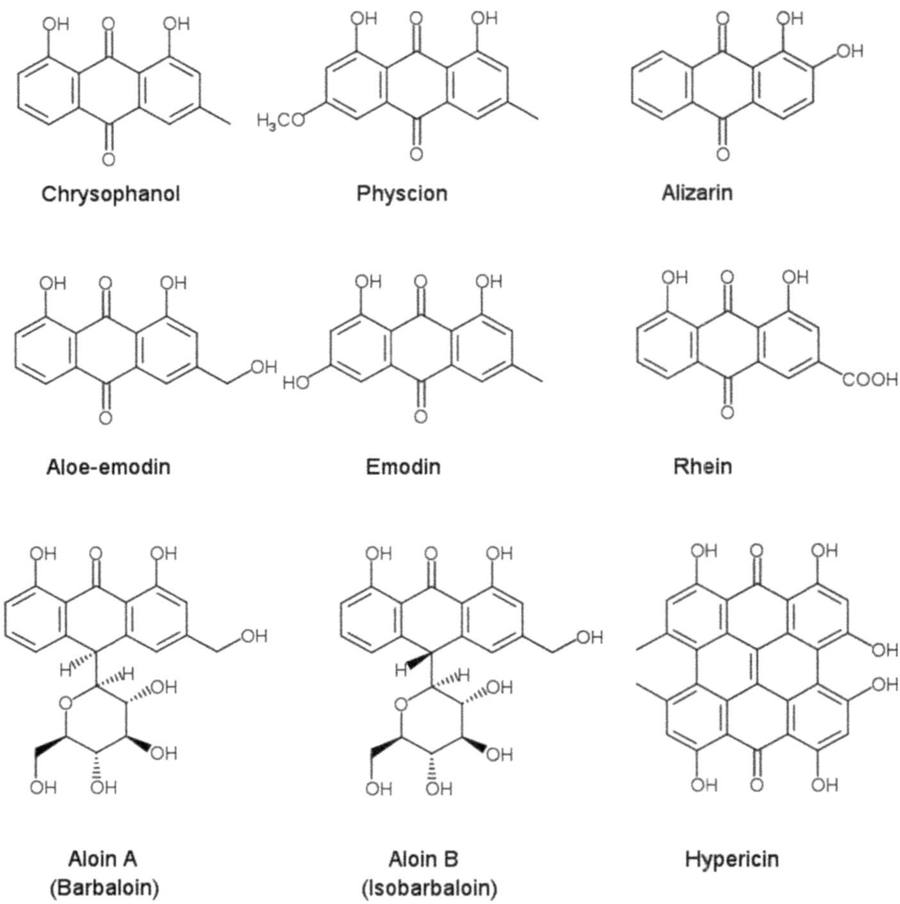

Figure 5.43 Different anthraquinones.

St. John's wort (*Hypericum perforatum*, Hypericaceae) is used for treating depression and other mood disorders. Products are available in the form of tablets, capsules, teas, and tinctures. The drug is standardized in terms of the anthraquinone hypericin. Two major constituents identified are hypericin (an anthraquinone derived naphthodianthrone) and hyperforin (a phloroglucinol derivative). Hypericin is a pigment responsible for the red color of St. John's wort oils. The compound is highly photoreactive.[84]

Anthraquinones in higher plants are derived from different precursors and pathways. In general, the isochorismate and plastidial MEP pathways are involved in the biosynthesis of anthraquinones. Anthraquinone structures may also be produced through polyketide formation[81] (Figure 5.44).

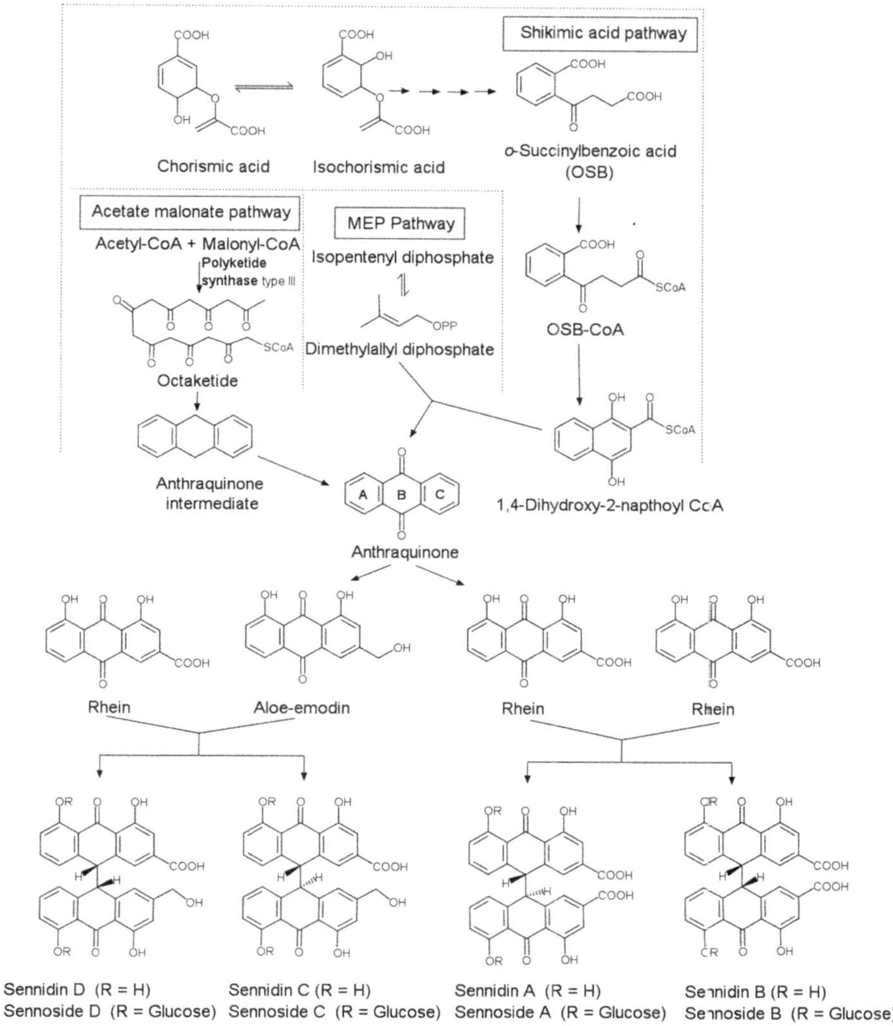

Figure 5.44 Biosynthesis of anthraquinones in senna.

5.11 Stilbenes

Stilbenes are C14 polyphenols composed of two benzene rings linked by an ethylene bridge. Naturally occurring stilbenes are present in *trans*-form. They have C6–C2–C6 structure. Stilbenes are reported in 72 plant species belonging to 31 genera belonging to 12 distantly related families such as Fabaceae, Cyperaceae, Dipterocarpaceae, Moraceae, Pinaceae, Polygonaceae, Poaceae, Vitaceae and Gnetaceae.[85,86]

Stilbenes are biosynthesized *via* a phenylalanine/polymalonate route. Coenzyme A-esters of cinnamic acid derivatives (generated from the phenylpropanoids) and three units of malonyl CoA react to produce the monomeric stilbenes pinosylvin, piceatannol, and resveratrol (Figure 5.45) with intermediate formation of tetraketides. The last step is catalyzed by stilbene synthase that belongs to type III of the polyketide synthase enzyme and carries out the condensation reactions with malonyl CoA. Stilbene synthase is a defense protein against stresses.[85–87]

Stilbenes have glycosylated, methylated, and prenylated derivatives. Resveratroloside and polydatin are glycosides of resveratrol. Piceatannol-3-*O*-glucoside and astringin are glycosylated derivatives of piceatannol (Figure 5.46). Pinosylvin methyl ether and pinosylvin dimethyl ether are methylated derivatives of pinosylvin. Pterostilbene and rhapontigenin are methylated derivatives of resveratrol and piceatannol, respectively. Arachidin 3 and arachidin 1 are prenylated derivatives of resveratrol and piceatannol respectively (Figure 5.47). Glycosylation may protect resveratrol from oxidation and enzymatic degradation. Oligomeric stilbenes are produced by oxidative coupling of homogeneous or heterogeneous monomers[87] (Figure 5.48).

Stilbenes are defense molecules in plants. They act as phytoalexins particularly in peanut, pine, and grapevine. Stilbene phytoalexins are derived from pinosylvin in pine and resveratrol in peanut and grapevine. ε-Viniferin and δ-viniferin are oligomers of resveratrol, formed in grapevine in response to infections and stress. Pinosylvin 3-*O*-methyl ether acts as a constitutive metabolite in the resistance of pines to nematodes. Constitutive stilbenes can also act as allelochemicals. Abiotic stresses (heavy metals, UV

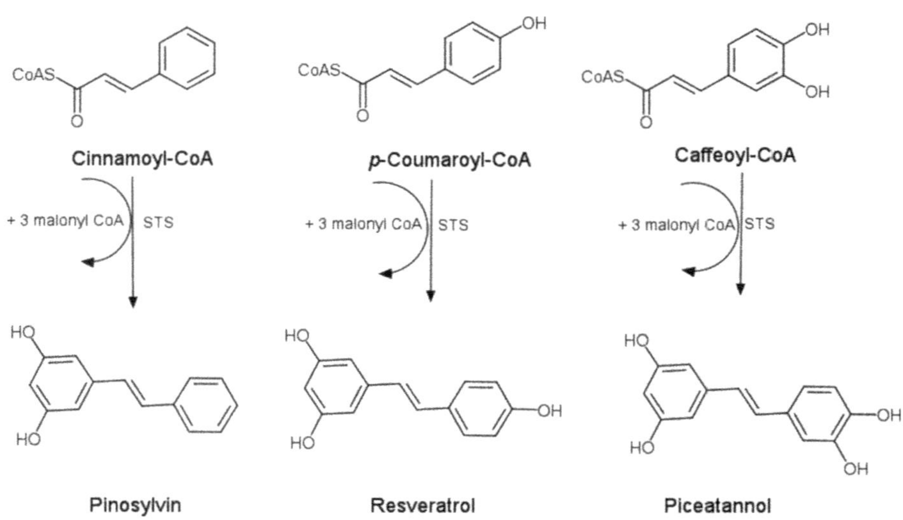

Figure 5.45 Biosynthesis of stilbene monomers: STS: Stilbene synthase.

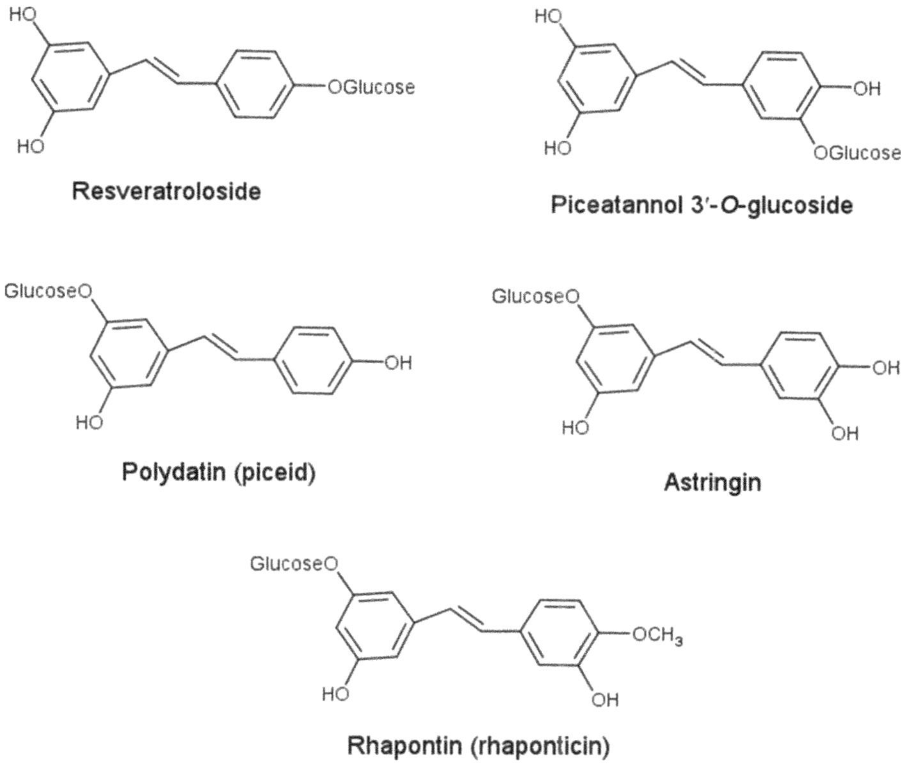

Figure 5.46 Glycosylated derivatives of stilbene monomers.

radiation, salt stress, drought, and tropospheric ozone) also induce stilbene biosynthesis.[85–88]

Stilbenes also have promising health benefits because of their antioxidant, antimutagenic, cardiovascular (French paradox), antiobesity, antidiabetic, anti-aging, neuroprotective and chemopreventive properties in experimental models.[86–88] In recent years, more attention has been paid to the monomeric stilbenes (pterostilbene, pinosylvin, piceatannol) and the oligomeric stilbene viniferin which showed more health promoting properties than resveratrol.[85]

5.12 Xanthones

Xanthones are oxygenated tricyclic polyphenols consisting of two benzenoid and one pyranoid rings.[89] Natural xanthones are found in higher plants, lichens, algae, and fungi, from terrestrial and marine origins. Xanthones are mainly found in Anacardiaceae (*e.g.*, mangiferin from *Mangifera indica*), Gentianaceae (*e.g.*, swerchirin and mangiferin from *Swertia chirayita*) and Clusiaceae, (*e.g.*, mangostin from *Garcinia mangostana*).[89,90]

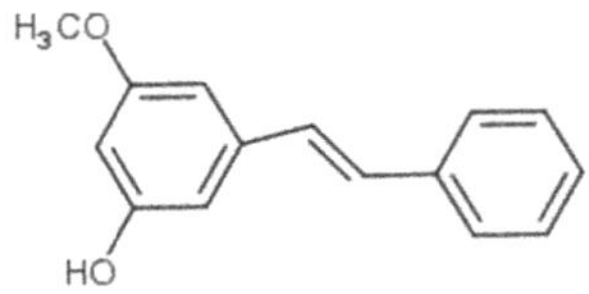

Pinosylvin methyl ether

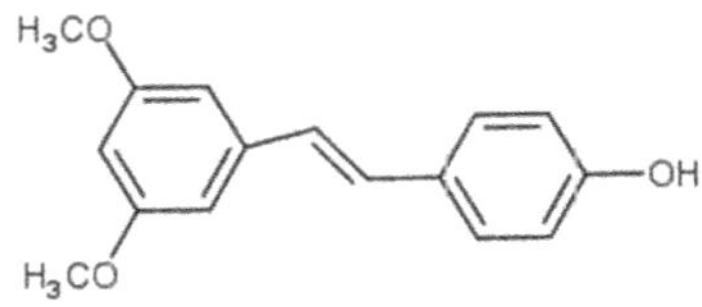

Pterostilbene

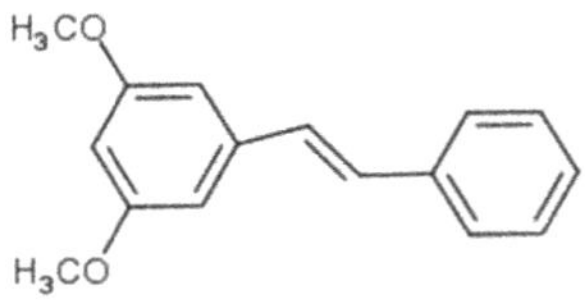

Pinosylvin dimethyl ether

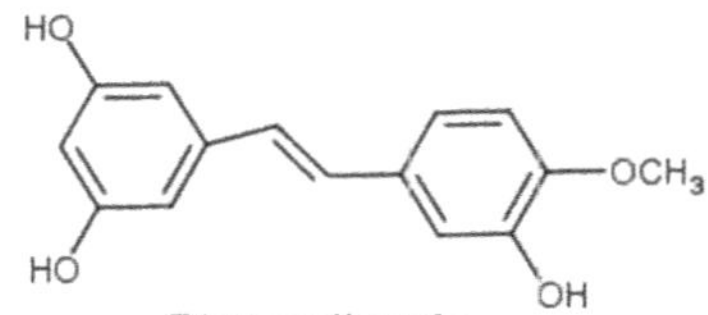

Rhapontigenin

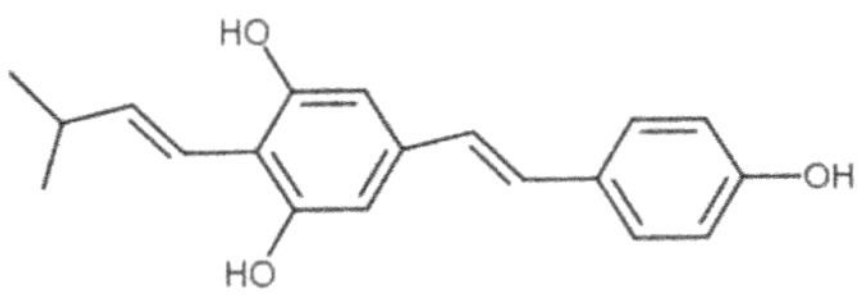

Arachidin 3

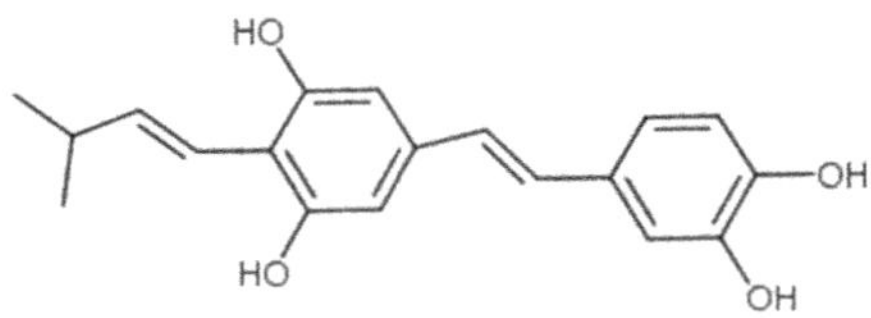

Arachidin 1

Figure 5.47 Methylated and prenylated derivatives of stilbene monomers.

In higher plants, the A-ring (carbons 1–4) of xanthone is acetate derived. The B-ring (carbons 5–8) is derived from the shikimic acid pathway.[89] The basic structure of xanthone and a few examples are shown in Figure 5.49.

Two distinct xanthone biosynthetic pathways are known. In fungi and lichen eight acetyl units yield anthraquinone which is cleaved and recyclized to yield xanthone. Plants use shikimate pathway derived benzoic acid or 3-hydroxybenzoic acid to form respective CoA esters. Condensation of benzoyl CoA and three molecules of malonyl CoA is catalyzed by benzophenone synthase (a type III polyketide synthase) to form 2,4,6-trihydroxybenzophenone. Alternatively, benzophenone synthase catalyzes condensation of 3-hydroxybenzoyl CoA and three molecules of malonyl CoA to form 2,3′,4,6-tetrahydroxybenzophenone. C–O phenol couplings yield 1,3,5- and 1,3,7-trihydroxyxanthones, respectively by intramolecular cyclizations. The two hydroxyxanthones are likely to be the precursors of all higher plant xanthones[91,92] (Figure 5.50).

Xanthones are finding roles in new drug development. Many xanthones have been found to have free radical scavenging, antimicrobial, and antiviral activities. Cancer cytotoxicity has been evaluated with many xanthones.

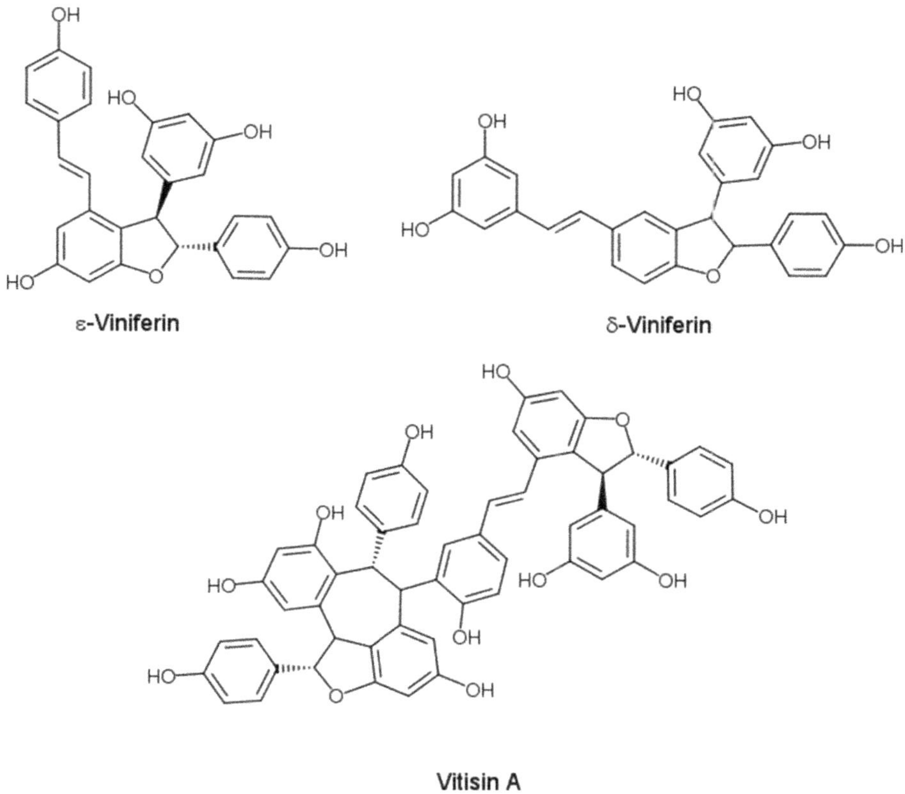

Figure 5.48 Oligomers of stilbene monomers.

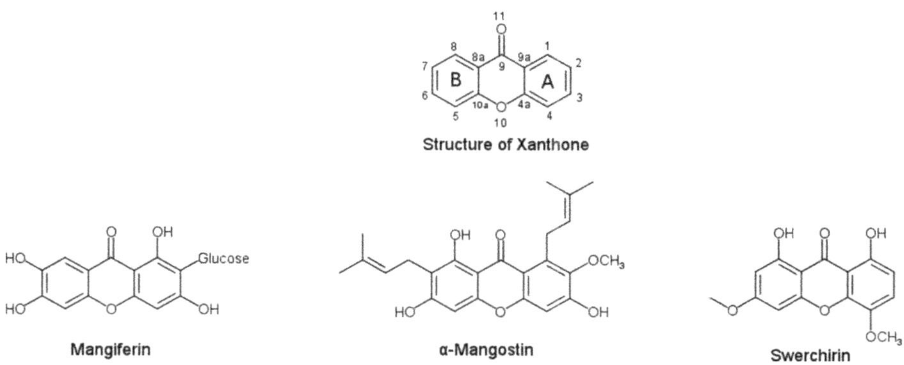

Figure 5.49 Examples of xanthones.

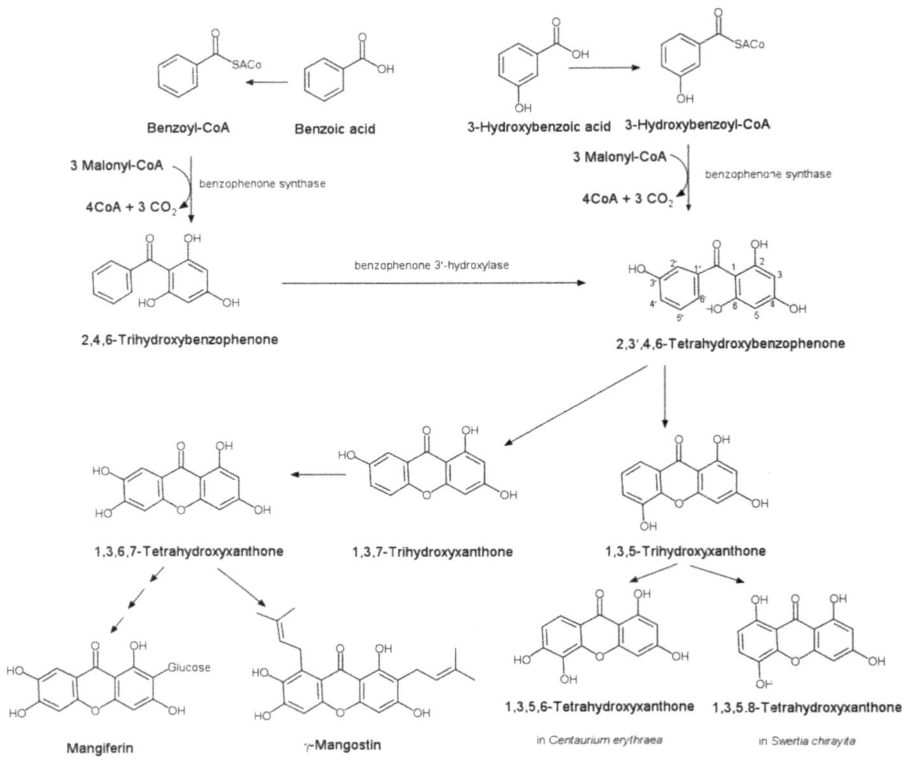

Figure 5.50 Biosynthesis of xanthones.

Other activities reported are anti-inflammatory and antidiabetic activities.[93] The increasing number of hydroxyl groups in xanthone enhance the antiproliferative activity.

5.13 Curcuminoids

Curcuminoids are polyphenolic yellow pigments present in the rhizome of *Curcuma longa* (Zingiberaceae), also known as turmeric. Curcuminoids consist of mainly three components **curcumin**, **demethoxycurcumin**, and **bisdemethoxycurcumin**.[94]

Condensation of feruloyl CoA and malonyl CoA by the type III polyketide synthase named diketide-CoA synthase produces feruloyldiketide CoA. Curcumin synthase which also belong to the polyketide synthase family (type III) ultimately forms the curcuminoids (Figure 5.51). The molecule curcumin is interconverted between its keto and enol tautomers in aqueous solution. The predominant form of curcumin in the solid state and in alkaline conditions is the enol form. In neutral and acidic solutions, it exhibits the keto form[95–97] (Figure 5.52)

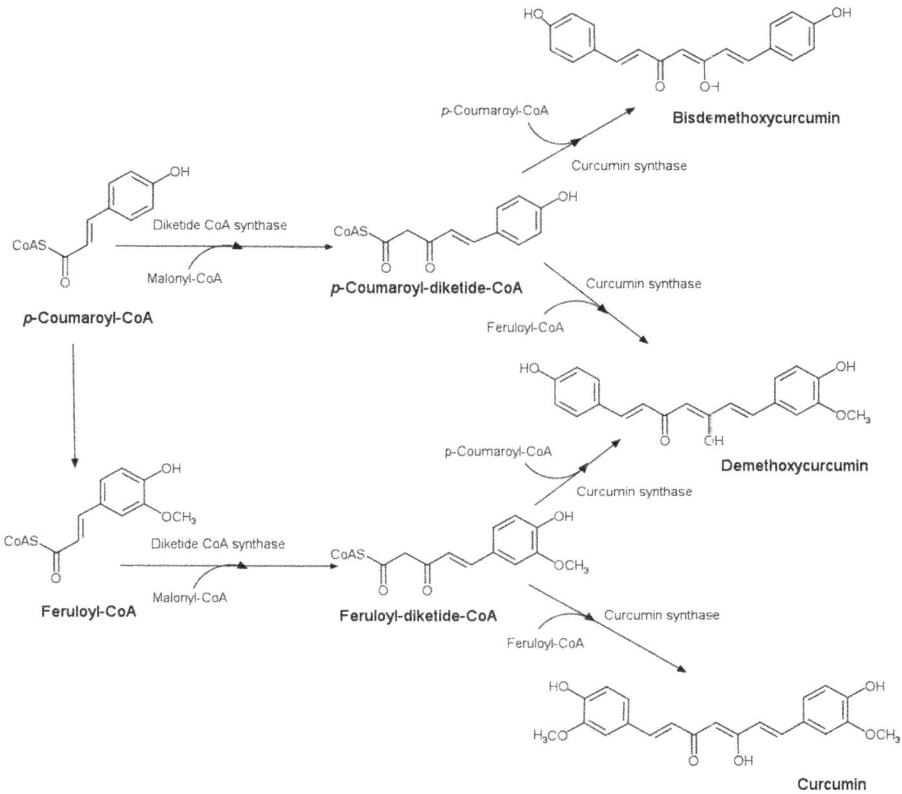

Figure 5.51 Biosynthesis of curcuminoids.

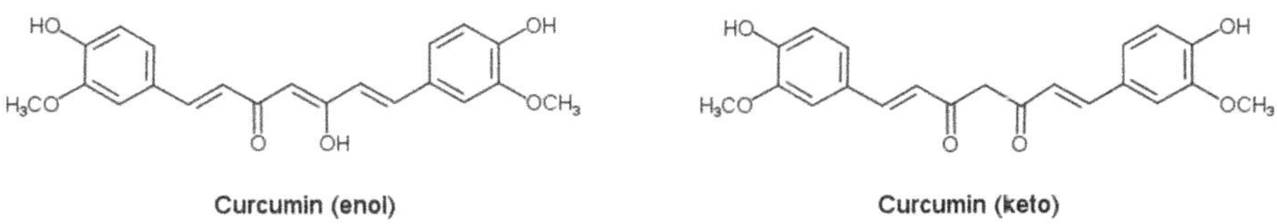

Figure 5.52 Tautomers of curcumin.

Turmeric is used as a spice in curry and in cosmetics. Turmeric is known for its insecticidal and repellent effect on insect pests. But curcuminoid pigments were not found to be active against cabbage looper. ar-Turmerone, the sesquiterpene essential oil constituent of turmeric, was found to be active.[98]

Of all the curcuminoids, curcumin was found to be most important. Various reports on curcumin and turmeric extracts show different biological activities such as antioxidant, anti-inflammatory, cardioprotective, neuroprotective, anticancer, antirheumatic, hepatoprotective, immunomodulating, wound healing, and antimicrobial activities as reported from different

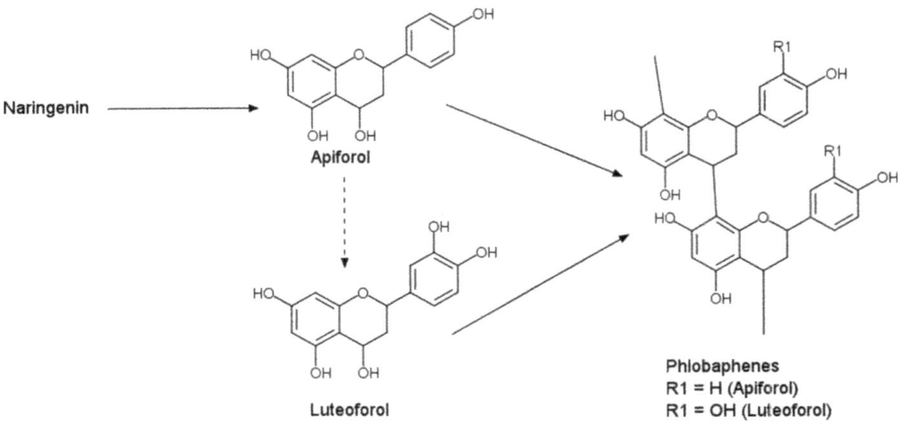

Figure 5.53 Biosynthesis of phlobaphenes.

laboratories. Despite the long-established safety record of turmeric, curcumin has been reported to have side effects. Curcumin is noted to have poor pharmacokinetic/pharmacodynamic properties, low efficacy, and toxic effects in several disease models. Double blinded, placebo controlled clinical trials were not successful.[94,98,99]

5.14 Phlobaphenes

Phlobaphenes are red-brown phenolic pigments found in maize (*Zea mays*) seed pericarp, husks, tassel glumes and certain lines of *Sorghum bicolor* (Poaceae). The flavanone naringenin is converted to the 3-deoxyflavonoids (flavan-4-ols) **apiforol** and **luteoforol**. Apiforol and luteoforol are then polymerized into phlobaphenes. Phlobaphenes are believed to arise by non-enzymatic polymerization (Figure 5.53).[100–102]

Phlobaphenes serve as potential UV-protectants in plants. This pigment provides resistance to plants against various pathogens. They have beneficial effects to human health due to their antioxidant properties.[100–102]

References

1. R. Marchiosi, W. D. dos Santos, R. P. Constantin, R. B. de Lima, A. R. Soars, A. Finger-Teixeira, T. R. Mota, D. M. de Oliveira, M. P. Foletto-Felipe, J. Abrahao and O. Ferrarese-Filho, Biosynthesis and metabolic actions of simple phenolic acids in plants, *Phytochem. Rev.*, 2020, **19**, 865–906.
2. J. R. Widhalm and N. Dudareva, A familiar ring to it: Biosynthesis of plant benzoic acids, *Mol. Plant*, 2015, **8**, 83–97.
3. H. Yoo, J. R. Widhalm, Y. Qian, H. Maeda, B. R. Cooper, A. S. Jannasch, I. Gonda, E. Lewinsohn, D. Rhodes and N. Dudareva, An alternate pathway

contributes to phenylalanine biosynthesis in plants *via* a cytosolic tyrosine: phenylpyruvate aminotransferase, *Nat. Commun.*, 2013, **4**, 2833.
4. Y. Qian, J. H. Lynch, L. Guo, D. Rhodes, J. A. Morgan and N. Dudareva, Completion of the cytosolic post-chorismate phenylalanine biosynthetic pathway in plants, *Nat. Commun.*, 2019, **10**, 15.
5. W. Biala and M. Jasinski, The phenylpropanoid case – it is transport that matters, *Front. Plant Sci.*, 2018, **9**, 1610.
6. V. Yadav, Z. Wang, C. Wei, A. Amo, B. Ahmed, X. Yang and X. Zhang, Phenylpropanoid pathway engineering: an emerging approach towards plant defense, *Pathogens*, 2020, **9**, 312.
7. D. A. Dempsey and D. F. Klessig, How does the multifaceted plant hormone salicylic acid combat disease in plants and are similar mechanisms utilized in humans?, *BMC Biol.*, 2017, **15**, 23.
8. E. Valanciene, I. Jonuskiene, M. Syrpas, E. Augustiniene, P. Matulis, A. Simonavicius and N. Malys, Advances and prospects of phenolic acids production, biorefinery and analysis, *Biomolecules*, 2020, **10**, 874.
9. Y. Deng and S. Lu, Biosynthesis and regulation of phenylpropanoids in plants, *Crit. Rev. Plant Sci.*, 2017, **36**, 257–290.
10. J. Chen, J. Yang, L. Ma, J. Li, N. Shahzad and C. K. Kim, Structure-antioxidant activity relationship of methoxy, phenolic hydroxyl, and carboxylic acid groups of phenolic acids, *Sci. Rep.*, 2020, **10**, 2611.
11. K. W. Singletary, Vanilla, *Nutr. Today*, 2020, **55**, 186–196.
12. W. Boerjan, J. Ralph and M. Baucher, Lignin Biosynthesis, *Annu. Rev. Plant Biol.*, 2003, **54**, 519–546.
13. R. Vanholme, B. D. Meester, J. Ralph and W. Boerjan, Lignin biosynthesis and its integration into metabolism, *Curr. Opin. Biotechnol.*, 2019, **56**, 230–239.
14. J. Barros, L. Escamilla-Trevino, L. Song, X. Rao, J. C. Serrani-Yarce, M. D. Palacios, N. Engle, F. K. Choudhury, T. J. Tschaplinsiko, B. J. Venables, R. Mittler and R. A. Dixon, 4-Coumarate 3-hydroxylase in the lignin biosynthesis pathway is a cytosolic ascorbate peroxidase, *Nat. Commun.*, 2019, **10**, 1994.
15. R. A. Dixon and J. Barros, Lignin Biosynthesis: old roads revisited and new road explored, *Open Biol.*, 2019, **9**, 190215.
16. M. Xie, J. Zhang, T. J. Tschaplinski, G. A. Tuskan, J. Chen and W. Muchero, Regulation of lignin biosynthesis and its role in growth-defense tradeoffs, *Front. Plant Sci.*, 2018, **9**, 1427.
17. Y. Tobimatsu and M. Schuetz, Lignin polymerization: how do plants manage the chemistry so well?, *Curr. Opin. Biotechnol.*, 2019, **56**, 75–81.
18. O. Yu and K. H. Kim, Lignin to materials: A focused review on recent novel lignin applications, *Appl. Sci.*, 2020, **10**, 4626.
19. M. Mariana, T. Alfatah, A. H. P. S. Khalil, E. B. Yahya, N. G. Olaiya, A. Nuryawan, E. M. Mistar, C. K. Abdullah, S. N. Abdulmadjid and H. Ismail, A current advancement on the role of lignin as sustainable reinforcement material in biopolymeric blends, *J. Mater. Res. Technol.*, 2021, **15**, 2287–2316.

20. C. Rodriguez-Garcia, C. Sanchez-Quesada, E. Toledo, M. Delgado-Rodriguez and J. J. Gaforio, Naturally lignan-rich foods: a dietary tool for health promotion?, *Molecules*, 2019, **24**, 917.
21. A. W. K. Yeung, N. T. Tzvetkov, A. A. Balacheva, M. G. Georgieva, R. Gan, A. Jozwik, B. Pyzel, J. O. Horbanczuk, E. Novellino, A. Durazzo, M. Lucarini, E. Camilli, A. B. Souto, A. G. Atanasov and A. Santini, Lignans: quantitative analysis of the research literature, *Front. Pharmacol.*, 2020, **11**, 37.
22. M. Nadeem, I. T. Khan, F. Khan, M. A. Shah and K. Niaz, Lignans and flavonolignans, in *Recent Advances in Natural Products Analysis*, ed. A. S. Silva, S. F. Nabazi. M. Saeedi and S. M. Nabavi, Elsevier, 2020, pp. 98–116.
23. R. B. Teponno, S. Kusari and M. Spiteller, Recent advances in research on lignans and neolignans, *Nat. Prod. Rep.*, 2016, **33**, 1044–1092.
24. A. I. Smeds, P. C. Eklund and S. M. Willfor, Chemical characterization of high-molar-mass fractions in a Norway spruce knotwood ethanol extract, *Phytochemistry*, 2016, **130**, 207–217.
25. A. Toure and X. Xueming, Flaxseed lignans: source, biosynthesis, metabolism, antioxidant activity, bioactive components, and health benefits, *Compr. Rev. Food Sci. Food Saf.*, 2010, **9**, 261–269.
26. Z. Shah, U. F. Gohar, I. Jamsed, A. Mushtaq, H. Mukhtar, M. Zia-Ul-Haq, S. I. Toma, R. Manea, M. Moga and B. Popovici, Podophyllotoxin: History, recent advances and future prospects, *Biomolecules*, 2021, **11**, 603.
27. S. C. Yoder, S. M. Lancaster, M. A. J. Hullar and J. W. Lampe, Gut microbial metabolism of plant lignans: influence on human health, in *Diet-Microbe Interactions in the Gut*, Elsevier Inc., 2015, pp. 103–117.
28. M. Gordaliza, P. A. Garcia, J. M. M. del Corral, M. A. Castro and M. A. Gomez-Zurita, Podophyllotoxin: distribution, sources, applications and new cytotoxic derivatives, *Toxicon*, 2004, **44**, 441–459.
29. K. Buzun, A. Bielawska, K. Bielawski and A. Gornowicz, DNA topoisomerases as molecular targets for anticancer drugs, *J. Enzyme Inhib. Med. Chem.*, 2020, **35**, 1781–1799.
30. X. Zhang, K. P. Rakesh, C. S. Shantharam, H. M. Manukumar, A. M. Asiri, H. M. Marwani and H. Qin, Podophyllotoxin derivatives as an excellent anticancer aspirant for future chemotherapy: a key current imminent needs, *Bioorg. Med. Chem.*, 2018, **26**, 340–355.
31. Y. Lin, Y. Li, Y. Zeng, B. Tian, X. Qu, Q. Yuan and Y. Song, Pharmacology, toxicity, bioavailability, and formulation of magnolol: an update, *Front. Pharmacol.*, 2021, **12**, 632767.
32. A. AbouZid and O. M. Ahmed, Silymarin flavonolignans: structure–activity relationship and biosynthesis, in *Studies in Natural Products Chemistry*, ed. Atta-ur-Rahman, Elsevier B.V, 2013, vol. 40, pp. 469–484.
33. C. Hellerbrand, J. M. Schattenberg, P. Peterburs, A. Lechner and R. Brignoli, The potential of silymarin for the treatment of hepatic disorders, *Clin. Phytosci.*, 2017, **2**, 7.

34. A. Gillesses and H. H. Schmidt, Silymarin as supportive treatment in liver diseases: A narrative review, *Adv. Ther.*, 2020, **37**, 1279–1301.
35. M. Bule, F. Khan, M. F. Nisar and K. Niaz, Tannins (hydrolysable tannins, condensed tannins, phlorotannins, flavono-ellagitannins), in *Recent Advances in Natural Product Analysis*, Elsevier, 2020, pp. 132–146.
36. Z. Chang, Q. Zhang, W. Liang, K. Zhou, P. Jian, G. She and L. Zhang, A comprehensive review of the structure elucidation of tannins from *Terminalia* Linn, *J. Evidence-Based Complementary Altern. Med.*, 2019, 8623909.
37. S. Soares, E. Brandao, C. Guerreiro, S. Soares, N. Mateus and V. de Freitas, Tannins in food: Insights into the molecular perception of astringency and bitter taste, *Molecules*, 2020, **25**, 2590.
38. K. Macakova, V. Koleckar, L. Cahlikova, J. Chlebek, A. Hostalkova, K. Kuca, D. Jun and L. Opletal, Tannins and their influence on health, in *Recent Advances in Medicinal Chemistry*, ed. Atta-ur-Rahman, M. I. Choudhury, and G. Perry, Bentham Science Publishers Ltd. Published by Elsevier Inc, 2014, vol. 1, pp. 159–208.
39. K. Khanbabaee and T. van Ree, Tannins: Classification and definition, *Nat. Prod. Rep.*, 2001, **18**, 641–649.
40. V. U. Kavitha and B. Kandasubramanian, Tannins for freshwater treatment, *SN Appl. Sci.*, 2020, **2**, 1081.
41. Y. Wu, X. Li, W. Xiang, C. Zhu, Z. Lin, Y. Wu, J. Li, S. Pandravada, D. D. Ridder, G. Bai, M. L. Wang, H. N. Trick, S. R. Bean, M. R. Tuinstra, T. T. Tesso and J. Yu, Presence of tannins in sorghum grains is conditioned by different natural alleles of Tannin1, *Proc. Natl. Acad. Sci. U. S. A.*, 2012, **109**, 1028–10286.
42. M. Samitya, R. E. Aluko and T. Dhewa, Plant food antinutritional factors and their reduction strategies: an overview, *Food Prod., Process. Nutr.*, 2020, **2**, 6.
43. D. R. Gang, J. Wang, N. Dudareva, K. H. Nam, J. E. Simon, E. Lewinsohn and E. Pichersky, An investigation of the storage and biosynthesis of phenylpropenes in sweet basil, *Plant Physiol.*, 2001, **125**, 539–555.
44. I. M. Garcia, V. C. B. Leitune, M. S. Ibrahim, M. A. S. Melo, V. F. Matoses, S. Sauro and F. M. Collares, Determining the effects of eugenol on bond strength of resin-based restorative materials to dentin: a meta-analysis of the literature, *Appl. Sci.*, 2020, **10**, 1070.
45. P. Kemprai, B. P. Mahanta, D. Sut, R. Barman, D. Banik, M. Lal, S. P. Saikia and S. Haldar, Review on safrole: identity shift of the 'candy shop' aroma to a carcinogen and deforester, *Flavour Fragrance J.*, 2020, **35**, 5–23.
46. I. B. Slimen, T. Najar and M. Abderrabba, Chemical and antioxidant properties of betalains, *J. Agric. Food Chem.*, 2017, **65**, 675–689.
47. D. Strack, T. Vogt and W. Schliemann, Recent advances in betalain research, *Phytochemistry*, 2003, **62**, 247–269.
48. M. G. Miguel, Betalains in some species of the Amaranthaceae family: A review, *Antioxidants,* 2018, **7**, 53.

49. S. Lopez-Nieves, Y. Yang, A. Timoneda, M. Wang, T. Feng, S. A. Smith, S. F. Brockington and H. A. Maeda, Relaxation of tyrosine pathway regulation underlies the evolution of betalain pigmentation in Caryophyllales, *New Phytol.*, 2018, **217**, 896–908.
50. G. Polturak and A. Aharoni, "La Vie en Rose": biosynthesis, sources, and applications of betalain pigments, *Mol. Plant*, 2018, **11**, 7–22.
51. F. Annunziata, C. Pinna, S. Dallavalla, L. Tamborini and A. Pinto, An overview of coumarin as a versatile and readily accessible scaffold with broad-ranging biological activities, *Int. J. Mol. Sci.*, 2020, **21**, 4618.
52. K. N. Venugopala, V. Rashmi and B. Odhav, Review on natural coumarin lead compounds for their pharmacological activity, *BioMed. Res. Int.*, 2013, 963248.
53. E. K. Akkol, Y. Genc, B. Karpuz, E. Sobarzo-Sanchez and R. Capasso, Coumarins and coumarin-related compounds in pharmacotherapy of cancer, *Cancers*, 2020, **12**, 1959.
54. J. Rajniak, R. F. H. Giehl, E. Chang, I. Murgia, N. von Wiren and E. Sattely, Biosynthesis of redox-active metabolites in response to iron deficiency in plants, *Nat. Chem. Biol.*, 2018, **14**, 442–450.
55. J. C. Amaral, M. M. da Silva, M. F. G. F. da Silva, T. C. Alves, A. G. Ferreira, M. R. Forim, J. B. Fernandes, E. S. Pina, A. A. Lopes, A. M. S. Pereira and V. M. Novelli, Advances in the biosynthesis of pyranocoumarins: isolation and 13C-incorporation analysis by high-performance liquid chromatography-ultraviolet-solid-phase extraction-nuclear magnetic resonance data, *J. Nat. Prod.*, 2020, **83**, 1409–1415.
56. R. Vanholme, L. Sundin, K. C. Seetso, H. Kim, X. Liu, J. Li, B. D. Meester, L. Hoengenaert, G. Goeminne, K. Moreel, J. Haustraete, H. Tsai, W. Schmidt, B. Vanholme, J. Ralph and W. Boerjan, COSY catalyses *trans–cis* isomerization and lactonization in the biosynthesis of coumarins, *Nat. Plants*, 2019, **5**, 1066–1075.
57. K. Yonekura-sakakibara, Y. Higashi and R. Nakabayashi, The origin and evolution of plant flavonoid metabolism, *Front. Plant Sci.*, 2019, **10**, 943.
58. A. Rani, A. Anand, K. Kumar and V. Kumar, Recent developments in biological aspects of chalcones: the odyssey continues, *Expert Opin. Drug Discovery*, 2019, **14**, 249–288.
59. B. Salehi, C. Quispe, I. Chamkhi, N. E. Omari, A. Balahbib, J. Sharafi-Rad, A. Bouyahya, M. Akram, M. Iqbal, A. O. Docea, C. Caruntu, G. Leyva-Gomez, A. Dey, M. Martorell, D. Calina, V. Lopez and F. Les, Pharmacological properties of chalcones: a review of preclinical including molecular mechanisms and clinical evidence, *Front. Pharmacol.*, 2021, **11**, 592654.
60. K. M. Davies, S. J. Bloor, G. B. Spiller and S. Deroles, Production of yellow colour in flowers: redirection of flavonoid biosynthesis in Petunia, *Plant J.*, 1998, **13**, 259–266.
61. J. Ferrer, J. M. Jez, M. E. Bowman, R. A. Dixon and J. P. Noel, Structure of chalcone synthase and the molecular basis of plant polyketide biosynthesis, *Nat. Struct. Biol.*, 1999, **6**, 775–784.

62. T. T. H. Dao, H. J. M. Linthorst and R. Verpoorte, Chalcone synthase and its functions in plant resistance, *Phytochem. Rev.*, 2011, **10**, 397–412.
63. S. Ayabe, H. Uchiyama, T. Aoki and T. Akashi, Plant Phenolics: Phenylpropanoids, in *Comprehensive Natural Products II*, Elsevier, 2010, vol. 1, pp. 929–976.
64. B. Boucherle, M. Peuchmaur, A. Boumendjel and R. Haudecoeur, Occurrences, biosynthesis and properties of aurones as high-end evolutionary products, *Phytochemistry*, 2017, **142**, 92–111.
65. T. Nakayama, Biochemistry and regulation of aurone biosynthesis, *Biosci., Biotechnol., Biochem.*, 2022, **86**, 557–573.
66. A. Boumendjel, Aurones: a subclass of flavones with promising biological potential, *Curr. Med. Chem.*, 2003, **10**, 2621–2630.
67. E. Ono, M. Fukuchi-Mizutani, N. Nakamura, Y. Fukui, K. Yonekura-Sakakibara, M. Yamaguchi, T. Nakayama, T. Tanaka, T. Kusumi and Y. Tanaka, Yellow flowers generated by expression of the aurone biosynthetic pathway, *Proc. Natl. Acad. Sci. U. S. A.*, 2006, **103**, 11075–11080.
68. A. N. Panche, A. D. Diwan and S. R. Chandra, Flavonoids: an overview, *J. Nutr. Sci.*, 2016, **5**, e47.
69. S. K. Kumar and A. K. Pandey, Chemistry and biological activities of flavonoids: an overview, *Sci. World J.*, 2013, 162750.
70. R. K. Singla, A. K. Dubey, A. Garg, R. K. Sharma, M. Fiorino, S. M. Ameen, M. A. Haddad and M. Al-Hiary, Natural polyphenols: chemical classification, definition of classes, subcategories, and structures, *J. AOAC Int.*, 2019, **102**, 1397–1400.
71. H. E. Khoo, A. Azlan, S. T. Tang and S. M. Lim, Anthocyanidins and anthocyanins: colored pigments as food, pharmaceutical ingredients, and the potential health benefits, *Food Nutr. Res.*, 2017, **61**, 1361779.
72. T. C. Wallace and M. M. Giusti, Anthocyanins – nature's bold, and health-promoting colors, *Foods*, 2019, **8**, 550.
73. J. M. Awika, L. W. Rooney and R. D. Waniska, Properties of 3-deoxyanthocyanins from sorghum, *J. Agric. Food Chem.*, 2004, **52**, 4388–4394.
74. L. Krizova, K. Dadakova, J. Kasparovska and T. Kasparovsky, Isoflavones, *Molecules* 2019, **24**, 1076.
75. H. B. Patisaul and W. Jefferson, The pros and cons of phytoestrogens, *Front. Neuroendocrinol.*, 2010, **31**, 400–419.
76. A. Laohavisit, T. Wakatake, N. Ishihama, H. Mulvey, K. Takizawa, T. Suzuki and K. Shirasu, Quinone perception plants *via* leucine-rich-repeat receptor-like kinases, *Nature*, 2020, **587**, 92–97.
77. M. Liu and S. Lu, Plastoquinone and ubiquinone in plants: biosynthesis, physiological function and metabolic engineering, *Front. Plant Sci.*, 2016, **7**, 1898.
78. I. Hook, C. Mills and H. Sheridan, Bioactive napthaquinones from higher plants, in *Studies in Natural Products Chemistry*, Elsevier BV, 2014, vol. 41, pp. 119–160.
79. S. P. Devi, S. Kumaria, S. R. Rao and P. Tandon, Carnivorous plants as a source of potent bioactive compound: napthoquinones, *Trop. Plant Biol.*, 2016, **9**, 267–279.

80. J. R. Widhalm and D. Rhodes, Biosynthesis and molecular actions of specialized 1,4-napthaquinone natural products produced by horticultural plants, *Hortic. Res.*, 2016, **3**, 16046.
81. N. R. R. Reddy, R. H. Mehta, P. H. Sonin, J. Makasana, N. A. Gajbhiye, M. Ponnuchamy and J. Kumar, Next generation sequencing and transcriptome analysis predicts biosynthetic pathway of sennosides from senna (*Cassia angustifolia* Vahl.), a non-model plant with potent laxative properties, *PLoS One*, 2015, 0129422.
82. R. Mandrioli, L. Mercolini, A. Ferranti, S. Fanali and M. A. Raggi, Determination of aloe emodin in Aloe vera extracts and commercial formulations by HPLC with tandem UV absorption and fluorescence detection, *Food Chem.*, 2011, **126**, 387–393.
83. D. Kline, V. Ritruthal and S. Babajanian, Quantitative analysis of aloins and aloin-emodin in Aloe vera raw materials and finished products using high-performance liquid chromatography: single laboratory validation. First action 2016.09, *J. AOAC Int.*, 2017, **100**, 661–670.
84. V. Ivetic, S. Trivic, M. K. Pogancev, M. Popovic and J. Zlinska, Effects of St John's Wort (*Hypericum perforatum* L.) extracts on epileptogenesis, *Molecules*, 2011, **16**, 8062–8075.
85. A. Valletta, L. M. Iozia and F. Leonelli, Impact of environmental factors on stilbene biosynthesis, *Plants*, 2021, **10**, 90.
86. P. Jeandet, B. Delaunois, A. Conreux, D. Donnez, V. Nuzzo, S. Cordelier, C. Clement and E. Courot, Biosynthesis, metabolism, molecular engineering, and biological functions of stilbene phytoalexins in plants, *Biofactors*, 2010, **36**, 331–341.
87. A. S. Dubrovina and K. V. Kiselev, Regulation of stilbene biosynthesis in plants, *Planta*, 2017, **246**, 597–623.
88. M. Reinisalo, A. Karlund, A. Koskela, K. Kaarniranta and R. O. Karjalainen, Polyphenol stilbenes: Molecular mechanisms of defence against oxidative stress and aging-related diseases, *Oxid. Med. Cell. Longevity*, 2015, 340520.
89. M. M. M. Pinto, A. Palmeira, C. Fernandes, D. I. S. P. Resende, E. Sousa, H. Cidade, M. E. Tiritan, M. Correia-da-Silva and S. Cravo, From natural products to new synthetic small molecules: a journey through the world of xanthones, *Molecules*, 2021, **26**, 431.
90. N. Tocci, M. Gaid, F. Kaftan, A. K. Belkheir, I. Belhadj, B. Liu, A. Svatos, R. Hansch, G. Pasqua and L. Beerhues, Exodermis and endodermis are the sites of xanthone biosynthesis in *Hypericum perforatum* roots, *New Phytol.*, 2018, **217**, 1099–1112.
91. J. Remali, I. Sahidin and W. M. Aizat, Xanthone biosynthetic pathway in plants: A review, *Front. Plant Sci.*, 2022, **13**, 809497.
92. I. Ei-Awaad, M. Bocola, T. Beuerle, B. Liu and L. Beerhues, Bifunctional CYP81AA Proteins catalyse identical hydroxylations but alternative regioselective phenol couplings in plant xanthone biosynthesis, *Nat. Commun.*, 2016, **7**, 11472.

93. J. Ruan, C. Zheng, Y. Liu, L. Qu, H. Yu, L. Han, Y. Zhang and T. Wang, Chemical and biological research on herbal medicines rich in xanthones, *Molecules*, 2017, 1698.
94. J. Sharifi-Rad, Y. E. I. Rayess, A. A. Rizk, C. Sadaka, R. Zgheib, W. Zam, S. Sestito, S. Raposelli, K. Neffe-Skocinska, D. Zielinska, B. Salehi, W. N. Stzer, N. S. Dosoky, Y. Taheri, M. E. I. Beyrouthy, M. Martorell, E. A. Ostrander, H. A. R. Suleria, W. C. Cho, A. Maroyi and N. Martins, Turmeric and its major compound curcumin on health: bioactive effects and safety profiles for food, pharmaceutical, biotechnological and medicinal applications, *Front. Pharmacol.*, 2020, **11**, 01021.
95. T. Kita, S. Imai, H. Sawada, H. Kumagai and H. Seto, The biosynthetic pathway of curcuminoid in turmeric (*Curcuma longa*) as revealed by 13C-labelled precursors, *Biosci., Biotechnol., Biochem.*, 2008, **72**, 1789–1798.
96. Y. Katsuyama, T. Kita, N. Funa and S. Horonouchi, Curcuminoid biosynthesis by two type III polyketide synthases in the herb *Curcuma longa*, *J. Biol. Chem.*, 2009, **284**, 11160–11170.
97. J. L. Rodrigues, K. L. J. Prather, L. D. Kluskens and L. R. Rodrigues, Heterologous production of curcuminoids, *Microbiol. Mol. Biol. Rev.*, 2015, **79**, 39–59.
98. W. S. Tavares, Y. Akhtar, G. L. P. Goncalves, J. C. Zanucio and M. B. Isman, Turmeric powder and its derivatives from *Curcuma longa* rhizomes: insecticidal effects on cabbage looper and the role of synergists, *Sci. Rep.*, 2016, **6**, 34093.
99. K. M. Nelson, J. L. Dahlin, J. Bisson, J. Graham, G. F. Pauli and M. A. Walters, The essential medicinal chemistry of curcumin, *J. Med. Chem.*, 2017, **60**, 1620–1637.
100. M. Landoni, D. Puglisi, E. Cassani, G. Borlini, G. Brunoldi, C. Comaschi and R. Pilu, Phlobaphenes modify pericarp thickness in maize and accumulation of the fumonisin mycotoxins, *Sci. Rep.*, 2020, **10**, 1417.
101. B. Wu, H. Chang, R. Marini, S. Chopra and L. Reddivari, Characterization of maize near-isogenic lines with enhanced flavonoid expression to be used as tools in diet-health complexity, *Front. Plant Sci.*, 2021, **11**, 619598.
102. J. Boddu, C. Svabek, F. Ibraheem, A. D. Jones and S. Chopra, Characterization of deletion allele of a sorghum Myb gene yellow seed1 showing loss of 3-deoxyflavonoids, *Plant Sci.*, 2005, **169**, 542–552.

CHAPTER 6

Alkaloids

Alkaloids are a structurally diverse group of organic compounds containing **nitrogen** usually in the **heterocyclic ring** and are chemically basic in nature. They are toxic and have pharmacological activities. They are mainly produced by plants. However, these metabolites are also produced by some microorganisms and animals. Amino acids, peptides, nucleosides, amino sugars, and antibiotics are not considered as alkaloids. Most alkaloids are derived from different amino acids except the purine, terpenoid and steroidal alkaloids that are biosynthesized from other precursors and the main skeleton is aminated. More than 21 000 alkaloids have been reported from plants.

6.1 Classification

Based on the molecular precursors and the structure of the molecules, alkaloids are classified into three main types: (1) true alkaloids, (2) protoalkaloids, and (3) pseudoalkaloids.[1] **True alkaloids** are derived from amino acids (Figure 6.1) and have nitrogen in the heterocyclic ring. They are bitter in taste, white (except berberine which is yellow in colour and the salt of sanguinarine which is copper red in colour), solid (except nicotine which is a brown liquid and oxygen free). Solubilities of true alkaloids vary because they are extremely diverse in structure. They may occur in free state/as salts/as *N*-oxides. **Protoalkaloids** are also derived from amino acids, but the nitrogen atom is not present in the heterocyclic ring. **Pseudoalkaloids** are nitrogen containing specialized metabolites in which the nitrogen is introduced at a late stage by transamination after the main skeleton has been synthesized. The classification of alkaloids is shown in Table 6.1. Betalains (discussed in Chapter 5) are structurally related to alkaloids and are often known as chromoalkaloids.

Chemical Diversity of Plant Specialized Metabolites: A Biosynthetic Approach
By Bratati De

Published by the Royal Society of Chemistry, www.rsc.org

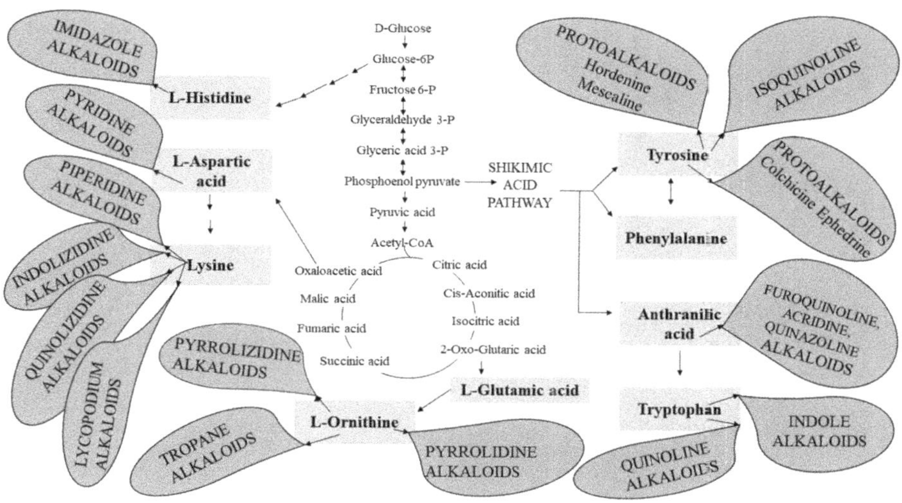

Figure 6.1 Origin of true alkaloids and some protoalkaloids.

Generally, there are four key steps in the biosynthesis of true alkaloids.[2]

- Amine accumulation
- Aldehyde accumulation
- Formation of an iminium cation
- A Mannich-like reaction and scaffold formation.

The reactions may be intramolecular or intermolecular depending on the amino acid precursor/type of alkaloid produced.

6.2 Ornithine Derived Alkaloids

6.2.1 Tropane Alkaloids

Three groups of tropane alkaloids identified from plants are (i) hyoscine, hyoscyamine, (ii) cocaine, and (iii) calystegines.[3]

Hyoscyamine, atropine, and **scopolamine (hyoscine)** are examples of tropane alkaloids. These tropane alkaloids are produced by species of Solanaceae such as *Atropa*, *Datura*, *Hyoscyamus*, *Duboisia*, *Mandragora*, and *Scopolia*. Medicinally important species are *Atropa belladonna*, *Datura stramonium*, *Datura metel*, *Hyoscyamus niger*, and *Scopolia japonica*. Roots and leaves of *Atropa belladonna* (commonly known as belladonna), leaves of different species of *Datura* and *Hyoscyamus niger*, and roots of *Scopolia japonica* are used as medicine and the sources of the important tropane alkaloids. Species of *Duboisia* particularly *Duboisia myoporoides* from Australia are also a major source of scopolamine and hyoscyamine type tropane alkaloids. Atropine

Table 6.1 Classes of alkaloids.

Precursor compound	Alkaloid type	Skeletal structure	Examples	Plant sources
True alkaloids				
Ornithine	Tropane alkaloids	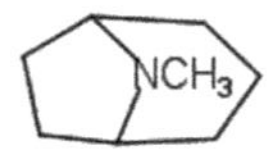	Atropine, hyoscine, hyoscyamine	Species of *Atropa*, *Datura*, *Hyoscyamus, Duboisia, Mandragora*, and *Scopolia* (Solanaceae)
			Cocaine	*Erythroxylum coca* (Erythroxylaceae)
			Calystegines	Species of *Convolvulus, Ipomoea,* and others
	Pyrrole and pyrrolidine alkaloids	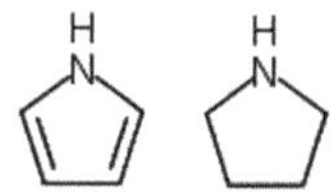	Hygrine, cuscohygrine	*Erythroxylum coca* (Erythroxylaceae)
			Stachydrine	*Stachys tuberifera* (Lamiaceae)
	Pyrrolizidine alkaloids	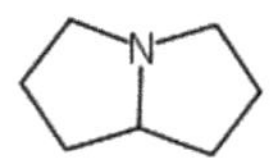	Symphitine	*Symphytum* spp. (Boraginaceae)
			Seneciphylline	*Senecio* spp. (Asteraceae)
			Senecionine	
			Indicine-*N*-oxide	*Heliotropium indicum* (Boraginaceae)
Aspartic acid	Pyridine alkaloids	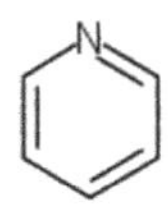	Nicotine	*Nicotiana tabacum*
			Anabasine	*Nicotiana tabacum* (Solanaceae)
			Trigonelline	*Trigonella foenum-graecum* (Fabaceae)
			Ricinine	*Ricinus communis* (Euphorbiaceae)
			Arecoline	*Areca catechu* (Arecaceae)

Lysine	Piperidine alkaloids	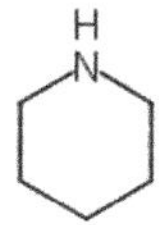	Lobeline Pelletierine Piperine	*Lobelia* spp. (Campanulaceae) *Punica granatum* (Lythraceae) *Piper* spp. (Piperaceae)
	Quinolizidine alkaloids	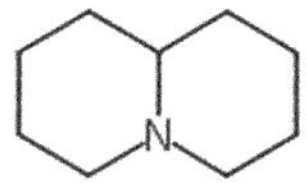	Sparteine, cytisine Lupanine	*Cytisus scoparius* (Fabaceae) *Lupinus* spp. (Fabaceae)
	Indolizidine alkaloids	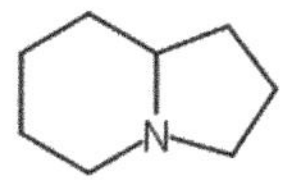	Castanospermine Swainsonine	*Castanospermum australe* (Fabaceae) *Swainsona* spp. (Fabaceae)
	Lycopodium alkaloids		Lycopodine, lycodine, huperzine A	*Lycopodium* species, *Huperzia* species (Lycopodiaceae)
Tyrosine	Isoquinoline alkaloids	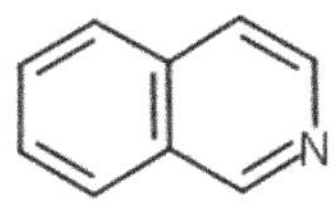	Morphine, codeine, papaverine, thebaine	*Papaver somniferum* (Papaveraceae)
			Tubocurarine	*Chondrodendron tomentosum* (Menispermaceae)
			Berberine	*Berberis* spp. (Berberidaceae) *Coptis japonica* (Ranunculaceae)
			Emetine, cephaeline	*Cephaelis* ipecacuanha (Rubiaceae)
			Galanthamine	Plants of Amaryllidaceae
Histidine	Imidazole alkaloids		Pilocarpine, pilosine	*Pilocarpus* spp. (Rutaceae)

(*continued*)

Table 6.1 (*Continued*)

Precursor compound	Alkaloid type	Skeletal structure	Examples	Plant sources
Tryptophan	Indole/benzopyrrole alkaloids	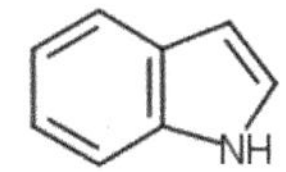	Reserpine, ajmalicine, catharanthine	*Rauvolfia* spp. (Apocynaceae)
			Vinblastine, vincristine	*Catharanthus roseus* (Apocynaceae)
			Physostigmine	*Physostigma venenosum* (Fabaceae)
			Strychnine, brucine	*Strychnos nux-vomica* (Loganiaceae)
			Ergometrine, ergotamine	Ergot of rye
	Quinoline alkaloids	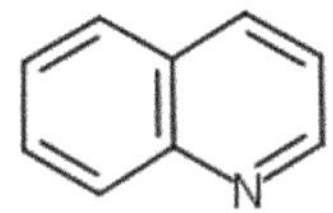	Quinine, quinidine, cinchonine, cinchonidine	Species of *Cinchona* and *Remijia* (Rubiaceae)
			Camptothecin	*Camptotheca acuminata* (Nyssaceae)
Anthranilic acid	Furoquinoline alkaloids		Skimmianine, dictamine	*Skimmia japonica* (Rutaceae)
	Quinazoline alkaloids		Peganin (vasicine)	*Peganum harmala* (Nitrariaceae) *Adhatoda vasica* (Acanthaceae)
Protoalkaloids				
Tyrosine			Hordenine	Germinated barley
			Mescaline	*Lophophora williamsii* (Cactaceae)
Tyrosine and phenylalanine			Colchicine	*Colchicum autumnale* (Colchicaceae)
Phenylalanine			Ephedrine	*Ephedra* spp. (Ephedraceae)
Pseudoalkaloids				
Acetate	Piperidine alkaloids		Coniine	*Conium maculatum* (Apiaceae)
Geraniol	Terpenoid alkaloids		Aconitine	*Aconitum* spp. (Ranunculaceae)
Cholesterol	Steroidal alkaloids		Conessine	*Holarrhena antidysenterica* (Apocynaceae)
			Jervine, protoveratrine A, protoveratrine B	*Veratrum* spp. (Melanthiaceae)
			Solanidine, solasodine, tomatidine	Some species of *Solanum* (Solanaceae)
Purine nucleotides	Purine alkaloids	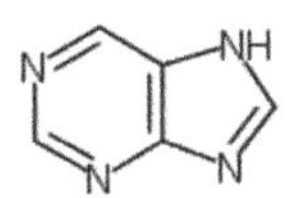	Caffeine, theophylline, theobromine	Tea (*Camellia sinensis*, Theaceae) Coffee (*Coffea arabica*, Rubiaceae)

(a racemate of hyoscyamine) and scopolamine are suppressors of the parasympathetic nervous system. Atropine dilates the pupil of the eye and is widely used in ophthalmology for examination of eyes. Scopolamine is used in the treatment of motion sickness. Scopolamine ingestion may lead to hallucinations and psychoactive effects. To overcome this drawback scopolamine derivatives have been developed. *N*-Butylscopolamine is used to treat abdominal pain, cramping, renal colic, and bladder spasms.[3,4]

Cocaine (benzoylmethylecgonine), obtained from *Erythroxylum coca* (Erythroxylaceae) leaves, is a diester of ecgonine and benzoyl CoA. Cocaine is one of the frequently consumed illicit drugs because of its toxic and addictive properties. Cocaine consumption also causes hallucinations and psychoactive effects. Cocaine has local anaesthetic properties. It is occasionally used for ophthalmic, ear, nose, and throat surgeries.[4]

Calystegines are polyhydroxylated nortropane alkaloids occurring mainly in Convolvulaceae, Solanaceae, Moraceae and Brassicaceae. Several calystegines have been reported from plant sources. Examples are calystegines A_3, A_5, A_6, A_7, B_1, B_2, B_3, B_4, B_5, C_1, C_2, N_1, *N*-methyl-calystegine B_2, and *N*-methyl-calystegine C_1. Calystegines are not absorbed by the central nervous system due to their hydrophilicity and exhibit no psychoactive effects. But research shows that calystegines inhibit mammalian α-glucosidases.[5]

Biosynthesis of tropane alkaloids starts from ornithine which is derived from glutamic acid. The carbon atom of the carboxylic acid is lost to produce a putrescine intermediate. An *N*-methyl-Δ-pyrrolinium cation is formed through formation of *N*-methyl putrescine and amino aldehyde. Two molecules of malonyl-CoA react with the *N*-methyl-Δ-pyrrolinium ion. This is the branch point for biosynthesis of atropine/scopolamine, cocaine, calystegines as well as nicotine, and hygrine. Ultimately tropine is formed after hydrolysis and decarboxylation. On the other hand, cocaine is produced through formation of ecgonine from the combined product of *N*-methyl-Δ-pyrrolinium and malonyl-CoA (Figure 6.2).[6] Recent investigation shows that the *N*-methyl-Δ-pyrrolinium cation produces the intermediate 4-(1-methyl-2-pyrrolidinyl)-3-oxobutanoic acid in presence of an atypical type III polyketide synthase (PKS).[7]

The tropane moiety is originated from ornithine. Condensation of tropine with and phenyl-lactyl-CoA yields littorine. Further rearrangement in the structure forms hyoscyamine. Epoxidation of hyoscyamine yields scopolamine. The reaction is catalyzed by 2-oxoglutarate dependent dioxygenase and hyoscyamine 6β-hydroxylase (Figure 6.2).[6,8]

Tropane alkaloid biosynthesis exhibits **compartmentation** of metabolites and enzyme activities at the level of tissues, cell types, and within the intracellular organelles. Biosynthesis of tropane alkaloids in Solanaceae takes place in the root from where the alkaloids are transported to the aerial parts for storage. Cell specific compartmentalization has been observed in the root (in the region 1–3 cm from the root apex and younger lateral roots). Expression of the genes of putrescine *N*-methyl transferase in *Atropa belladonna* and hyoscyamine 6β-hydroxylase, catalyzing the last step of scopolamine

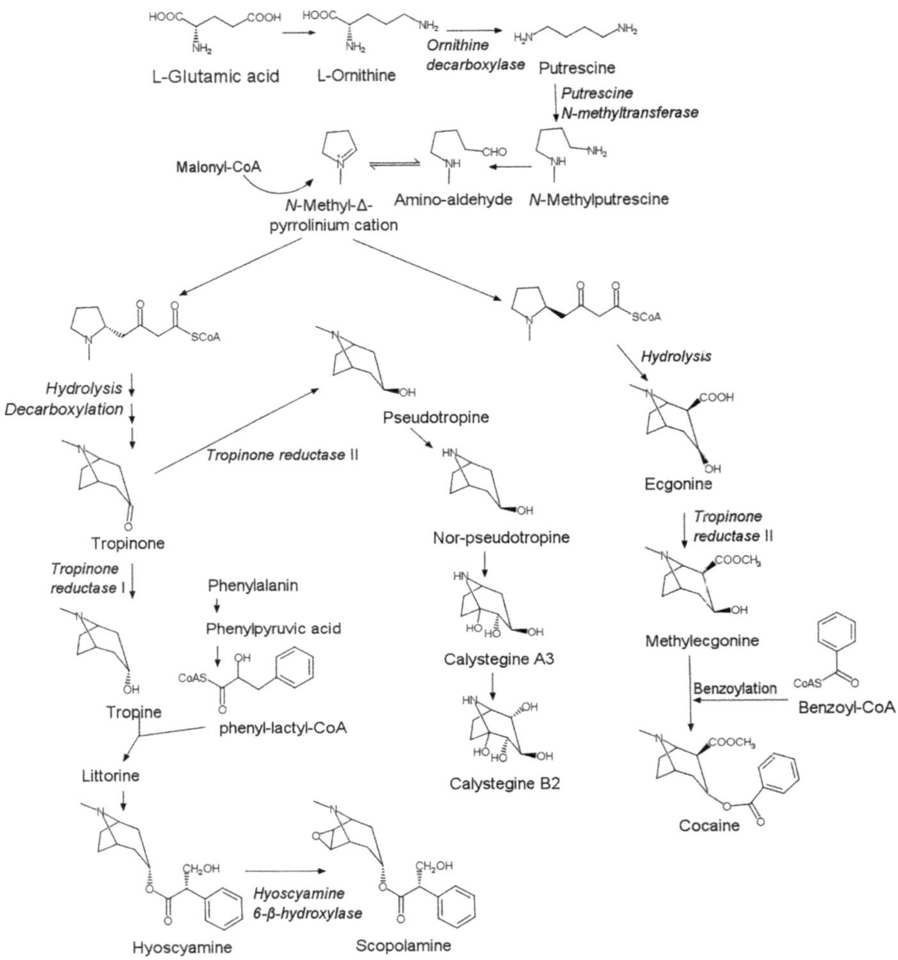

Figure 6.2 Biogenesis of tropane alkaloids.

biosynthesis, in *Hyoscyamus niger* was detected in the root pericycle. The amino acid ornithine is transported through the vascular tissues and is delivered to the pericycle. The enzyme tropinone reductase I, an intermediate enzyme for production of hyoscyamine and scopolamine, resides in the endodermis and nearby cortical cells in *Hyoscyamus niger*. Tropinone reductase II for the formation of calystegines was found to be localized in the companion cells of the sieve elements in the phloem of potato. The different compartmentation of the biosynthetic enzymes suggests that the biosynthetic intermediates must shuttle between the cellular pericycle and endodermis compartments for completion of the biosynthetic process leading to the formation of scopolamine (Figure 6.3). These tropane alkaloids are accumulated mostly within vacuoles.[3,8,9]

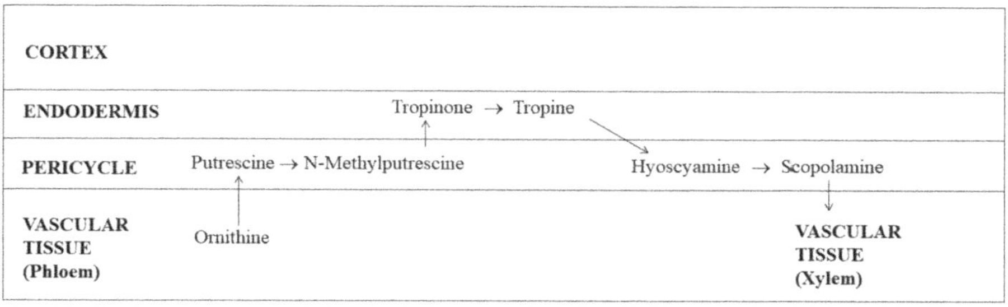

Figure 6.3 Shuttling of intermediate metabolites in scopolamine biosynthesis in different zones.

6.2.2 Pyrrolidine Alkaloids

Hygrine and cuscohygrine are pyrrolidine alkaloids found in the leaves of coca (*Erythroxylum coca*). Formation of the *N*-methyl-Δ-pyrrolinium cation from L-ornithine *via* the formation of putrescine and *N*-methylputrescine is involved in the biosynthesis of these two pyrrolidine alkaloids. After Claisen condensation with malonyl-CoA, hygrine and cuscohygrine are formed (Figure 6.4).[6]

Hygrine and cuscohygrine could be used as markers to distinguish between chewing coca leaves and cocaine abuse.[10] These two alkaloids are less toxic than cocaine and lack the euphoric effects of cocaine.[11]

6.2.3 Pyrrolizidine Alkaloids

This group of alkaloids are reported from several plants of Asteraceae (*Senecio vulgaris, Senecio pierotii* and other species of *Senecio*), Boraginaceae, Fabaceae, Apocynaceae and some genera of Orchidaceae.

Indicine-*N*-oxide (Figure 6.5) from *Heliotropium indicum* (Boraginaceae) gained importance because it was found that the alkaloid was active in the treatment of acute lymphoblastic leukaemia in children. However, the alkaloid showed liver toxicity. The liver is the main target of toxicity of pyrrolizidine alkaloids although the lungs may also be a target of injury.[12,13] It is of great concern when pyrrolizidine alkaloids are detected in food stuff. This group of alkaloids from *Senecio* and some species of Boraginaceae are sometimes found in milk. **Senecionine, seneciphylline** (Figure 6.5), **retrorsine, senecivernine-*N*-oxides** and their free bases were detected in herbal teas. Many food supplements and honey were also reported to contain pyrrolizidine alkaloids.[12]

Most of the pyrrolizidine alkaloids are mono- or di-**esters** of an aminoalcohol called **necine** that is a pyrrolizidine derivative, and straight or branched chain mono- or dicarboxylic aliphatic acids called **necic acids** (Figure 6.6).

Biosynthesis of pyrrolizidine alkaloids starts from L-ornithine followed by production of putrescines. Condensation of two molecules of putrescine

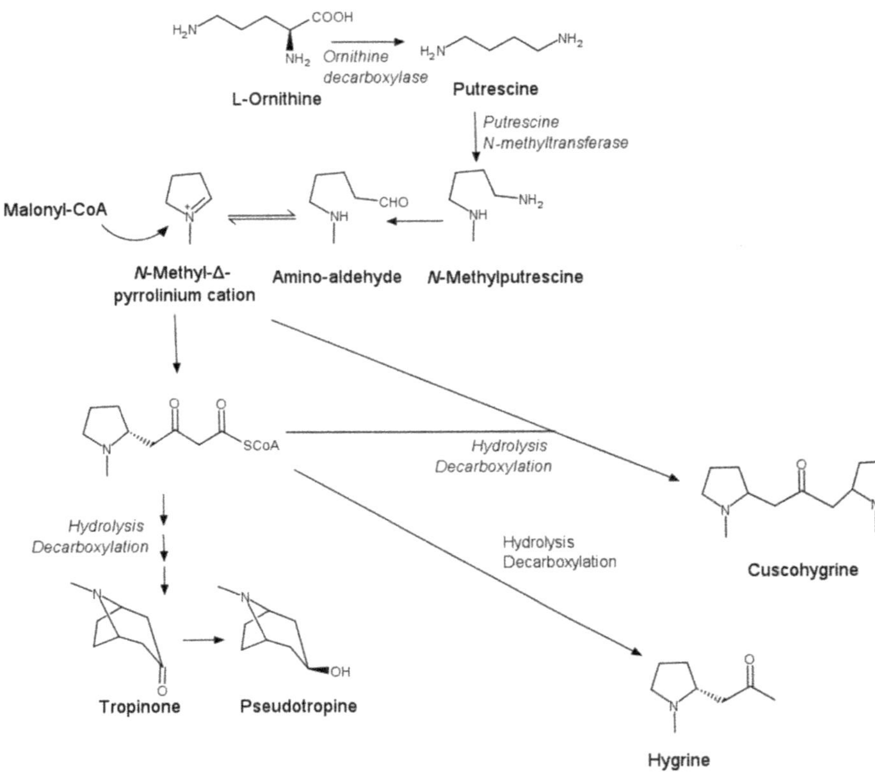

Figure 6.4 Biosynthesis of pyrrolidine alkaloids.

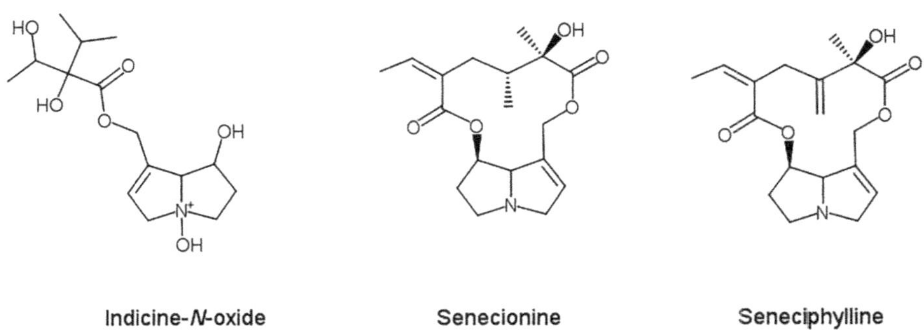

Figure 6.5 Examples of some pyrrolizidine alkaloids.

initiates the process. However, there is dispute. According to some authors one molecule of putrescine and one molecule of spermidine are condensed. This condensation reaction yields an aminobutyl group with subsequent production of the corresponding iminium ion which is reduced and cyclized to trachelanthamidine (Figure 6.7). The necic acids are derived from L-valine, L-leucine, L-isoleucine, and L-threonine.[12]

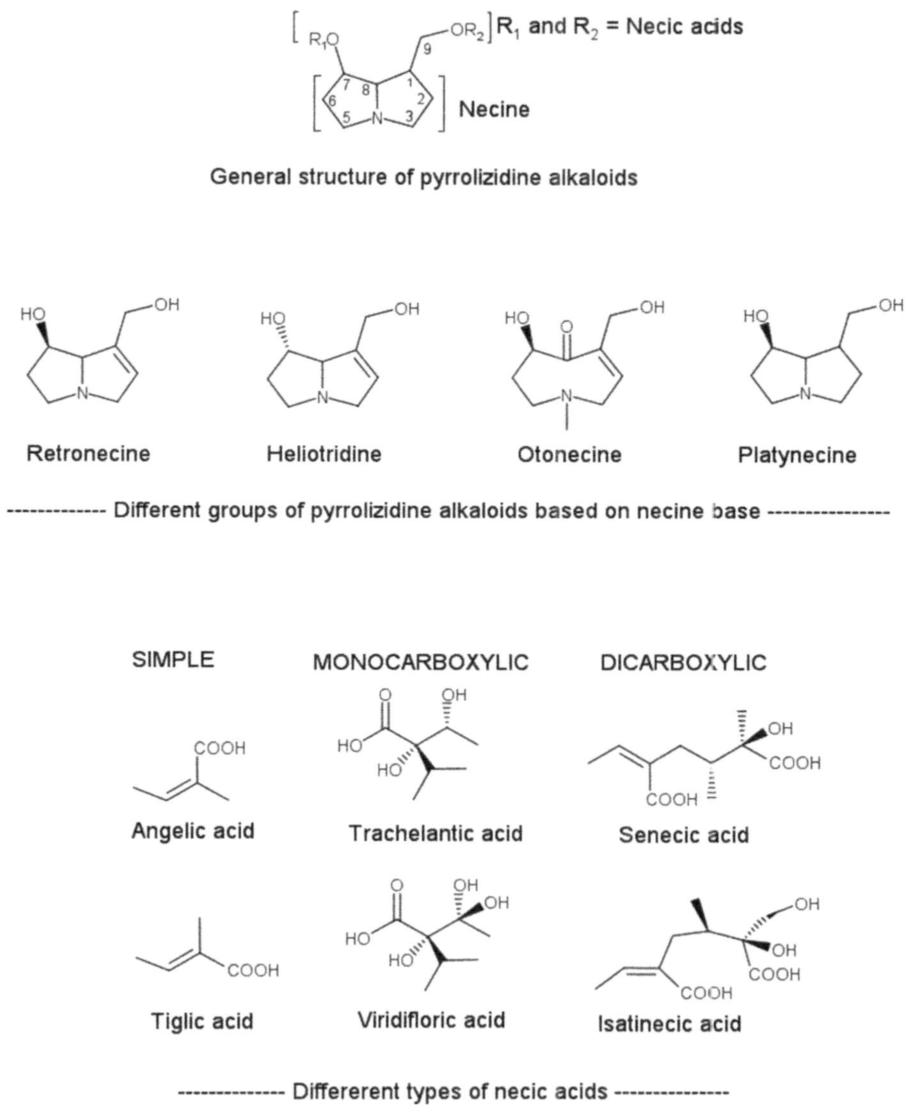

Figure 6.6 Structural components of pyrrolizidine alkaloids.

6.3 Aspartate/Lysine Derived Alkaloids

6.3.1 Pyridine Alkaloids

Nicotine, nornicotine, and **anabasine** (Figure 6.8) isolated from tobacco leaves (*Nicotiana tabacum*, Solanaceae) are examples of pyridine alkaloids. These three alkaloids possess insecticidal properties. Nicotine, as nicotine

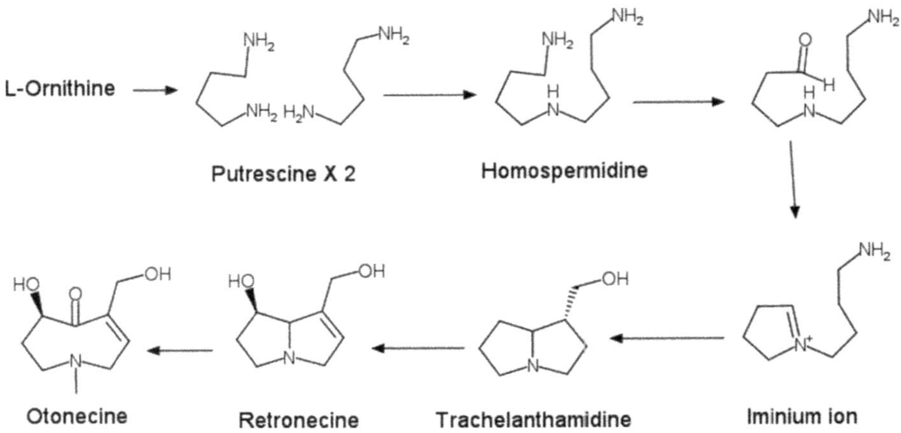

Figure 6.7 Biosynthesis of necines.

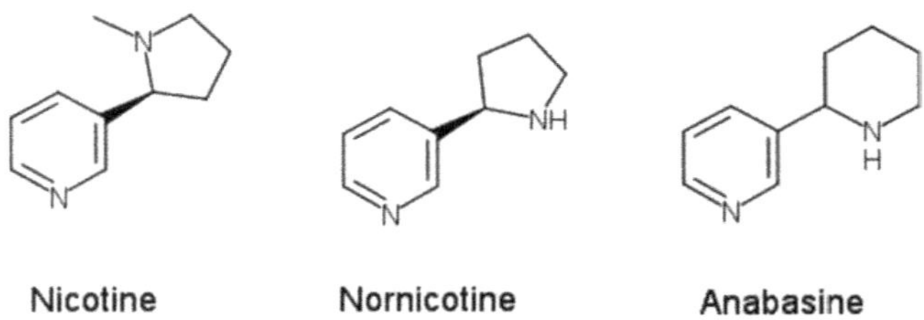

Figure 6.8 Pyridine alkaloids.

sulphate, is used in agriculture as an insecticide. **Anatabine** is present in small quantity.

The pyridine rings of these alkaloids are derived from nicotinic acid which is biosynthesized from aspartic acid. The pyrrolizidine ring of nicotine is biosynthesized from ornithine *via* the *N*-methyl-Δ-pyrrolinium cation. The piperidine moiety of anabasine is derived from lysine which is biosynthesized from aspartic acid (Figure 6.9).[8,14,15]

Nicotine is biosynthesized in the root cortical cells and translocated to aerial parts *via* xylem. Four intracellular compartments are involved in nicotine biosynthesis in the root cortical cells. These are cytosol, mitochondria, peroxisome, and vacuole (Figure 6.10). Nicotine accumulates in the vacuoles of the leaves in high concentration and plays an important role in deterring herbivores. Long distance transport of nicotine from root to leaf through xylem and unloading of nicotine from xylem into leaf cells should include several transporters which are not known.[16]

There are some more pyridine alkaloids derived from nicotinic acid including trigonelline, ricinine and arecoline (Figures 6.9 and 6.11).

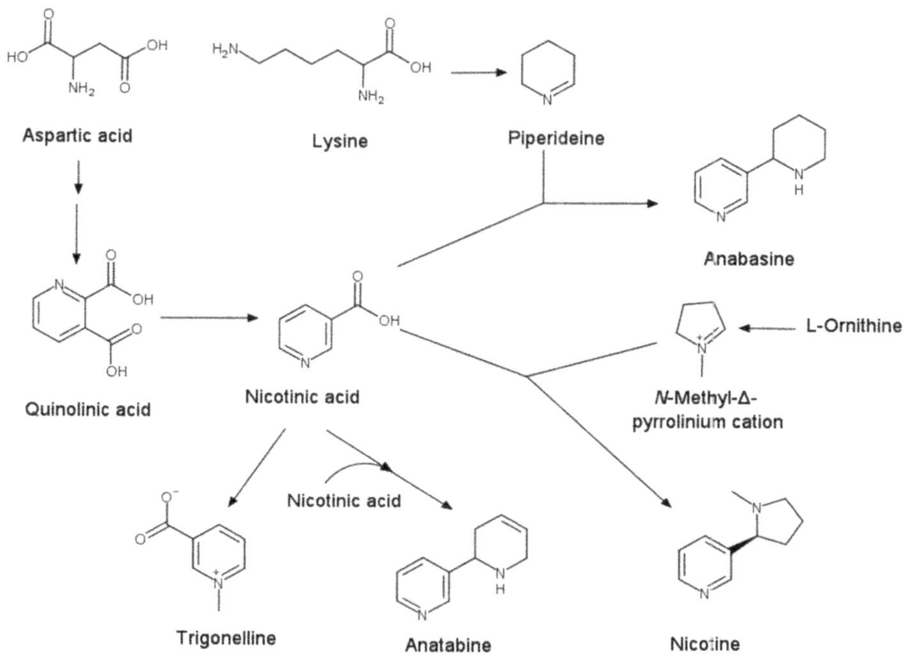

Figure 6.9 Biosynthesis of pyridine alkaloids.

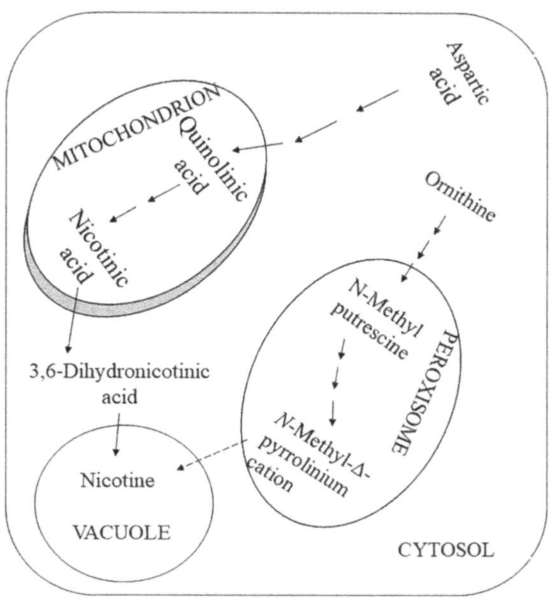

Figure 6.10 Nicotine biosynthesis in root cortical cells in *Nicotiana*.

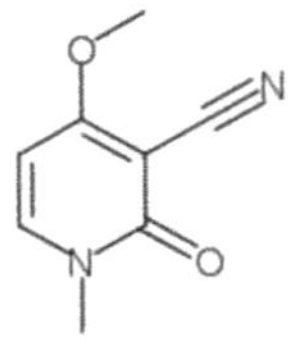

Ricinine

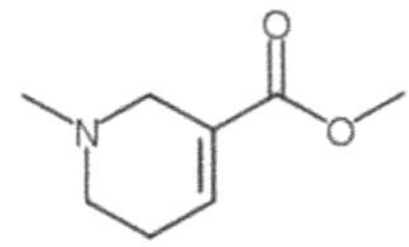

Arecoline

Figure 6.11 Ricinine and arecoline: the two other pyridine alkaloids.

Trigonelline (*N*-methylnicotinic acid) is a pyridine alkaloid produced from nicotinic acid. This alkaloid has attracted the attention of plant biologists. It is a plant hormone and regulates diverse functions like plant cell cycle, nodulation *etc.* The alkaloid is very effective in inducing G2 arrest in the root apices of many plants. The alkaloid may act as a cell cycle regulator by lengthening both S-phase and cell cycle, and function as a signal transmitter in the response of the plants to oxidative stress. The alkaloid has been detected from many plant species such as *Trigonella foenum-graecum, Pisum sativum*, *Glycine max* (Fabaceae), *Coffea* sp. (Rubiaceae), *Lycopersicon esculentum* (Solanaceae) *etc.* Coffee beverage is the major food containing trigonelline. Trigonelline has been reported to have many biological properties such as the protection of heart and liver and to act against hypercholesterolemia, hyperglycemia, nervous and hormonal disorders, and cancers.[14,15,17]

Ricinine (3-cyano-4-methoxy-*N*-methyl-2-pyridone) is a toxic alkaloid found in the leaves and seeds of castor plant *Ricinus communis* (Euphorbiaceae). It may cause different toxic reactions like liver and kidney damage, convulsions, hypotension, and even may lead to death. Ricinine has insecticidal activity against some Hymenoptera and Lepidoptera insects.[18]

Arecoline was isolated from the seeds of *Areca catechu* (Arecaceae) also known as betel nut palm. It is a cholinergic alkaloid.[19]

6.3.2 Piperidine Alkaloids

Piperidine alkaloids are present in many plants. *Lobelia inflata* (Campanulaceae) is used in spasmodic asthma and chronic bronchitis. Many piperidine alkaloids were isolated from the plant. These are **lobeline, lobelanine**, and **lobelanidine**. Lobeline has a stimulatory effect on the respiratory centre. It is frequently used in antismoking preparations as the alkaloid is a competitive nicotinic receptor antagonist. *Lobelia nicotianifolia*, known as Indian lobelia, grows at altitude 700–2200 m.

The piperidine part of *Lobelia* alkaloids are derived from lysine *via* cadaverine or 5-aminopentanal.[6] The side chain is possibly derived from benzoylacetic acid (Figure 6.12).

Piper nigrum (Piperaceae), commonly known as pepper, is used to produce black pepper and white pepper. Black pepper is prepared from the unripe fruits of *Piper nigrum*. White pepper is prepared from the ripe fruits of the

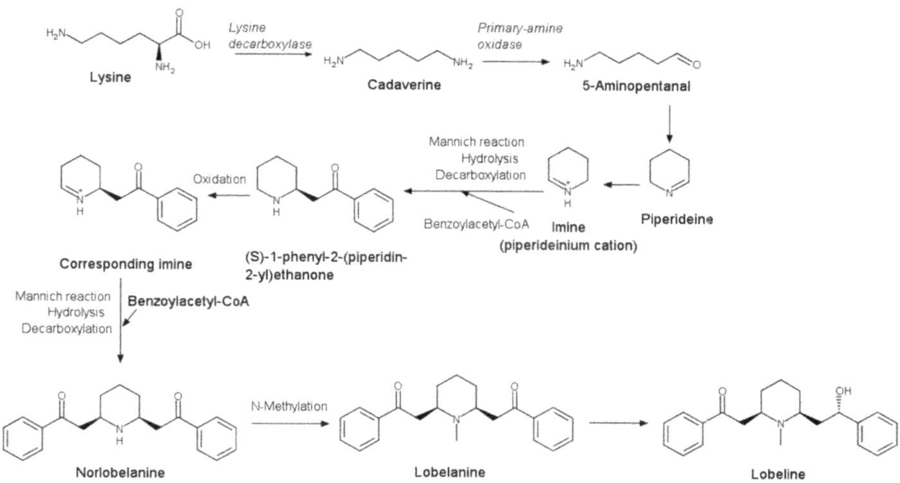

Figure 6.12 Biosynthesis of *Lobelia* alkaloids.

plant after removal of the pericarp. The predominant pungent metabolite of pepper is the piperidine alkaloid **piperine (1-piperoyl-piperidine)**, a component of pepper oleoresin. Fruits from other *Piper* species such as long pepper (*Piper longum*) and Balinese long pepper (*Piper retrofractum*) also contain piperine. Pepper is an extensively used spice worldwide. It is also used as medicine. In the Indian system of medicine pepper is used for alleviating and treating pain, chills, influenza, fever, headache, strep throat, digestion, *etc.*[20]

The piperidine part of piperine is derived from lysine. Piperic acid and piperoyl-CoA are derived from ferulic acid. Piperidine and piperoyl-CoA form piperine by piperine synthase[21] (Figure 6.13).

The root bark of *Punica granatum* (Lythraceae), earlier used to treat worm infestations, contains **pelletierine** and **isopelletierine**. The anthelmintic activity was mainly due to isopelletierine. This use has now been abandoned due to harmful effects.[22]

6.3.3 Quinolizidine Alkaloids

Lupinine is the bitter-tasting alkaloid of *Lupinus* spp. (Fabaceae). The plant also contains another quinolizidine alkaloid **lupanine**. *Cytisus scoparius* (Fabaceae) contains the biosynthetically related alkaloid **cytosine**. Cytosine has an antiarrhythmic property. However, it is not approved by FDA because of severe toxicity.

The biosynthetic pathway is not fully explored yet. From the knowledge presently available, quinolizidine alkaloids are produced from lysine *via* cyclization of the cadaverine unit.[23] (Figure 6.14).

The quinolizidine alkaloids show compartmentalization of the metabolites and enzymes involved in biosynthesis. These alkaloids are synthesized

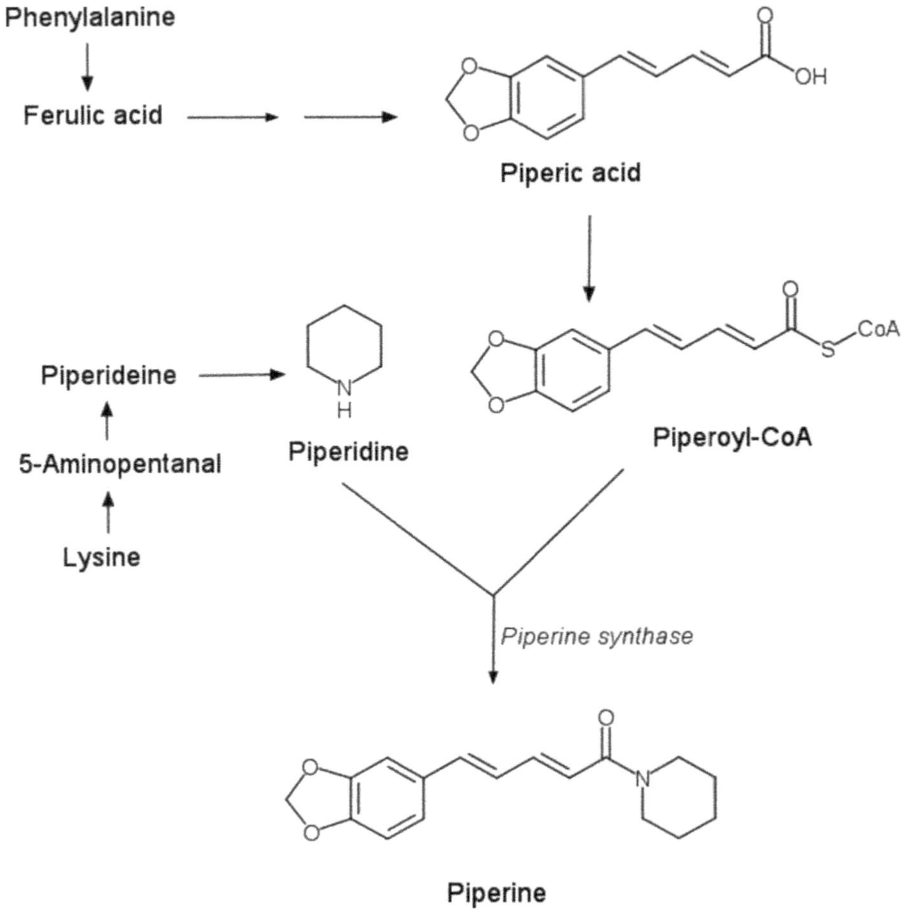

Figure 6.13 Pathway of piperine biosynthesis.

in the green parts of the shoot tissues where lysine, cadaverine and enzymes for lupanine synthesis are present. They do also accumulate in seeds where about 50% of the alkaloids are translocated by phloem and about 50% are produced *in situ*. Quinolizidine alkaloid content decreases during germination. Again, the level of the alkaloids increases with formation of new leaves in the seedlings. Lysine decarboxylase is localized in the chloroplasts. Enzymes for ester formation are present in the mitochondria.[23]

6.3.4 Indolizidine Alkaloids

Many indolizidine alkaloids have been reported from marine and terrestrial plants, fungi, ants, and frogs. These alkaloids have wide range of biological activities. Indolizidine alkaloids have been identified in plants of the families

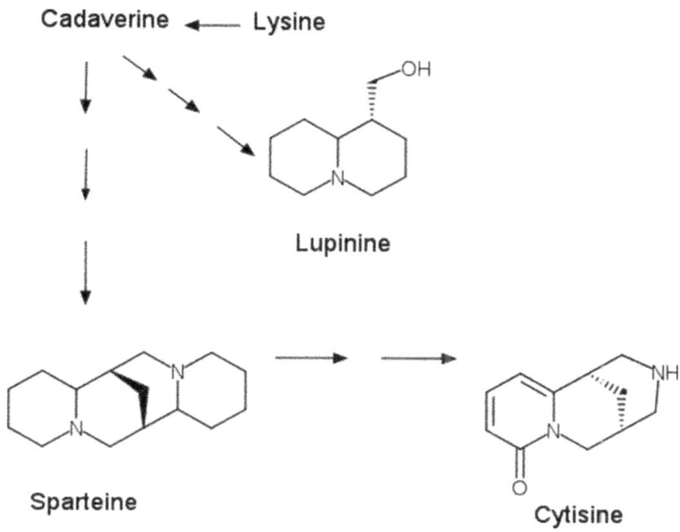

Figure 6.14 Biosynthesis of quinolizidine alkaloids from lysine.

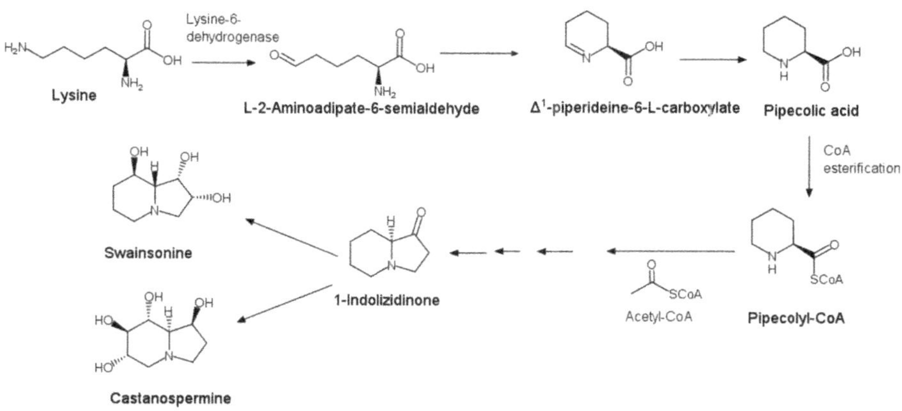

Figure 6.15 Biosynthesis of two indolizidine alkaloids.

Apocynaceae, Convolvulaceae, Fabaceae, Orchidaceae and Pandanaceae. Indolizidine alkaloids have a 5-membered ring fused with a 6-membered ring sharing a common nitrogen atom. Apart from this simple structure, indolizidine alkaloids may form complex multicyclic structures. A few of these alkaloids are mentioned here.

Swainsonine (Figure 6.15) is one of the most active indolizidine alkaloids obtained from *Swainsona canescens* and *Astragalus lentiginosus*, both belonging to the family Fabaceae. The alkaloid is toxic to livestock. Swainsonine inhibited metastasis of cancer cells and is a novel anticancer drug candidate.

Securine, a tetracyclic indolizidine alkaloid, was isolated from *Securinega suffruticosa* (Phyllanthaceae). The alkaloid has shown anticancer, and neuroprotective activities in many studies. It is a promising compound with potential for further research. **Castanospermine** (Figure 6.15) was isolated from the seeds of *Castanospermum australe* and seven other species of *Castanospermum* (Fabaceae). This alkaloid inhibits α-glucosidase activity involved in glycoprotein biosynthesis and alters the glycogen distribution in animals. It interferes with the glycoprotein envelope of the human immunodeficiency virus and inhibits viral replication. Thus, castanospermine has potential in the treatment of AIDS because of its relative non-toxicity and new mode of action. Castanospermine also displayed significant effect against the Moloney murine leukaemia virus.[24,25]

Indolizidine alkaloids are biosynthesized from lysine but this follows a pathway other than formation of cadaverine. Lysine is first deaminated and then forms pipecolic acid *via* formation of Δ^1-piperidine-6-L-carboxylate.[6,23] The biosynthesis of swainsonine and castanospermine is shown in Figure 6.15.

6.3.5 Lycopodium Alkaloids

Hundreds of Lycopodium alkaloids from species of *Lycopodium* and *Huperzia* (Lycopodiaceae) have been reported. Lycopodium alkaloids are divided into four major groups based on structure and biogenetic relationship. Representatives of such groups are lycopodine, lycodine, fawcettimine and phlegmarine. The Lycopodium alkaloid skeleton usually contains 16 carbons. Sometimes there are as many as 32 carbons apparently due to formation of dimers or less than 16 carbons probably due to bond cleavage. Huperzine A was isolated from *Huperzia serrata* (synonym of *Lycopodium serratum*). Huperzine A is a competitive and reversible inhibitor of acetylcholinesterase and a promising candidate drug with clinical interest in the treatment of Alzheimer's disease, myasthenia gravis and other neurological diseases. The Lycopodium alkaloids contain diverse groups of lysine derived compounds that are of quinolizine or pyridine type[26–31] (Figure 6.16).

6.4 Tyrosine Derived Alkaloids

6.4.1 Isoquinoline Alkaloids

Isoquinoline alkaloids form one of the largest and diverse groups of alkaloids. More than 4000 isoquinoline alkaloids have been detected from plant families such as Alangiaceae, Amaryllidaceae, Annonaceae, Apocynaceae, Aristolochiaceae, Berberidaceae, Fabaceae, Lauraceae, Menispermaceae, Papaveraceae, Ranunculaceae, and Rutaceae. The isoquinoline moiety consists of a benzene and pyridine ring fused at C3/C4 of the pyridine ring. Isoquinoline or tetrahydroisoquinoline rings (Figure 6.17) are derived biogenetically from tyrosine.[32–34]

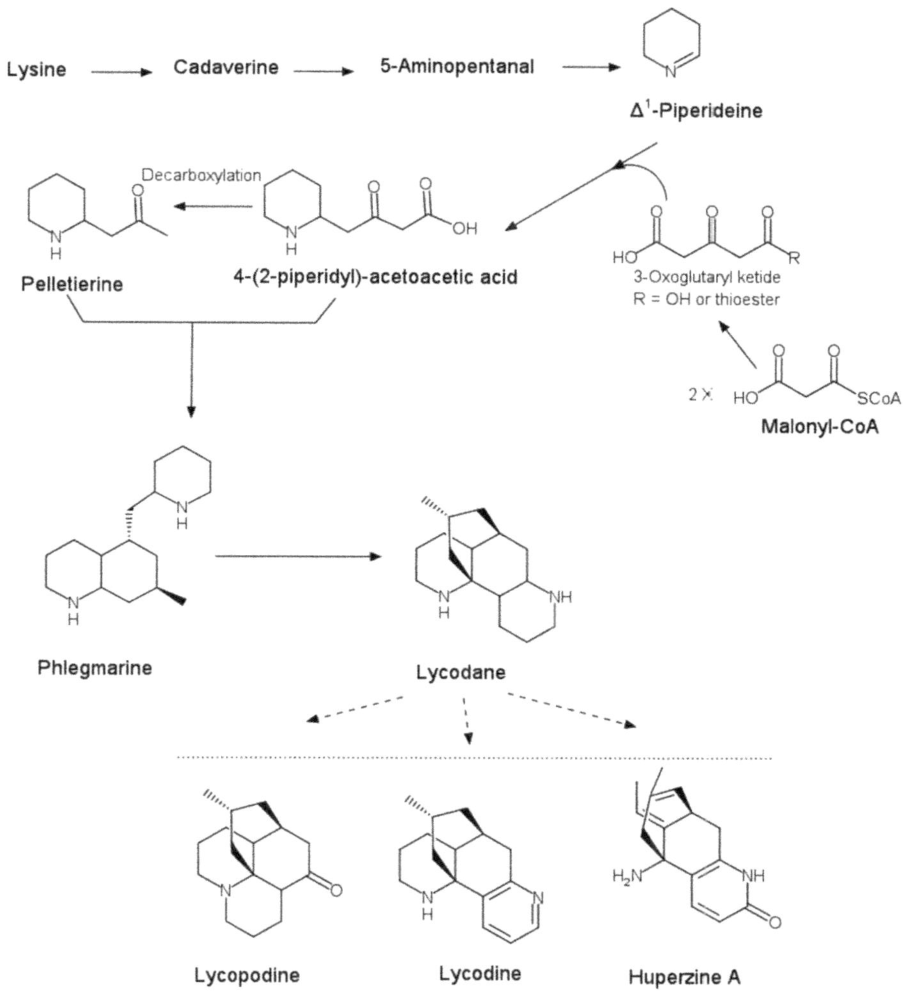

Figure 6.16 Biosynthesis of Lycopodium alkaloids.

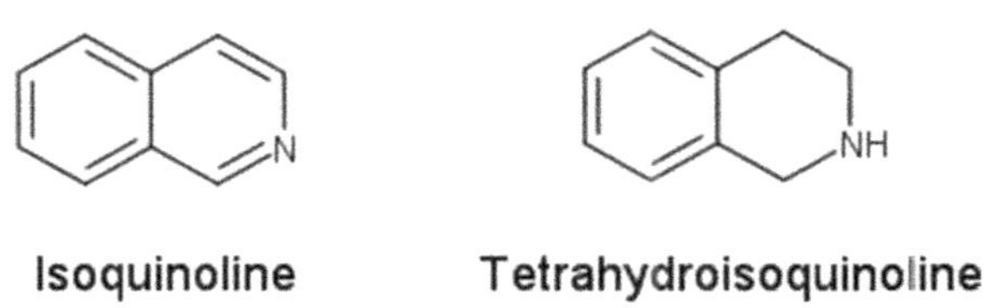

Figure 6.17 General structure of isoquinoline alkaloids.

Table 6.2 Different types of isoquinoline alkaloids.

Type	Examples
1. Simple isoquinoline/ tetrahydroisoquinoline	Anhalonine, anhalonidine, anhalamine
2. Benzyltetrahydroisoquinoline/ benzylisoquinoline	Papaverine
Bis-benzyltetrahydroisoquinoline	Tetrandrine, berbamunine, tubocurarine
Morphinan	Morphine, codeine, thebaine
Phthalideisoquinoline	Noscapine, hydrastine
Protopine	Protopine
Benzophenanthridine	Sanguinarine, chelerythrine, chelidonine
Protoberberine	Berberine, palmatine, coralyne, coptisine, jatrorhizine
Aporphine	Glaucine, boldine, isoboldine
3. Terpenoid tetrahydroisoquinoline alkaloids	Emetine, cephaeline
4. Amaryllidaceae alkaloids	Galanthamine
5. Phenethylisoquinoline alkaloids	Autumnaline, floramultine, kreysigine

Depending on the diversity in structure, the isoquinoline alkaloids are classified into different subgroups[6,35] (Table 6.2).

Tyrosine is converted to dopamine by hydroxylation and decarboxylation, and to 4-hydroxyphenylacetaldehyde by transamination and decarboxylation. Dopamine and 4-hydroxyphenyacetaldehyde are condensed to form norcoclaurine. Norcoclaurine is converted to reticuline [(*S*)-reticuline and (*R*)-reticuline] through formation of coclaurine, and derivatives. Reticuline may also be formed through formation of norlaudanosine (Figure 6.18). Reticuline is the precursor of many isoquinoline alkaloids. However, many isoquinoline alkaloids are also derived from dopamine.

Simple tetrahydroisoquinoline alkaloids: These are structurally the simplest isoquinoline alkaloids of this group. They are usually bicyclic with a tertiary nitrogen atom. Most of the simple isoquinolines are reported from the family Cactaceae but many are also reported from other families. Anhalamine, anhalonidine and anhalonine, representatives of tetrahydroisoquinoline derivatives, are reported from peyote cactus (*Lophophora williamsii*, Cactaceae). They are produced from dopamine (Figure 6.19). Peyote cactus also contains mescaline, the protoalkaloid.[36]

Benzylisoquinoline alkaloids: Benzylisoquinoline alkaloids are diverse in structure. Approximately 2500 such alkaloids are known from plant families Papaveraceae, Ranunculaceae, Berberidaceae, and Menispermaceae. Based on structural diversity and substitution pattern, this group of alkaloids are classified into many subgroups.[37] Some of these subgroups are mentioned in Figure 6.20. **Papaverine** is a benzylisoquinoline alkaloid of opium, the dried latex of *Papaver somniferum.* It is formed from norreticuline by *O*-methylation (Figure 6.18). It is a weak base.

Bisbenzylisoquinoline alkaloids are reported from Berberidaceae, Ranunculaceae, Lauraceae, Menispermaceae families. Bisbenzylisoquinoline

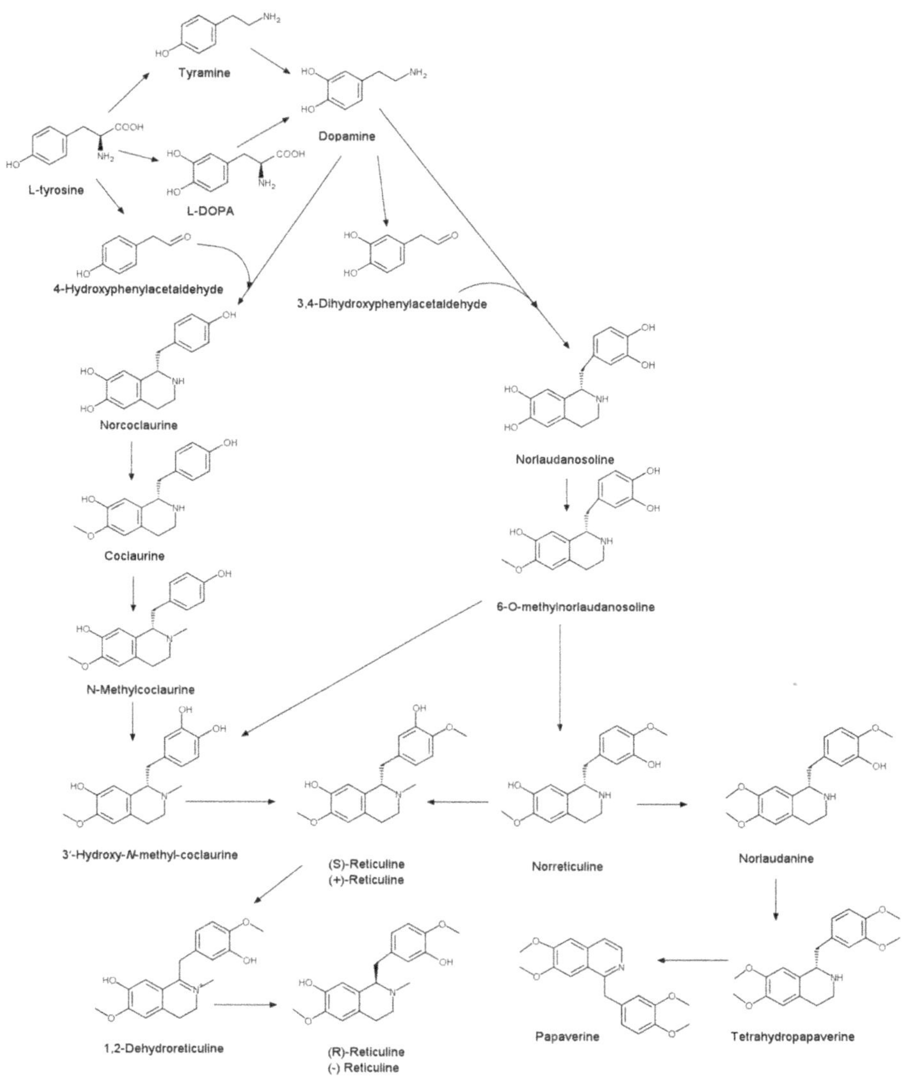

Figure 6.18 Biosynthesis of reticuline and some other benzylisoquinoline alkaloids.

alkaloids are dimers each consisting of two benzylisoquinoline parts. Mostly, these alkaloids are formed by condensation of two coclaurine units. Some of them can arise from the condensation of a coclaurine and a reticuline unit. The two units are linked *via* diphenyl ether, biphenyl, or benzyl phenyl ether bonds. Structurally, these types of alkaloids show variation depending on the number, site, orientation, and the nature of linkages, substitution of nitrogen atoms, and the degree of unsaturation. Over five

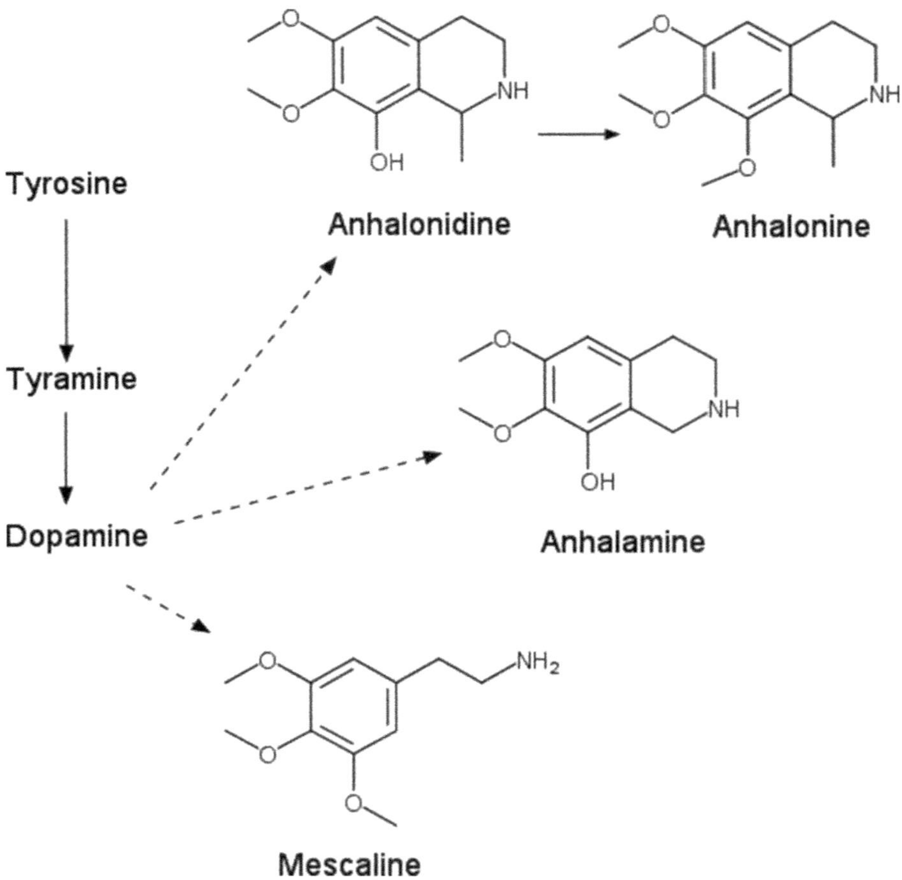

Figure 6.19 Biosynthesis of simple tetrahydroisoquinoline alkaloids of peyote cactus and mescaline (the protoalkaloid).

hundred bisbenzylisoquinoline alkaloids are in report.[38,39] Some examples are **tetrandrine** (from *Laureliopsis philippiana*, Monimiaceae), **berbamine** (*Berberis vulgaris*, Berberidaceae), **neferine** (from *Nelumbo nucifera*, Nelumbonaceae), and **tubocurarine** (from *Chondrodendron tomentosum*, Menispermaceae). Most of the bisbenzylisoquinoline alkaloids are bioactive molecules. The alkaloid **tubocurarine** is a cyclic molecule with two molecules of methylcoclaurine connected head to tail by diphenyl ether linkages[39] (Figure 6.21). Tubocurarine is a well-known muscle relaxant. It is one of the ingredients of the arrow poison 'curare' used by South American native tribes. Curare is made from different plants belonging to the families Menispermaceae (including *Chondrodendron tomentosum*) and Loganiaceae. Menispermaceous curare contained tubocurarine. (+)-Tubocurarine is more active than the (−)-tubocurarine. This alkaloid was used as a muscle relaxant during surgical operations. Safer drugs are now available.[40]

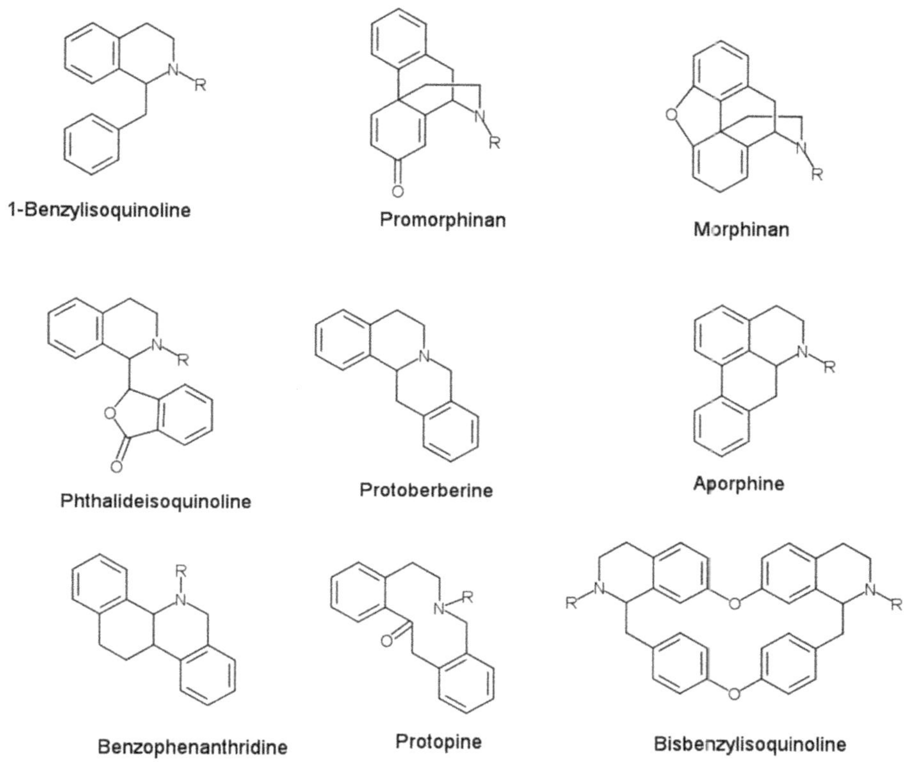

Figure 6.20 Diversity in structures of benzylisoquinoline alkaloids.

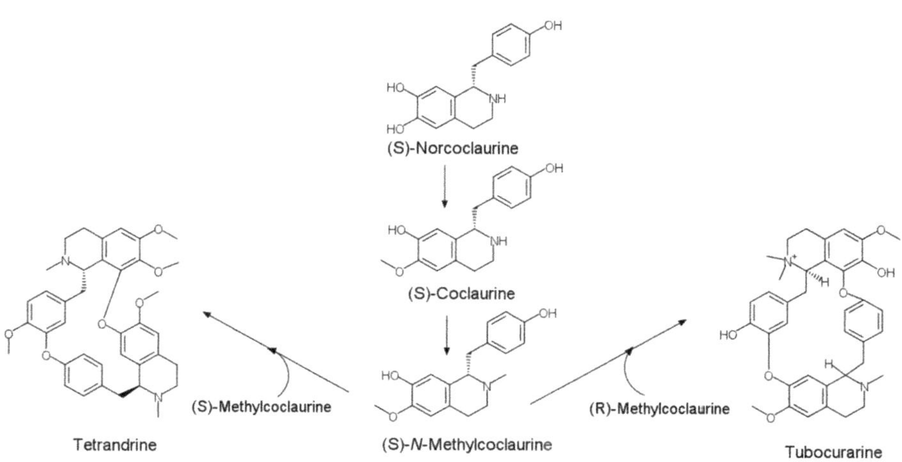

Figure 6.21 Biosynthesis of some bisbenzylisoquinoline alkaloids.

Morphinan type alkaloids are derived from *R*-reticuline. They include the important **opium** alkaloids **morphine, codeine,** and **thebaine** which are the strong bases. In addition to the previously mentioned benzylisoquinoline alkaloid **papaverine,** opium also contains two other feeble bases **noscapine** and **narceine** of **phthalideisoquinoline type** isoquinoline alkaloids. Opium, the dried latex obtained from the unripe fruits of *Papaver somniferum* (Papaveraceae) by incision, contains approximately 30 alkaloids including those mentioned above. Opium alkaloids also contain **protopine**. The biogenesis of opium alkaloids is detailed in Figures 6.22 and 6.23. Opium alkaloids are narcotic. The narcotic properties decrease in the order morphine, codeine, noscapine. Opium and morphine are used to relieve pain and are hypnotics. Regular users of opium smoke develop addiction. Codeine is a mild sedative and is used for the treatment of coughs. Thebaine is the precursor for synthesis of other drugs. Papaverine relaxes smooth muscles in blood vessels. Opium poppy seeds do not contain significant quantity of these alkaloids.[4,40]

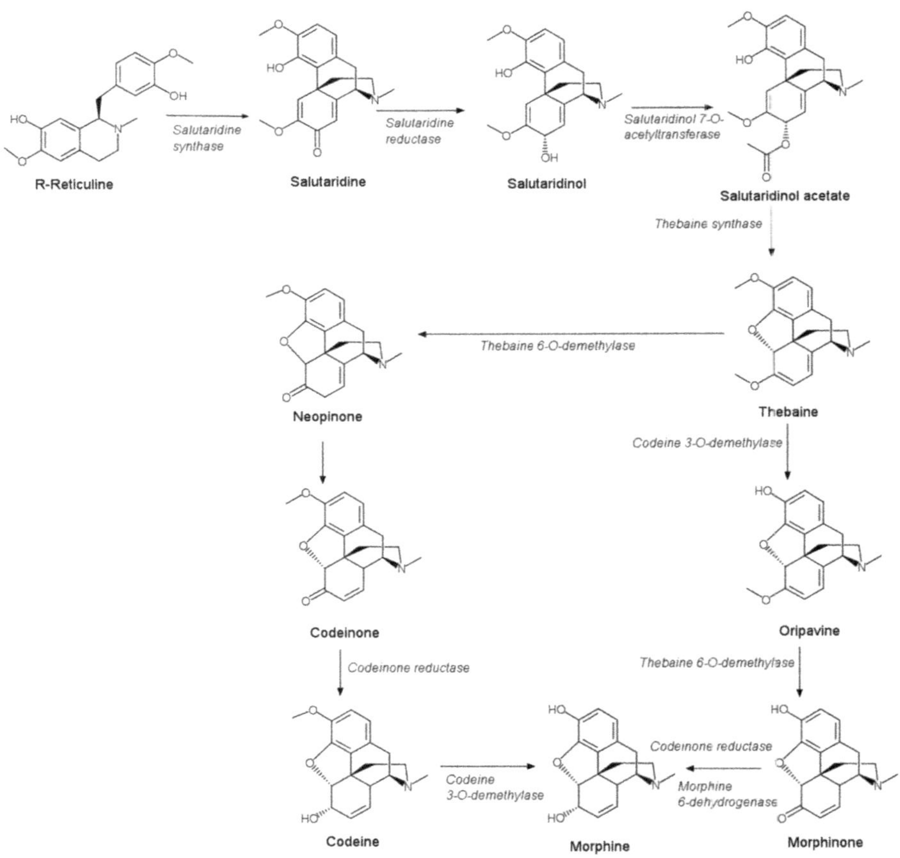

Figure 6.22 Biosynthesis of morphinan type alkaloids.

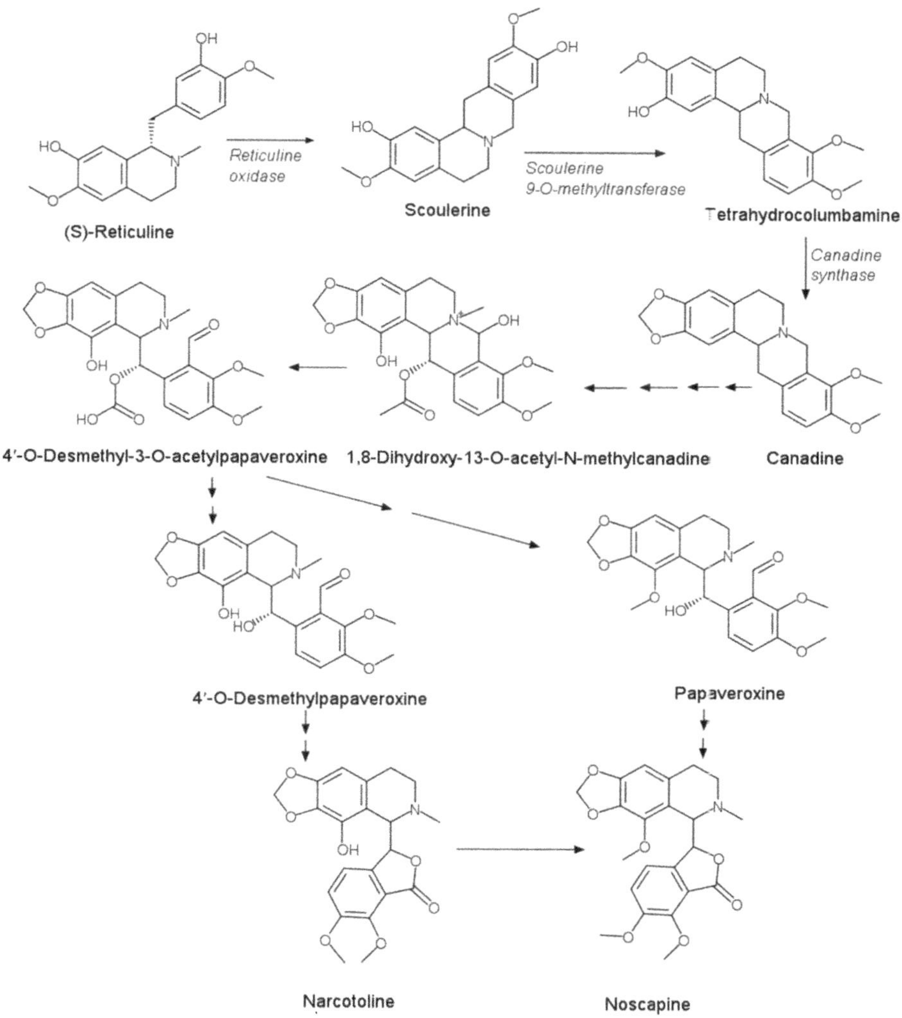

Figure 6.23 Biosynthesis of noscapine.

Synthesized **diamorphine** or **heroin** is a diacetate of morphine and causes illicit drug problems worldwide. But at the same time, heroin is used as analgesic and for suppression of coughing in cases of terminally ill cancer patients.[4]

Phthalide type isoquinoline alkaloids are **noscapine** and **narceine** as discussed under opium alkaloids. Their biosynthesis is mediated by (*S*)-reticuline (Figure 6.23). Noscapine lacks any analgesic activity but it is a cough suppressant. **Hydrastine** from *Hydrastis canadensis* (Ranunculaceae) is another example of a phthalide isoquinoline alkaloid. Hydrastine is a remedy traditionally used to control uterine bleeding.[4]

Protopine and benzophenanthridine type isoquinoline alkaloids are biosynthesized from (*S*)-reticuline. **Protopine** is obtained from *Chelidonium majus* (Papaveraceae). Protopine and **allocryptopine** are also reported from *Sanguinaria canadensis*. *Sanguinaria canadensis* (Papaveraceae) rhizome, known as blood root because of its red latex, has ethnobotanical importance because of its biological activities. The plant contains six quaternary **benzophenanthridine** alkaloids **sanguinarine, sanguilutine, sanguirubine, chelerythrine, chelilutine**, and **chelirubine**, in addition to the above mentioned two protopine alkaloids protopine and allocryptopine. These alkaloids show several bioactivities like preventing DNA break repair, inducing apoptosis, impairing cell division and tumour migration, cytotoxic effect *etc.* But more trials are required in human systems.[41] Protopine and benzophenanthridine type alkaloids are synthesized from (*S*)-reticuline (Figure 6.24).

Protoberberine type alkaloids are found in Berberidaceae, Ranunculaceae and some other families. Examples are **berberine, palmatine, jatrorrhizine,**

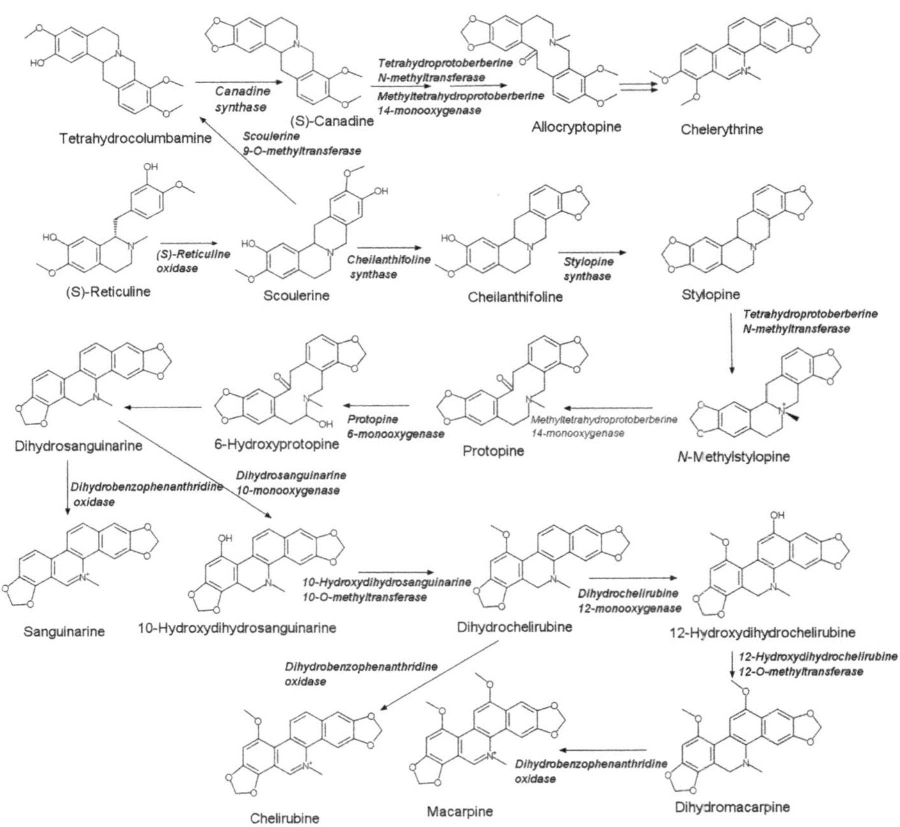

Figure 6.24 Biosynthesis of protopine and benzophenanthridine type isoquinoline alkaloids.

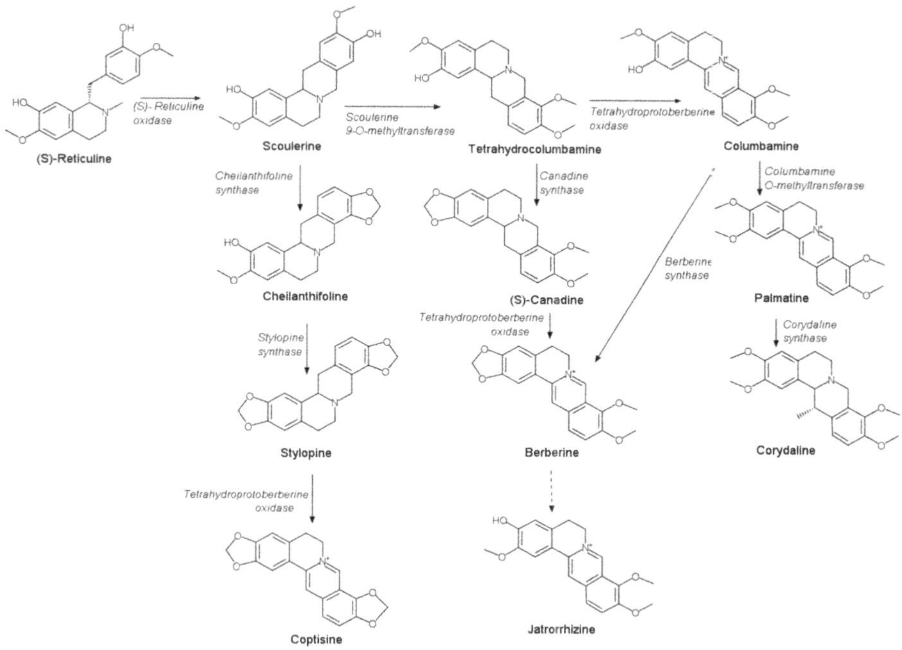

Figure 6.25 Biosynthesis of protoberberine type alkaloids.

coptisine, **corydaline** and many others. These alkaloids are produced in plants *via* (*S*)-reticuline (Figure 6.25).

Aporphine type isoquinoline alkaloids form one of the largest groups of isoquinoline alkaloids with more than 500 alkaloids reported so far. Some examples are **boldine**, **isoboldine**, **magnoflorine**, **isothebain**, **nuciferine**, **pronuciferine** and **stephanine**. Magnoflorine is synthesized from (*S*)-reticuline. Nuciferine and pronuciferine are produced from dopamine *via* *N*-methylcoclaurine[42,43] (Figure 6.26).

Different **compartments for biosynthesis** and storage of benzylisoquinoline alkaloids are reported. Benzylisoquinoline alkaloids are accumulated in the laticifers in all organs of opium poppy (*Papaver somniferum*). Morphinan, pthalideisoquinoline (*e.g.*, noscapine), and 1-benzylisoquinoline (papaverine) alkaloids are found in the latex (the cytoplasm of laticifers) of aerial organs (Figure 6.27). The benzophenanthridine alkaloid sanguinarine is predominantly found in the roots. Laticifers in *Papaver somniferum* are confined to the phloem. Laticifers and the phloem cells sieve elements and companion cells, that originate from the procambium and vascular cambium, are the three compartments for biosynthesis and storage of benzylisoquinoline alkaloids in this plant. The pathway enzymes are transcribed and translated in the companion cells. Enzymes are transported to the sieve elements or laticifers. Thebaine is synthesized in the sieve elements and transported

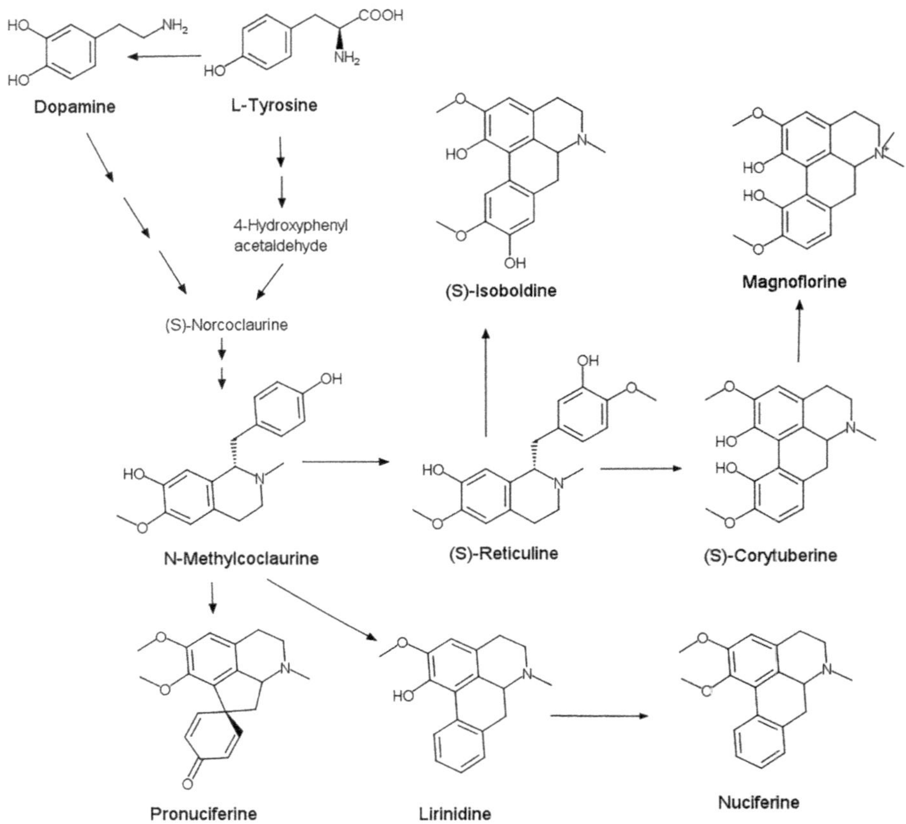

Figure 6.26 Aporphine type alkaloids.

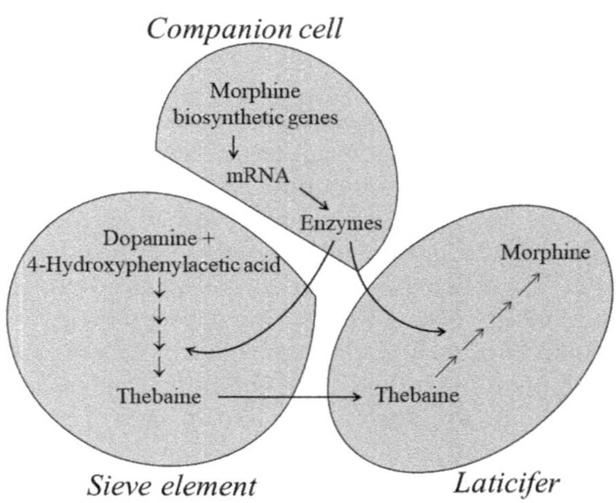

Figure 6.27 Compartmentation in morphine biosynthesis.

to laticifers. Morphine biosynthesis is completed in the laticifer. However, phloem tissues are not involved in the biosynthesis of benzylisoquinoline alkaloids in *Thalictrum flavum* (Ranumculaceae).[44,45]

Berberine is produced in the root of *Coptis japonica* (Ranunculaceae) and translocated to the rhizome which is used as the medicinal part of the plant in Asian countries.

Many experimental studies have suggested that the benzylisoquinoline alkaloids have defensive properties such as anti-herbivory, antifungal and antibacterial properties. The presence of defensive metabolites in the sugar rich phloem sap of *Papaver somniferum* deters sucking insects. Bismorphine produced in the cell walls of *Papaver somniferum* due to mechanical injury is suggested to serve a defensive function. Berberine has potent anti-herbivory effect against a variety of insects. Berberine has been commercialized as an insecticide because of its properties to reduce larval growth and survival rates of pests. Protoberberine, sanguinarine, and chelerythrine show antifungal and antibacterial activities against important pathogens. Due to bitterness of the alkaloids, benzylisoquinoline alkaloids have a defensive role against large animals.[45]

6.4.2 Terpenoid Tetrahydroisoquinoline Alkaloids

Principal alkaloids of this group are **emetine** and **cephaeline** produced in the roots and rhizomes of *Cephaelis ipecacuanha* (Rubiaceae), commonly known as ipecac. Other related alkaloids found in ipecac root are psychotrine, psychotrine methylether, ipecoside *etc.* These alkaloids possess tetrahydroisoquinoline structures added to part of the terpenoid secoiridoid **secologanin** (Figure 6.28). Secologanin condenses with dopamine by a Mannich like reaction. The resultant isoquinoline alkaloids are the glycosides deacetylipecoside and deacetylisoipecoside. **Alangiside** and **demethylalangiside** produced from deacetylipecoside are reported from the Cornaceae plant *Alangium lamarckii* (synonym: *Alangium salviifolium*). *Alangium lamarckii* also contains **tubulosine, cephaeline, isocephaeline, psychotrine, protoemetine** *etc.* Methylation of deacetylisoipecoside followed by removal of sugar unit by β-glucosidase and addition of dopamine results in the formation of emetine and cephaeline (Figure 6.28). Ipecac is used as an expectorant and emetic. Emetine has more expectorant activity and less emetic action than cephaeline.[46,47]

6.4.3 Amaryllidaceae Alkaloids of Isoquinoline Type

Combination of 3,4-dihydrobenzaldehyde with tyramine produces various alkaloids present in the family Amaryllidaceae. Approximately 600 alkaloids occur in Amaryllidaceae. **Galanthamine (or galantamine)** (Figure 6.29) is the most important alkaloid of all of them. This alkaloid has attracted the attention of scientists because of its property to inhibit acetylcholinesterase. An acetylcholinesterase inhibitor may be used to treat symptoms of neurodegenerative disorders like Alzheimer's disease. At present galanthamine has been approved by the United States Food and Drug Administration to treat

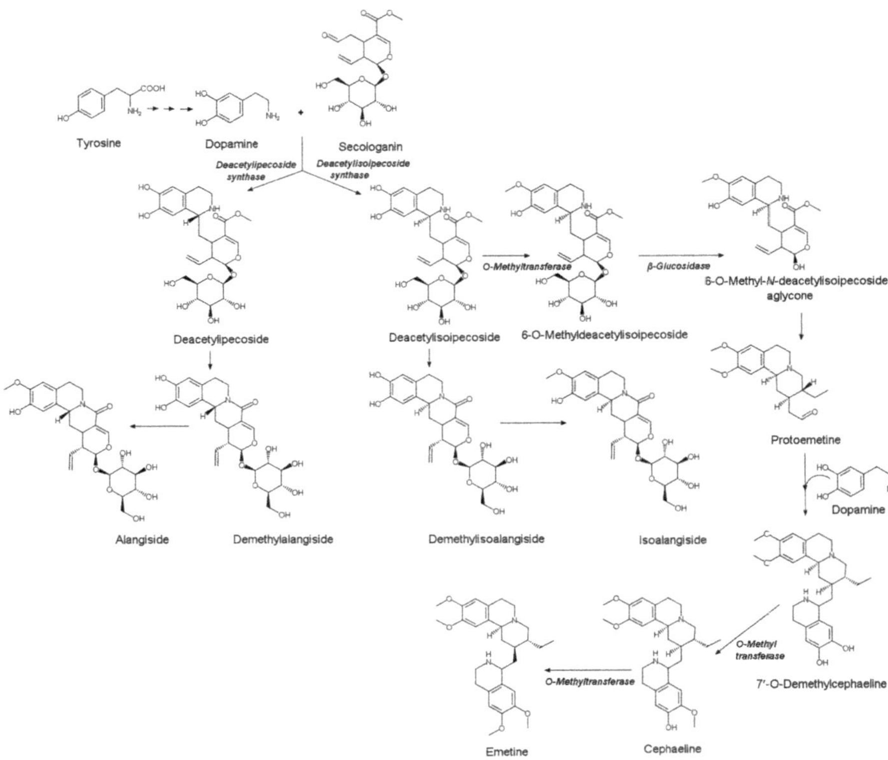

Figure 6.28 Biosynthesis of terpenoid isoquinoline alkaloids.

the symptoms of Alzheimer's Disease. Some other Amaryllidaceae alkaloids *e.g.*, **lycorine, haemanthamine** and **narciclasine** are the lead molecules in the areas of anticancer research.[48,49] Biosynthesis of these alkaloids is shown in Figure 6.29.

6.4.4 Phenethylisoquinoline Alkaloids

Autumnaline is a phenethylisoquinoline alkaloid found in some genera of Liliaceae. Structurally it is related to benzyltetrahydroisoquinoline alkaloids but with an extra carbon between the tetrahydroisoquinoline and the attached aromatic ring. Dopamine condenses with 4-hydroxydihydrocinnamaldehyde. Further reactions like hydroxylation and methylation produce autumnaline. From autumnaline, the phenethylisoquinolines **floramultine** and **kreysigine** are produced in one direction. In another pathway, autumnaline produces the protoalkaloid **colchicine**[4] (Figure 6.30).

Colchicine is obtained from the seeds and the corm of *Colchicum autumnale* (Colchicaceae). Colchicine is also obtained from other Colchicaceae plants *e.g.*, *Gloriosa* and *Sandersonia*. Colchicine is used in the treatment of gout and is approved by FDA. Colchicine is also used to treat Mediterranean

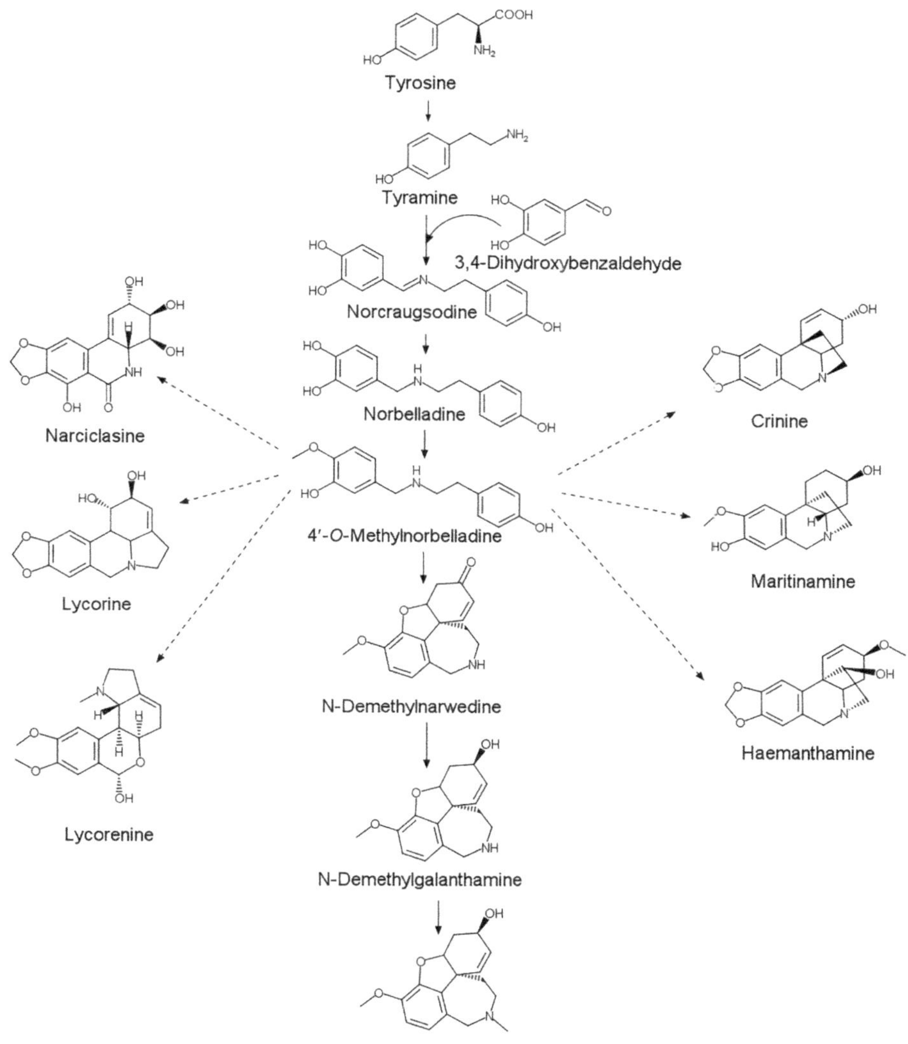

Figure 6.29 Biosynthesis of galanthamine and some other Amaryllidaceae alkaloids.

fever, pericarditis, and Behcet's disease. It is cytotoxic and binds to the tubulin dimers in the mitotic spindle and inhibit microtubule growth.[50]

6.5 Histidine Derived Alkaloids

Imidazole alkaloids are mostly important because of the therapeutic uses of **pilocarpine** (Figure 6.31). Commercially this alkaloid is mainly obtained from *Pilocarpus jaborandi* and *Pilocarpus microphyllus* (Rutaceae). This alkaloid is also present in other species such as *Pilocarpus pennatifolius*, *Pilocarpus*

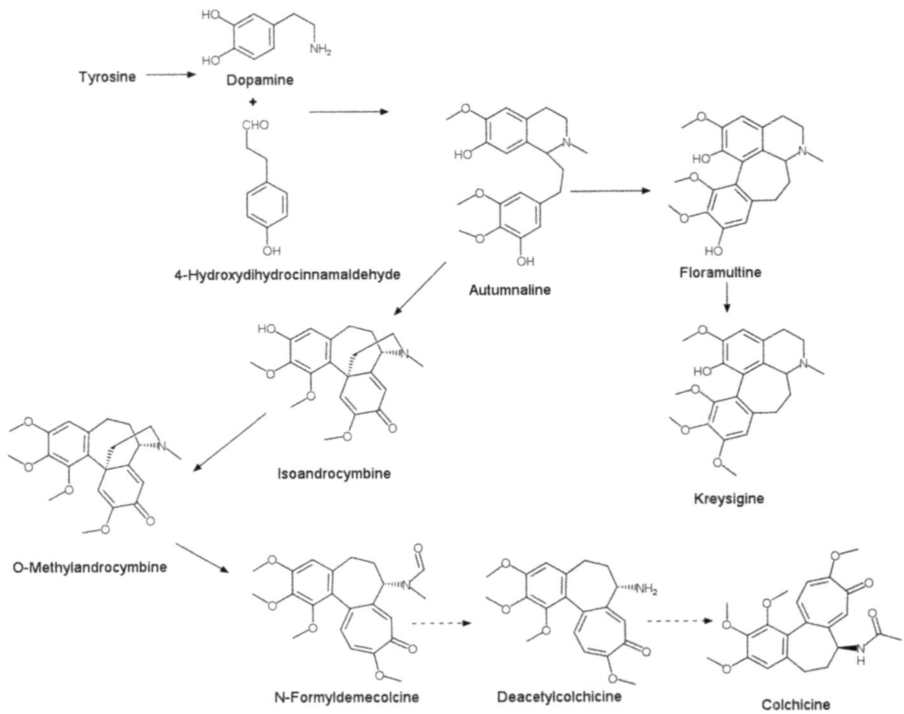

Figure 6.30 Biosynthesis of phenethylisoquinoline alkaloids and colchicine.

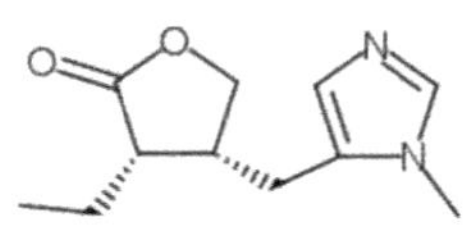

Pilocarpine

Figure 6.31 Pilocarpine.

racemosus, *Pilocarpus trachylophus*, *Pilocarpus carajaensis* and *Pilocarpus alatus*.[51,52] Pilocarpine is used in eye drops for the treatment of glaucoma and xerostomia (reduced production of saliva) in patients suffering from mouth and throat cancers undergoing radiotherapy. Pilocarpine causes constriction of the pupil of the eye. The alkaloid is a parasympathetic agonist and stimulates sweat and lachrymal glands increasing secretion.[53] The alkaloids **epiisopiloturine** and **epiisopilosine** are side products in the pharmaceutical industry. They have interested the scientific community because of their promising anti-inflammatory therapeutic potential.[54] Other alkaloids reported from different *Pilocarpus* species are **pilosinine, pilosine, isopilosine**, and **epipilosine** *etc.*[51]

Pilocarpine has an imidazole ring and a γ-lactone ring. Pilocarpine and structurally related compounds are assumed to be derived from L-histidine which also has an imidazole ring (glyoxaline).[51]

6.6 Tryptophan Derived Alkaloids

6.6.1 Indole Alkaloids

The amino acid L-tryptophan, produced *via* anthranilic acid in the shikimic acid pathway, is the precursor for many indole alkaloids. The indole alkaloids may be of different types: (1) simple indole alkaloids, (2) simple β-carboline alkaloids, (3) terpenoid indole alkaloids and (4) ergot alkaloids.

Simple indole alkaloids (Figure 6.32) are produced by plants and fungi. They have effects on plants as well as human beings. A few examples are mentioned here. **Serotonin and melatonin** are indoleamines produced from **tryptamine**, a decarboxylated form of tryptophan. **Serotonin (5-hydroxytryptamine, 5-HT)** is the hydroxylated derivative of tryptamine. In addition to its presence in human beings, plants also produce this simple alkaloid. The production of this compound has been found to be related to stress responses. Similarly, the animal hormone melatonin, produced by *N*-acetylation of serotonin followed by *O*-methylation, is also produced by plants. Serotonin has been found to be a plant growth regulator, and a biotic and abiotic stress defense molecule. Serotonin also plays an important role in long distance and rapid signaling in response to pathogen attack. In plants, melatonin functions in reducing oxidative stress, detoxifying free radicals

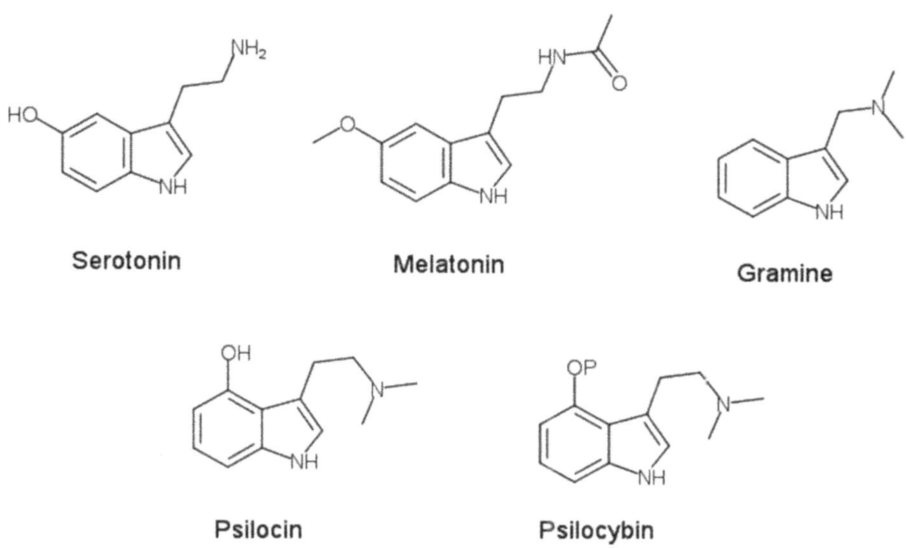

Figure 6.32 Simple indole alkaloids.

generated during photosynthesis and metabolism, improving resistance to both biotic and abiotic stresses including drought, cold, heat, osmotic stress, oxidative stress, herbicides, intense radiation, and bacterial pathogens. Melatonin promotes seed germination and controls stomatal closure. The auxin like effect of melatonin has also been documented.[55–57] **Gramine** is found in various species of Poaceae, Fabaceae, and other plant families. This indole alkaloid shows highest toxicity in barley (*Hordeum vulgare*, Poaceae). Gramine is responsible for its toxicity to ruminants. Gramine is an allelochemical. In addition to the phytotoxicity, gramine is toxic to mammals, insects, bacteria, and fungi. The toxicity could be due to its effect on energy metabolism. Enhancement of oxidative stress is probably responsible for its algicidal effect. Gramine accumulation is induced in barley with increase in temperature, drought, attacks of aphids and fungi. Gramine has a defensive role in response to biotic and abiotic stresses.[58–61]

About 30 species of the mushroom *Psilocybe* such as *Psilocybe mexicana* and *Psilocybe semilanceata* have been found to be hallucinogenic and psychoactive. The active components are **psilocybin** and **psilocin**. Psilocybin is produced by phosphorylation of the hydroxyl group of psilocin. These alkaloids have also been detected in some other fungi such as *Conocybe*, *Panaeolus* and *Stropharia*. Recently, effective improvement has been observed in psilocybin-assisted psychotherapy in the treatment of depression and anxiety in cancer patients. The FDA has approved psilocybin for treatment-resistant depression and post-traumatic stress disorder.[62–64]

Simple β-carboline alkaloids are reported to be formed endogenously in plants, in different human tissues including the human eye lens as well as in other living tissues of insects, and marine creatures. These tricyclic nitrogen heterocycles are formed by the Maillard reaction through the condensation of amino acids and reducing sugars or aldehydes. Common examples from plants are **harman** (isolated from the Rubiaceae plant *Sickingia rubra* bark); **harmine**, **harmaline**, and **harmalol** (isolated from the seeds of *Peganum harmala*, Nitrariaceae) and **elaeagnine** (from *Elaeagnus angustifolia*, Elaeagnaceae) (Figure 6.33). The psychoactive properties of *Peganum harmala*, used as hallucinogenic drinks or snuffs, are due to the presence of β-carboline alkaloids. Different fruits and fruit juices and crude drugs of plant origin are reported to be natural sources of various β-carboline alkaloids. Diet is a common exogenous source of β-carboline alkaloids in mammals and humans. β-Carboline alkaloids have recently attracted interest due to their potent antitumour, antiviral, antimicrobial, and antiparasitic activities. Harmine, an important metabolite detected in *Passiflora incarnata* and *Passiflora edulis* (Passifloraceae) fruits, has potential antitumour and cytotoxic effects. **Harmaline** and **harmine** possess acetylcholinesterase inhibitory activity, antioxidant and anti-inflammatory properties thus having a great potential for treating Alzheimer's disease. Some β-carboline alkaloids are reported to intercalate into DNA and inhibit topoisomerase and monoamine oxidase.[65–70]

Terpenoid indole alkaloids have two parts – the tryptamine derived indole moiety and the non-indole moiety which is monoterpenoid in origin. The

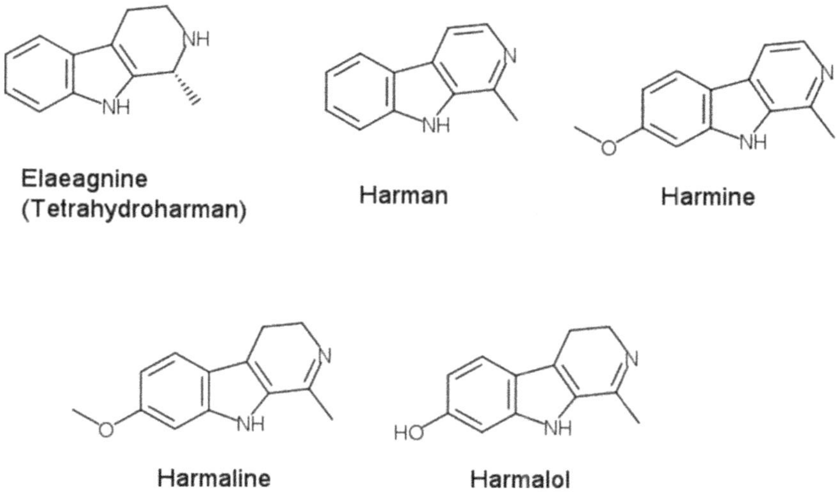

Figure 6.33 Simple β-carboline alkaloids.

ultimate terpenoid precursor is geraniol derived secologanin which constitutes the non-indole part or non-tryptamine part. The non-tryptamine unit consists of 9 or 10 skeletal carbon atoms. Tryptophan derived alkaloids are found in several families including Apocynaceae, Loganiaceae, Rubiaceae, and Nyssaceae. Tryptamine produced by decarboxylation of L-tryptophan is the direct precursor of the indole part. Condensation of tryptamine and secologanin results in the formation of strictosidine. Strictosidine is the precursor of major classes of terpenoid indole alkaloids. Subsequent transformation of strictosidine involves first the loss of the glucose moiety by strictosidine specific glucosidase, followed by formation of 4,21-dehydrogeissoschizine. 4,21-Dehydrogeissoschizine cyclizes to cathenamine in one direction and other terpenoid indole alkaloids also (Figure 6.34). Tabersonine is converted to 16-methoxytabersonine and then through a series of reactions to vindoline.[71–73]

According to the cyclic arrangement of the non-tryptamine part, three types (classes) of monoterpene units are constructed. These are corynanthe type, aspidosperma type and iboga type (Figure 6.35). Examples of **corynanthe type** of alkaloids are ajmalicine, serpentine, preakuammicine, strychnine, brucine *etc.* Corynanthe type of alkaloids are also termed the *Strychnos* type alkaloids. **Aspidosperma type** of alkaloids are tabersonine and vindoline. **Iboga type** of alkaloids are exemplified by catharanthine and ibogaine.[72]

Catharanthus roseus (Apocynaceae), commonly known as Madagascar periwinkle, is one of the most extensively investigated medicinal plants containing pharmaceutically important alkaloids. About 100 alkaloids are produced by *Catharanthus roseus*. This plant is important because of the presence of two bisindole antitumour alkaloids in the leaves. These are **vinblastine** and **vincristine**. Vinblastine is produced in the leaves by the condensation of

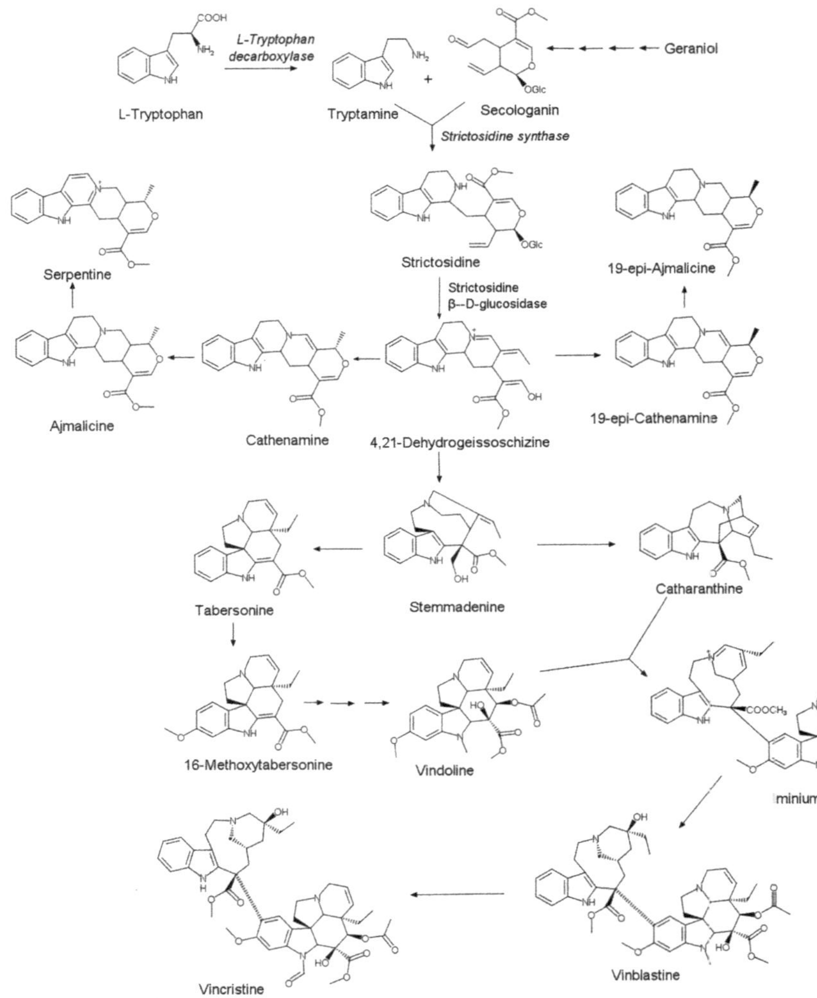

Figure 6.34 Biosynthesis of important alkaloids in *Catharanthus roseus.*

catharanthine and **vindoline** through formation of **3,4-anhydrovinblastine.** Vinblastine and vincristine are structurally closely related. Vincristine has an *N*-formyl group on the indole nitrogen of the vindoline fragment, while vinblastine has an *N*-methyl group (Figure 6.34). **Vinblastine** is mainly used for the treatment of Hodgkin's disease, a cancer that affects lymph glands, liver, and spleen. **Vincristine** is used in the treatment of childhood leukaemia and some other cancers such as lymphomas, small cell lung cancer, cervical and breast cancers. These alkaloids inhibit microtubule formation during mitosis thus preventing cells from reproducing. A few other semisynthetic derivatives including **vindesine** and **vinorelbine** are also used for the treatment of cancers. Vindesine is used to treat lymphoid leukaemia in children.

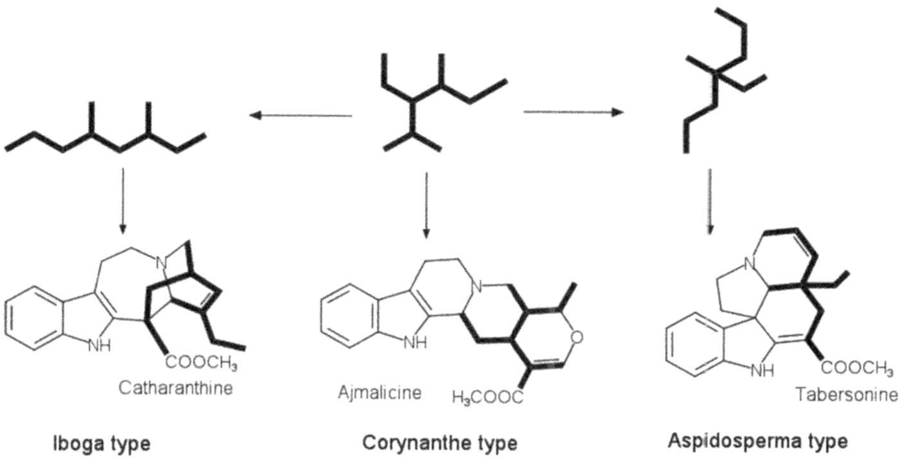

Figure 6.35 Arrangement of C9–C10 units in terpenoid indole alkaloids.

Vinorelbine is used against non-small-cell lung cancer and small-cell lung cancer. The yield of vinblastine and vincristine is very low (0.0002%) in the plant. The vinblastine content is higher in the plant than the vincristine content, whereas vincristine is more important than vinblastine. It is now possible to convert vinblastine to vincristine by controlled oxidation or by *N*-demethylation by *Streptomyces albogriseolus*. Apart from these dimeric alkaloids, the plant also contains other different terpenoid indole alkaloids.[4] Another important alkaloid detected in the roots is **ajmalicine**. Both *Catharanthus roseus* and *Rauvolfia serpentina* (Apocynaceae) are the commercial sources of ajmalicine that is used to treat circulatory disorders. The plant *Catharanthus roseus* contains corynanthe type (**ajmalicine, serpentine, tetrahydroalstonine, alstonine**), iboga type (**catharanthine**), and aspidosperma type (**tabersonine, vindoline, vincamine**) alkaloids.[4] The biosynthetic pathway[73] of terpenoid indole alkaloids have been studied in *Catharanthus roseus*. The pathway of the important alkaloids is depicted in Figure 6.34.

The terpenoid indole alkaloid biosynthesis often involves multiple organellar **compartments**. Immunocytochemical localization of different enzymes has confirmed that intracellular and intercellular translocation is involved in the biosynthesis of some major alkaloids.[74] **Intercellular compartments** for transport of metabolites during biosynthesis of terpenoid indole alkaloids involve four types of cells. These are **internal phloem-associated parenchyma, leaf epidermis, laticifers,** and **idioblasts** (Figure 6.36). **Internal phloem-associated parenchyma** existing in the periphery of stem pith or interxylary parenchyma on the upper part of vascular bundles in leaves are full of chloroplasts with plastid located enzymes. The MEP pathway happens in the green tissues. Early steps in the iridoid secologanin formation *i.e.*, conversion of geraniol to 10-oxogeraniol and formation of loganic acid also happen in the internal phloem-associated parenchyma cells of young tissues of leaves and roots. The **leaf epidermis** is the

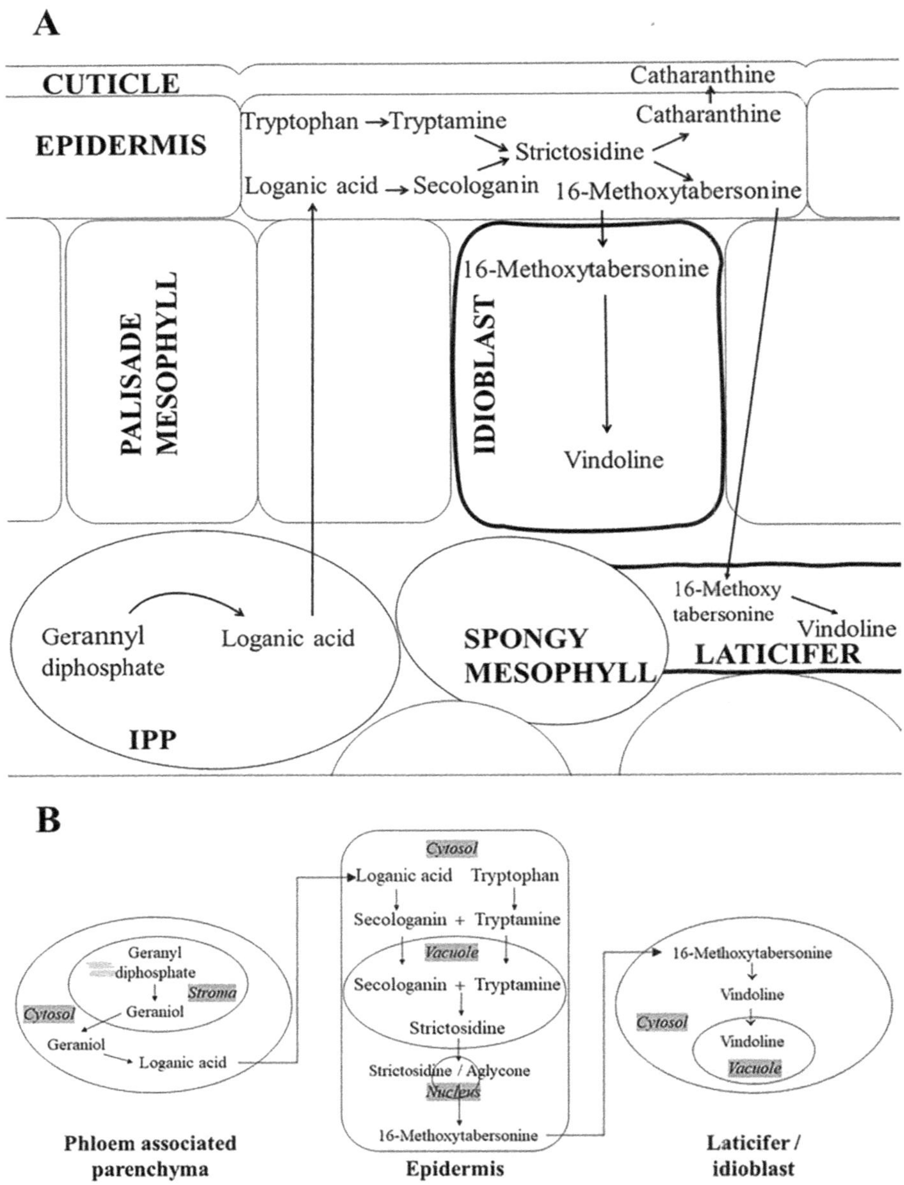

Figure 6.36 A: Intercellular compartmentation of terpenoid indole alkaloid biosynthesis in a leaf; B: intracellular compartmentation of terpenoid indole alkaloid biosynthesis. IPP: Internal Phloem-associated Parenchyma.

site where loganic acid is assumed to move from internal phloem-associated parenchyma to be converted into secologanin. Condensation of tryptamine and secologanin, resulting in the formation of strictosidine, takes place in the epidermal cells of young shoots and leaves. Catharanthine formed in the leaf epidermis is secreted to the leaf surface in the wax exudates. 16-Methoxytabersonine, produced from tabersonine and the precursor of vindoline, is also formed in the epidermis. **Laticifers** and **idioblasts** of leaves, stems, and flowers are the cellular compartments where 16-methoxytabersonine and other accessory products are transported from the epidermis. Vindoline is produced from these precursors in these two compartments (Figure 6.36).

Intracellular compartments and transport of metabolites within these compartments are required for different steps of terpenoid indole alkaloid biosynthesis. Geraniol, the precursor of loganin, is biosynthesized in the stroma of plastids or stromules in the internal phloem-associated parenchyma. Geraniol is exported from stroma to the cytosol where it is converted to loganic acid. Secologanin from loganic acid is synthesized in the cytosol of epidermal cells. Tryptamine is formed from tryptophan in the cytosol of the epidermis where the enzyme tryptophan decarboxylase is available. Tryptamine and secologanin are transported to the vacuole where condensation of these two metabolites results in the formation of strictosidine. The enzyme strictosidine synthase is localized in the vacuole. However, the strictosidine glucosidase is present in the nucleus. So strictosidine from the vacuole is transported to the cytosol and then probably to the nucleus to form the strictosidine aglycone However, this pathway is not yet clearly understood. Strictosidine aglycone is converted to catharanthine and tabersonine in the cytosol. 16-Methoxytabersonine is produced from tabersonine in the cytosol of the epidermis. 16-Methoxytabersonine and derivatives are converted to vindoline in the cytosol of laticifers/idioblasts. Vindoline is transported to the vacuole.[44,75] Ajmaline is mainly accumulated in root epidermis.

Rauvolfia serpentina (Apocynaceae) contains more than 50 indole alkaloids. The alkaloids identified from the plant are **ajmaline**, **ajmalicine**, **ajmalinine**, **ajmalidine**, **corynanthine**, **reserpine**, **deserpidine**, **reserpinine**, **serpentine**, **raucaffricine**, **rauhimbine**, **rescinnamine**, **sarpagine**, **canescine**, **yohimbine** *etc.* Other related species of *Rauwolfia* containing similar alkaloids are *Rauvolfia vomitoria*, *Rauvolfia caffra*, *Rauvolfia mombasiana* and *Rauvolfia obscura* from Africa, and *Rauvolfia tetraphylla* from India. Reserpine and deserpidine are trimethoxybenzoyl esters of yohimbine like alkaloids and rescinnamine is a trimethoxycinnamoyl ester. *Rauvolfia* preparations and reserpine are used to treat hypertension and certain neuropsychiatric disorders. Ajmaline is often used to treat cardiac arrhythmias. Biosynthesis of ajmalicine and serpentine has been shown in Figure 6.34. Biosynthetically yohimbine arises from 4,21-dehydrogeissoschizine. Reserpine, deserpidine and rescinnamine are produced at the later stage. Some other *Rauvolfia* alkaloids are presented in Figure 6.37.

Species of *Alstonia*, including *Alstonia scholaris* (Apocynaceae) contain **alstonine**, and **echitamine**. **Yohimbine** is obtained from *Pausinystalia*

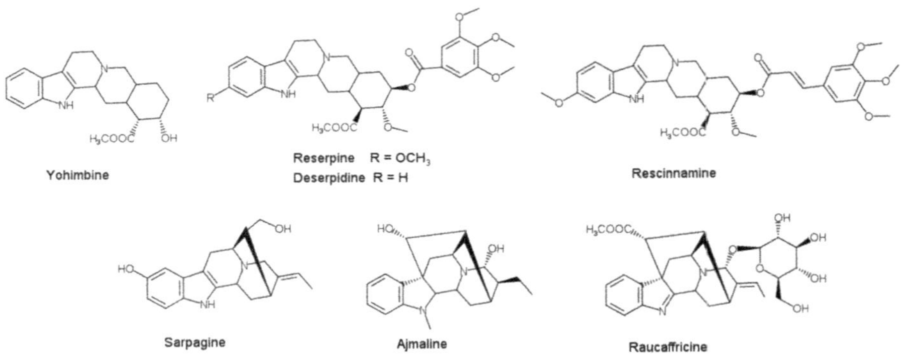

Figure 6.37 *Rauvolfia* alkaloids.

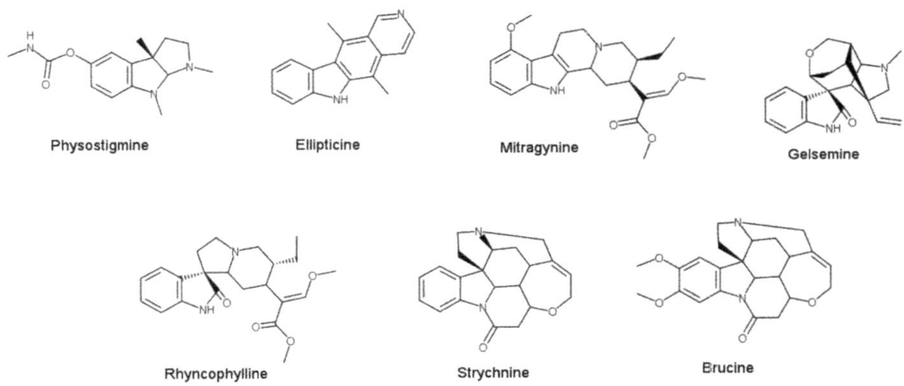

Figure 6.38 Other important terpenoid indole alkaloids.

yohimba (Rubiaceae) and *Aspidosperma* bark (Apocynaceae). Different species of *Mitragyna* (Rubiaceae) such as *Mitragyna speciosa* contain **mitragynine** (Figure 6.38), an indole alkaloid. Different *Uncaria* species of Rubiaceae such as *Uncaria gambir*, *Uncaria sinensis* and *Uncaria rhynchophylla* contain the indole alkaloid **rhyncophylline** (Figure 6.38), and its glycoside. *Gelsemium sempervirens* (Gelsemiaceae) contains the extremely toxic alkaloids **gelsemine** (Figure 6.38), and **gelsemicine**. **Ellipticine** (Figure 6.38), and **9-methoxyellipticine** have a pyridocarbazole skeleton. This alkaloid has been obtained from the bark of *Ochrosia elliptica* (Apocynaceae). It has potent anticancer activity but preclinical studies indicated side effects.[4,40]

Strychnine and **brucine** (corynanthe type) are well known indole alkaloids (Figure 6.38) obtained from the seeds and other parts of *Strychnos nux-vomica* (Loganiaceae). Other minor alkaloids reported from this plant are protostrychnine, α-colubrine, β-colubrine, icajine, vomicine, novacine, pseudostrychnine, pseudobrucine, isostrychnine *etc.* Strychnine is a very toxic alkaloid. Some species of *Strychnos e.g.*, *Strychnos toxifera* are used as

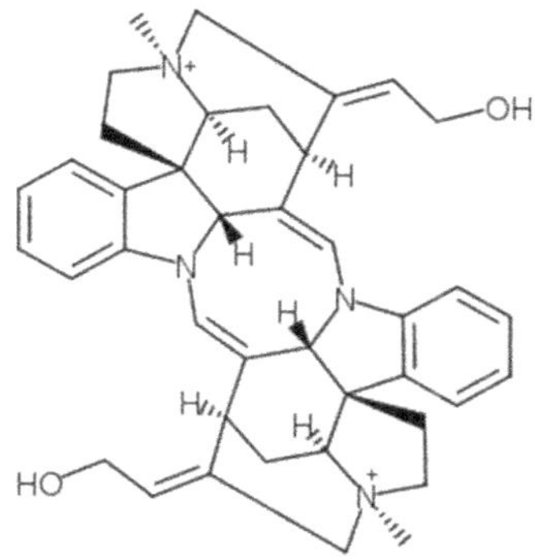

Toxiferine I

Figure 6.39 Toxiferine 1 from *Strychnos toxifera.*

one of the ingredients in the preparation of some curare type preparations, in addition to species of *Chondrodendron*. *Strychnos toxifera* contains dimeric indole alkaloid **toxiferine I**. The alkaloid is responsible for the activity of the arrow poison curare (Figure 6.39).

Pyrroloindole alkaloids are exemplified by **physostigmine**. The dried ripe seeds of *Physostigma venenosum* (Fabaceae), commonly known as calabar beans, contain the alkaloid **physostigmine** (or **eserine**) (Figure 6.38). Physostigmine is a known acetylcholinesterase inhibitor. But due to narrow therapeutic index it was not approved for the treatment of Alzheimer's Disease. Physostigmine salicylate is used for contracting the pupil of the eye. Rivastigmine, a drug developed from physostigmine, is used for treatment of early symptoms of Alzheimer's Disease. Neostigmine, pyridostigmine, and distigmine are related synthetic drugs used for the treatment of myasthenia gravis.[4,40]

Ergot Alkaloids

Ergot is the sclerotium growing on the infected ovary of members of the grass family. The fungi involved in the production of such sclerotia are different species of the parasitic fungal genera. Ergot of rye is the dried sclerotium of *Claviceps purpurea* growing on the ovary of rye (*Secale cereale*, Poaceae). The ergot may contaminate the flour obtained from such infected grains. Ergot infected rye, after consumption, results in a disease called ergotism. Ergot sclerotia contain biologically active alkaloids which are derived from tryptophan. All ergot alkaloids contain the tetracyclic ergoline ring structure.[76–80] Ergot alkaloids are divided into three classes based on substituents attached on the ergoline ring:

(1) **The clavines** (*e.g.*, chanoclavine, agroclavine) (none of this group is used medicinally).

(2) **Simple lysergic acid derivatives** are D-lysergic acid structures with a nonpeptidic amidation of the C8 carboxy group such as ergometrine (also known as ergonovine), ergometrinine, and paspalic acid (Figure 6.40). Lysergic acid derived amides are highly active.

The first step in the biosynthesis of ergot alkaloids is prenylation of L-tryptophan by dimethylallyl diphosphate (DMAPP) to produce 4-dimethylallyl-L-tryptophan (DMAT). *N*-Methylation of DMAT produces 4-dimethylallyl-L-abrine (*N*-Me-DMAT). Subsequent oxidation steps result in the cyclization of prenyl and indole moieties to produce chanoclavin-I, followed by chanoclavine-I-aldehyde, and then either agroclavine or festuclavine. Agroclavine is derivatized to form lysergic acid (Figure 6.41).

(3) **Peptide ergot alkaloids** (ergopeptines) consist of D-lysergic acid and proline containing cyclic tripeptide moieties *e.g.*, ergocristine, ergocornine and ergocryptine. The epimeric forms C8-(*R*) are known as ergopeptines and the C8-(*S*) isomers are known as ergopeptinines. Ergopeptinines are less toxic than ergopeptines.[77] Apart from proline, another two amino acids of the peptide chain are variable and form a diversity of peptide groups of ergot alkaloids. Based on the nature of the amino acids, the peptide ergot alkaloids can be divided into the ergotamine group and the ergotoxine group. The **ergotamine group** of alkaloids consists of **ergotamine**, **ergotaminine**, **ergosine**, and **ergosinine.** The **ergotoxine group** of alkaloids consists of **ergocristine, ergocristinine, ergocryptine, ergocryptinine, ergocornine** and **ergocorninine.**[79,81] Ergotamine incorporates alanine, phenylalanine, and proline in the peptide portion (Figure 6.42).

Ergot alkaloids are pharmacologically active alkaloids. **Lysergic acid diethylamide** or LSD is synthesized from lysergic acid. LSD is sold in the illicit market and exerts a hallucinogenic effect. **Ergometrine** (**ergonovine**) is a

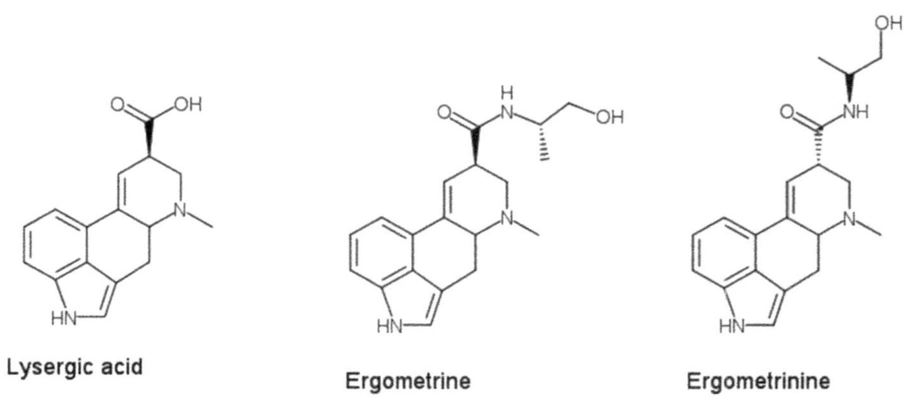

Figure 6.40 Simple lysergic acid derivatives.

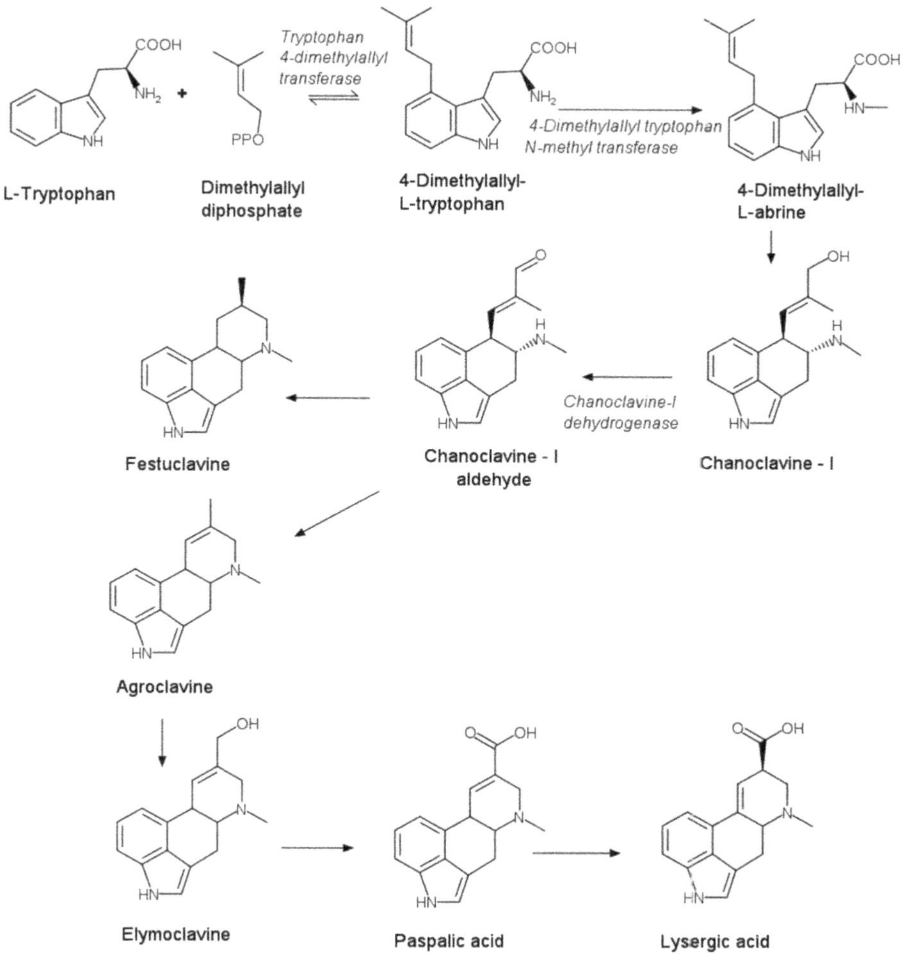

Figure 6.41 Early steps in ergot alkaloid biosynthesis: production of clavins and lysergic acid.

lysergamide having powerful action on the uterus. It is used in the management of the third stage of labour to prevent postpartum hemorrhage. Bleeding is reduced due to the vasoconstrictor effect of the alkaloid. **Ergotamine** has vasoconstrictor and analgesic effect. This alkaloid is used to treat acute migraine attacks. **Bromocriptine** (2-bromo-α-ergocryptine), a semisynthetic derivative of **ergocryptine**, promotes the release of dopamine. The FDA indicated its use for the treatment of hyperprolactinemia and pituitary prolactinoma. Bromocriptine is also used for patients with Parkinson's disease[4,76] who experience deteriorating response to L-dopa.

Ergot alkaloids have also been found in some members of Convolvulaceae. Plants like *Turbina corymbosa*, *Ipomoea violacea*, and *Ipomoea asarifolia* are

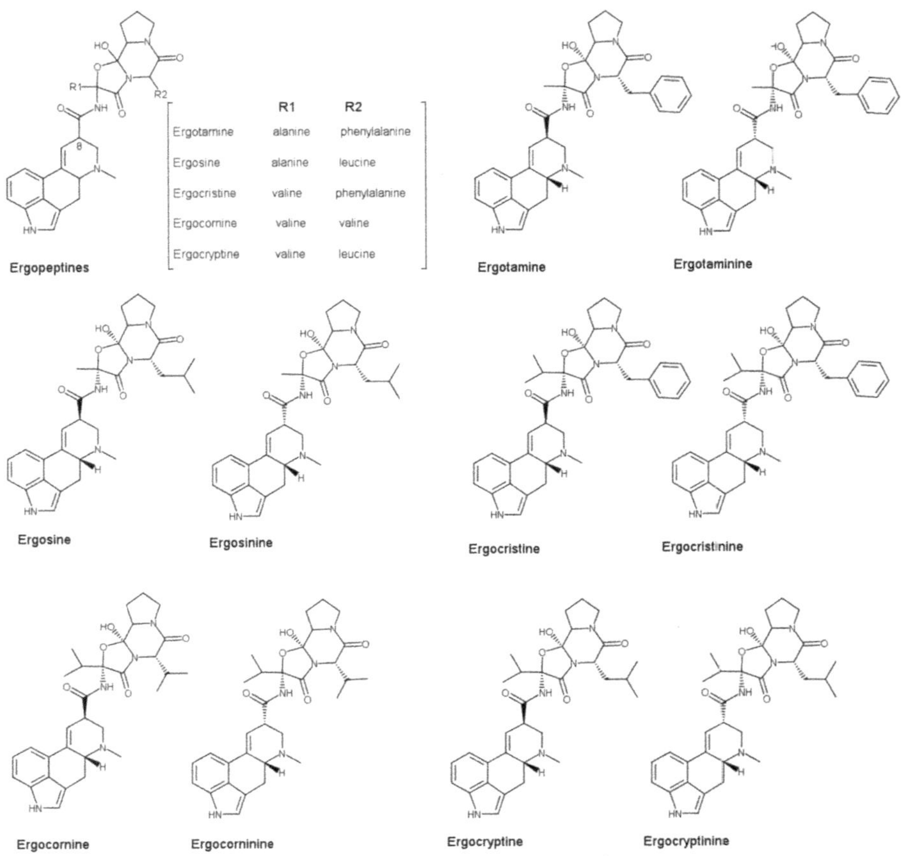

Figure 6.42 Ergot peptide alkaloids.

colonized by the ergot alkaloid producing *Periglandula* fungi living a symbiotic lifestyle. The fungi synthesize the ergot alkaloids and translocate the alkaloids *via* glandular trichomes into their host plant.[82]

6.6.2 Quinoline Alkaloids

Well-known representatives of quinoline alkaloids are **quinine**, **quinidine**, **cinchonidine**, **cinchonidinone**, and **cinchonine** produced in the root and stem bark of different species of *Cinchona* (Rubiaceae) such as *Cinchona ledgeriana* and *Cinchona succirubra.* The most important alkaloid quinine has long been used as antimalarial drug. The use of quinine was replaced by the synthetic compounds like chloroquine, primaquine, and mefloquine. Quinine is now being used as a drug for the treatment of *Plasmodium falciparum* infections which have developed resistance to chloroquine and other antimalarials. Quinidine, an optical isomer of quinine, is used for the treatment of cardiac arrythmias and atrial fibrillation. Quinidine is also detected in the bark of *Remijia pedunculata* (Rubiaceae). Another important quinoline alkaloid is **camptothecin** obtained from the seeds of *Camptotheca acuminata* (Nyssaceae), and some plants of Icacinaceae (*Nothapodytes foetida*, *Pyrenacantha klaineana*), Rubiaceae (*Ophiorrhiza pumila*), and Apocynaceae (*Ervatamia heyneana*). Camptothecin is a potential anticancer compound. Many structurally related synthetic drugs are undergoing clinical trials.[4,40,83]

Quinoline alkaloids are modified indole alkaloids. They originate from strictosidine. **Corynantheal**, and **cinchonamine** are the two indole alkaloids produced at the early stage of quinoline alkaloid biosynthesis. Corynantheal ultimately leads to the formation of other quinoline alkaloids of *Cinchona* species (Figure 6.43). Camptothecin is also formed from strictosidine *via* formation of **strictosamide** which is also an indole alkaloid (Figure 6.43).

6.7 Anthranilic Acid Derived Alkaloids

Anthranilic acid is the amino acid precursor of some quinoline, acridine, and quinazoline alkaloids (Figure 6.44).

The **furoquinoline alkaloids dictamnine** and **skimmianine**, obtained from both *Dictumnus albus* and *Skimmia japonica* (Rutaceae), are derived from anthraniloyl-CoA and malonyl-CoA. *N*-Methyl anthraniloyl-CoA, derived from anthranilic acid and 3 molecules of malonyl-CoA, produces the **acridine alkaloids melicopicine** (from *Melicope fareana*), **acronycine** (from *Acronychia baueri*) and **rutacridone** (from *Ruta graveolens*), all found in members of Rutaceae (Figure 6.45).[4,40]

Quinazoline alkaloids are formed from the amino acid anthranilic acid which is the precursor of tryptophan. **Peganin** from *Peganum harmala* (Nitrariaceae) is one example. This alkaloid is also known as **vasicine** and is responsible for bronchodilator activity of *Adhatoda vasica* (Syn: *Justicia adhatoda*) (Acanthaceae) where the alkaloid is formed from *N*-acetylanthranilic acid (Figure 6.46). Bromhexine and ambroxol are semisynthetic derivatives of vasicine, widely used as mucolytics. Peganin is associated with upper and lower respiratory tract infections and asthma. It is used as an expectorant.[4,40,83]

6.8 Pseudoalkaloids

Purine Alkaloids

Methylxanthines and methyluric acids are considered as purine alkaloids (Figure 6.47). Methylxanthines are 7-methylxanthine, caffeine, theophylline, theobromine, and paraxanthine. Methyluric acids are theacrine, liberine and methylliberine. Methyl groups are attached to the nitrogen atoms at 1, 3, 7, 9, positions or an oxygen atom at position 2.[84,85]

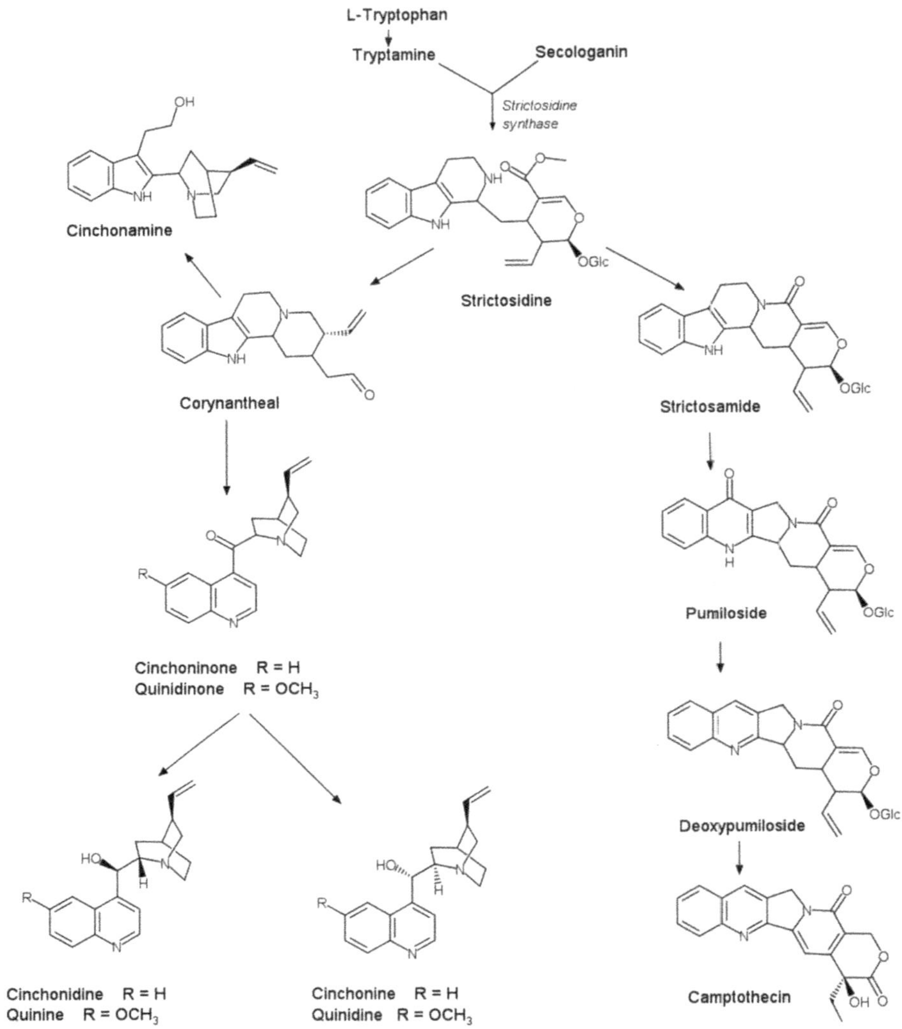

Figure 6.43 Biosynthesis of quinoline alkaloids.

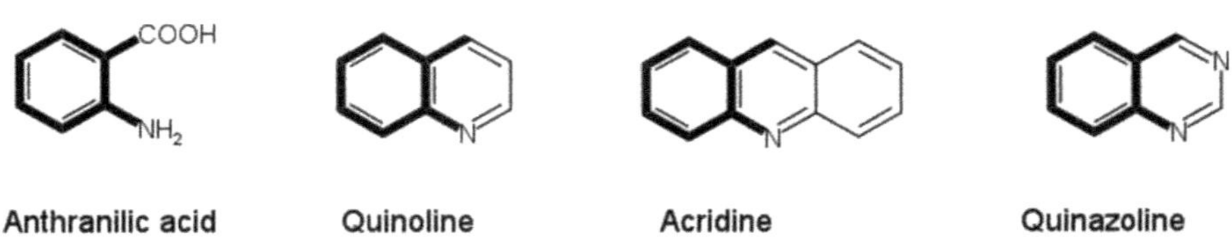

Figure 6.44 Alkaloids derived from anthranilic acid.

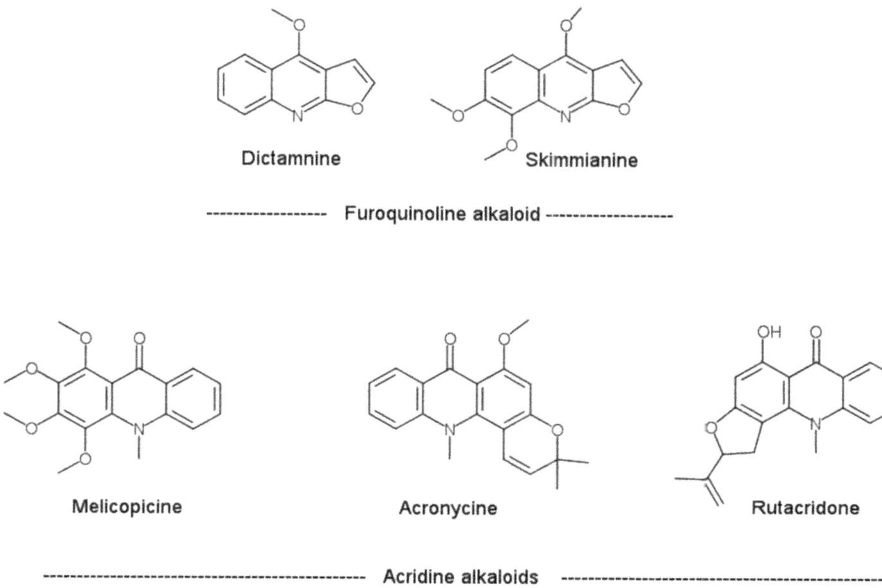

Figure 6.45 Examples of furoquinoline and acridine type alkaloids.

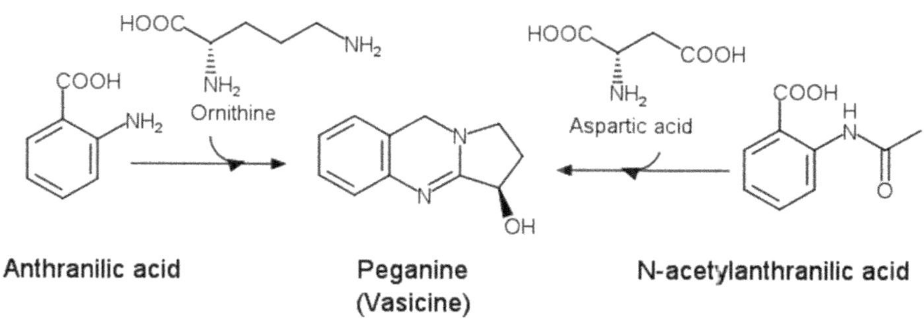

Figure 6.46 Biosynthesis of peganin (vasicine), a quinazoline alkaloid.

It is proposed that caffeine in young leaves, fruits, and flower buds of coffee and tea protect the tissues from pathogens and herbivores. Generally, short-hole borer (*Xyleborus fornicatus*) attack is followed by infection of the fungus *Monacrosporium ambrosium.* The beetle and the fungus live in symbiotic association. It has been proposed that accumulation of caffeine in tea stems after attack by beetles is a defense strategy. Caffeine also acts as allelopathic agent. Caffeine, in the seed coats and falling leaves, is released into the soil and inhibits germination of seeds around parent plants. Recently, it has also been observed that the honeybees remember the floral scent more

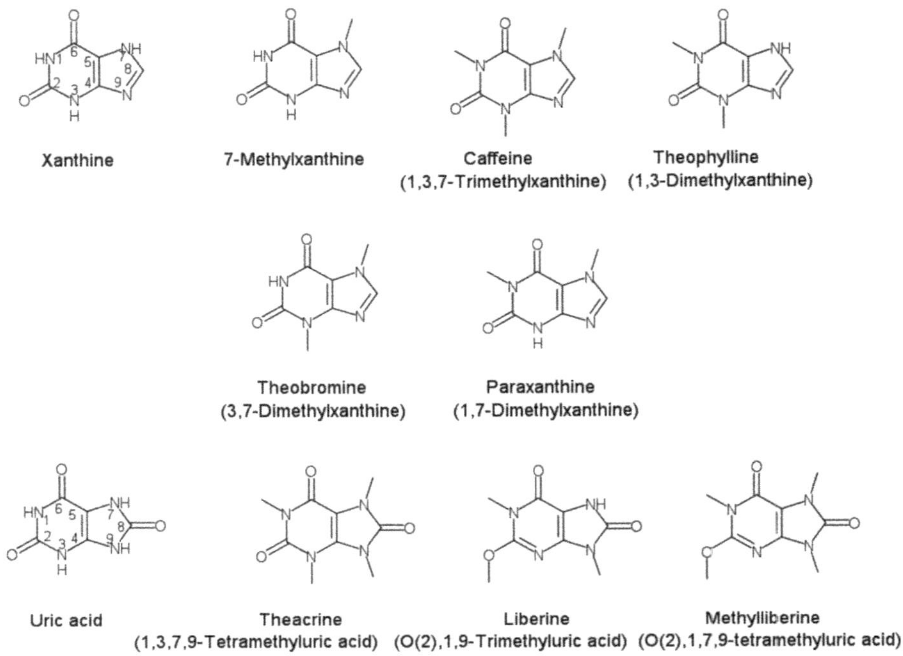

Figure 6.47 Different purine alkaloids.

when it is associated with caffeine present in the nectars of flowers of *Coffea* and *Citrus*. Caffeine appears to potentiate responses of neurons involved in olfactory learning and memory. Such behaviour leads to more effective pollination.

Purine alkaloids have been identified from at least 80 species in 13 orders of plants.[84,86] Some of the examples are shown in Table 6.3.

In the human system, caffeine is a central nervous system stimulant and has weak diuretic action. With other therapeutic agents, caffeine is used as analgesics. Theobromine is a smooth muscle relaxant and diuretic. Theophylline is a smooth muscle relaxant and is used for the relief of bronchospasm. Theacrine has been found to reduce inflammation, promote positive mood elevation, and has antioxidant activity.[40,87]

Purine alkaloids are derived from purine metabolites. The purine nucleotides are hydrolyzed by plant tissues forming nucleosides which then lead to the formation of xanthosine. The xanthine skeleton of methylxanthines is derived from xanthosine. Methylation of xanthosine produces 7-methylxanthosine. Hydrolysis and methylation of 7-methylxanthosine results in the formation of caffeine *via* theobromine or *via* paraxanthine (Figure 6.48).[84–86]

Methyluric acids are metabolic derivatives of methylxanthines. Caffeine, during catabolism is converted to theacrine with intermediate formation of 1,3,7-trimethyluric acid. Experimental evidence suggests that caffeine →

Table 6.3 Some sources of purine alkaloids[a].

Family	Plant	Common name (parts)	Alkaloids
Aquifoliaceae	*Ilex paraguariensis*	Mate (leaves)	**Caffeine**, theobromine
Malvaceae	*Cola acuminata*	Cola/Kola nut (cotyledons)	**Caffeine**, theobromine
	Theobroma cacao	Cocoa (seeds)	**Theobromine**, caffeine
	Theobroma grandiflorum	Cupu (cotyledons)	**Theacrine**
Rubiaceae	*Coffea arabica*	Coffee (seeds)	**Caffeine**, theophylline, theobromine
	Coffea liberica, Coffea canephora		Liberine, methylliberine (present at low levels)
Rutaceae	*Citrus paradisi*	Grapefruit	**Caffeine**, theophylline, theobromine, paraxanthine
	Citrus maxima	Pomelo, Chinese grapefruit (anthers)	**Caffeine**
	Citrus trifoliata	Trifoliate orange	**Caffeine**, theophylline, theobromine, paraxanthine
Sapindaceae	*Paullinia cupana*	Guarana (cotyledons)	**Caffeine**, theobromine, theophylline
Theaceae	*Camellia sinensis* var. sinensis	Tea (leaves)	**Caffeine**
	Camellia assamica	Assam tea (leaves)	**Caffeine**
	Camellia ptilophylla	Cocoa tea	**Theobromine**
	Camellia assamica var. kucha	Kucha (leaves)	**Theacrine**, methylliberine

[a]Bold names indicate predominance.

theacrine → methylliberine → liberine is the pathway occurring in species of *Coffea*.[84–86]

Immune-histochemical localization with caffeine antibodies revealed that caffeine was localized in the vascular bundles, mainly the precursor phloem in the young tea leaves. Palisade and spongy parenchyma also contained caffeine. The RNA *in situ* hybridization technique proved that the caffeine synthase gene was expressed in the palisade parenchyma, and less in the spongy parenchyma and hypoderm. At the subcellular level, caffeine is present in the vacuoles as a complex with chlorogenic acid in coffee. It was further shown that many of the enzymes were found in the purified chloroplast preparation from young tea leaves. This suggested that caffeine biosynthesis is localized in the chloroplasts, at least in tea leaves.[84]

Acetate derived alkaloids: See Chapter 3; Polyketides; Figure 3.35.
Steroidal alkaloids: See Chapter 4; Table 4.4 and Figures 4.51–4.54.
Terpenoid alkaloids: See Chapter 4, Diterpenes; Figure 4.33.

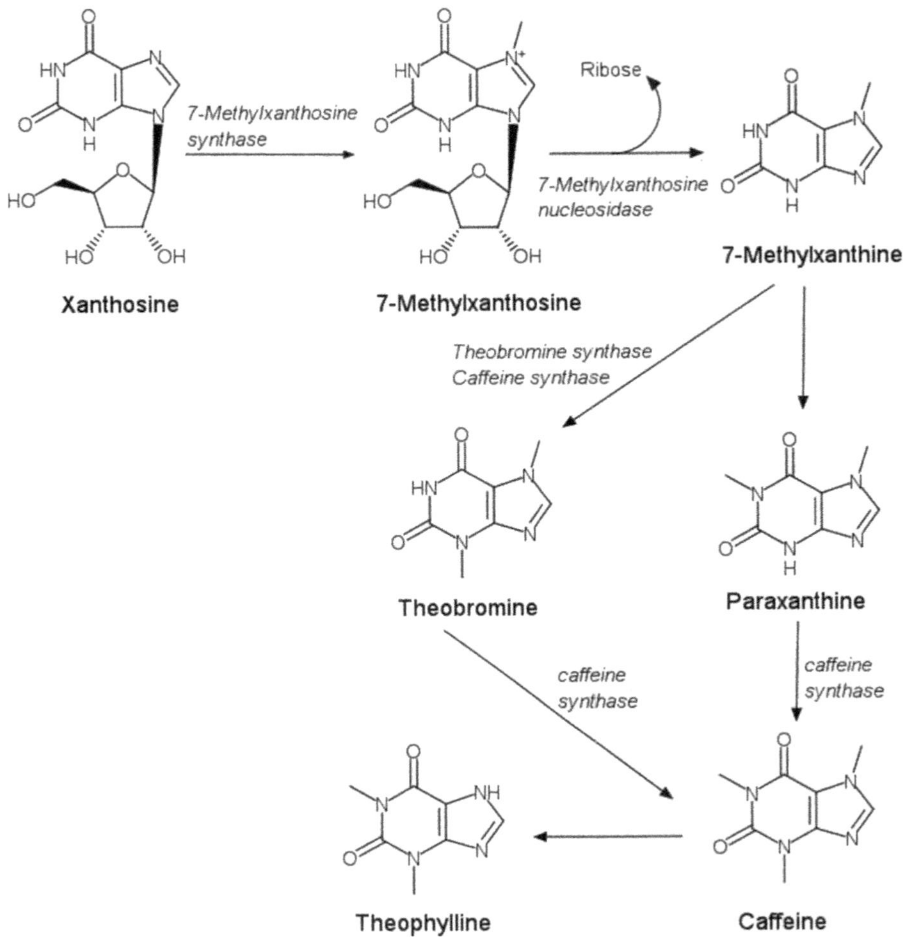

Figure 6.48 Biosynthesis of methylxanthines.

References

1. T. Aniszewski, Definition, typology and occurrence of alkaloids, in *Alkaloids: Chemistry, Biology, Ecology, and Applications*, Elsevier BV, 2015, pp. 2–97.
2. B. R. Lichman, The scaffold-forming steps of plant alkaloid biosynthesis, *Nat. Prod. Rep.*, 2021, **38**, 103–129.
3. K. L. Kohnen-Johannsen and O. Kayser, Tropane alkaloids: chemistry, pharmacology, biosynthesis and production, *Molecules*, 2019, **24**, 796.
4. P. M. Dewick, Alkaloids, in *Medicinal Natural Products: A Biosynthetic Approach*, John Wiley & Sons Ltd, 3rd edn, 2009.

5. B. Drager, Chemistry and biology of calystegines, *Nat. Prod. Rep.*, 2004, **21**, 211–223.
6. S. Brauch, W. S. Veldmate and F. P. J. T. Rutjes, Ornithine and lysine alkaloids, in *Biosynthesis to Total Synthesis: Strategies and Tactics for Natural Products*, ed. A. L. Zugrafos, John Wiley & Sons, Inc., 2016, pp. 385–427.
7. J. Huang, C. Fang, X. Ma, L. Wang, J. Yang, J. Luo, Y. Yan, Y. Zhang and S. Huang, Tropane alkaloids biosynthesis involves an unusual type III polyketide synthase and non-enzymatic condensation, *Nat. Commun.*, 2019, **10**, 4036.
8. M. F. Roberts, D. Strack and M. Wink, Biosynthesis of alkaloids and betalains, *Annu. Plant Rev.*, 2010, **40**, 20–91.
9. J. Zeigler and P. J. Facchini, Alkaloid biosynthesis: metabolism and trafficking, *Annu. Rev. Plant Biol.*, 2008, **59**, 735–769.
10. N. C. Rubio, S. Strano-Rossi, M. J. Tabernero, L. Anzillotti, M. Chiarotti and A. M. Bermejo, Hygrine and cuscohygrine as possible markers to distinguish coca chewing from cocaine abuse in workplace drug testing, *Forensic Sci. Int.*, 2014, **227**, 60–63.
11. A. S. Biondich and D. Joslin, Coca: The history and medicinal significance of an ancient Andean tradition, *Emerg. Med. Int.*, 2016, 4048764.
12. R. Moreira, D. M. Pereira, P. Valento and P. B. Andrade, Pyrrolizidine alkaloids: chemistry, pharmacology, toxicology and food safety, *Int. J. Mol. Sci.*, 2018, **19**, 1668.
13. X. Wei, W. Ruan and K. Vrieling, Current knowledge and perspectives of pyrrolizidine alkaloids in pharmacological applications: a mini review, *Molecules*, 2021, **26**, 1970.
14. H. Ashihara, I. A. Ludwig, R. Katahira, T. Yokota, T. Fujimura and A. Crozier, Trigonelline and related nicotinic acid metabolites: occurrence, biosynthesis, taxonomic considerations, and their roles in planta and in human health, *Phytochem. Rev.*, 2015, **14**, 765–798.
15. N. Mohamadib, F. Sharififar, M. Pournamdari and M. Ansari, A review on biosynthesis, analetical techniques, and pharmacological activities of trigonelline as a plant alkaloid, *J. Diet. Suppl.*, 2018, **15**, 207–222.
16. F. F. Zenker, M. Margis-Pinheiro and A. Cagliari, Nicotine biosynthesis in *Nicotiana*: a metabolic review, *Tob. Sci.*, 2019, **56**, 1–9.
17. P. V. Minorsky, The hot and classic, *Plant Physiol.*, 2002, **128**, 7–8.
18. P. Guo, D. Wei, J. Wang, G. Dong, Q. Zhang, M. Yang and l. Kong, Chronic toxicity of crude ricinine in rats assessed by 1H NMR metabolomics analysis, *RSC Adv.*, 2015, **5**, 27018–27028.
19. N. A. Horenstein, M. Quadri, C. Stokes, M. Shoaib and R. L. Papke, Cracking the betel nut: Cholinergic activity of *Areca* alkaloids and related compounds, *Nicotine Tob. Res.*, 2019, **21**, 805–812.
20. L. Gorgani, M. Mohammadi, G. D. Najafpour and M. Nikzad, Piperine – the bioactive compound of black pepper: from isolation to medicinal formulations, *Compr. Rev. Food Sci. Food Saf.*, 2016, **16**, 124–140.
21. A. Schnabel, B. Athmer, K. Manke, F. Schumacer, F. Cotinguiba and T. Vogt, Identification and characterization of piperine synthase from black pepper, Piper nigrum L., *Commun. Biol.*, 2021, **4**, 445.

22. E. Szoke, E. Lemberkovics and L. Kursinszki, Alkaloids derived from lysine: piperidine alkaloids, in *Natural Products*, ed. K. G. Ramawat and J. M. Merillon, Springer-Verlag Berlin Heidelberg, 2013, pp. 303–341.
23. S. Bunsupa, M. Yamazaki and K. Saito, Quinolizidine alkaloid biosynthesis: recent advances and future prospects, *Front. Plant Sci.*, 2012, **3**, 239.
24. J. Zhang, S. L. Morris-Natschke, D. Ma, X. Shang, C. Yang and Y. Liu, Biologically active indolizidine alkaloids, *Med. Res. Rev.*, 2021, **41**, 928–960.
25. S. Klochkov and M. Neganova, Unique indolizidine alkaloid securinine is a promising scaffold for development of neuroprotective and antitumour drugs, *RSC Adv.*, 2021, **11**, 19185–19195.
26. P. Siengalewicz, J. Mulzer and M. Rinner, *Lycopodium* alkaloids – synthetic highlights and recent developments, in *The Alkaloids*, Elsevier Inc., 2013, vol. 72.
27. S. Bunsupa, K. Hanada, A. Maruyama, K. Aoyagi, K. Komatso, H. Ueno, M. Yamashita, R. Sasaki, A. Oikawa, K. Saito and M. Yamazxaki, Molecular evolution and functional characterization of a bifunctional decarboxylase involved in Lycopodium alkaloid biosynthesis, *Plant Physiol.*, 2016, **171**, 2432–2444.
28. J. Wang, Z. Zhang, F. Jiang, B. Qi, N. Ding, S. Y. Y. Hnin, X. Liu, J. Li, X. Wang, P. Tu, I. Abe, H. Morita and S. Shi, Deciphering the biosynthetic mechanism of pelletierine in *Lycopodium* alkaloid biosynthesis, *Org. Lett.*, 2020, **22**, 8725–8729.
29. X. Kang, C. Liu, P. Shen, L. Hu, R. Lin, J. Ling, X. Xiong, B. Xie and D. Liu, Genomic characterization provides new insights into the biosynthesis of the secondary metabolite huperzine a in the endophyte Colletotrichum gloeosporoides Cg01, *Front. Microbiol.*, 2019, **9**, 3237.
30. X. Ma and D. R. Gang, The *Lycopodium* alkaloids, *Nat. Prod. Rep.*, 2004, **21**, 752–772.
31. R. S. Nett, Y. Dho, Y. Low and E. S. Sattely, A metabolic regulon reveals early and late acting enzymes in neuroactive Lycopodium alkaloid biosynthesis, *Proc. Natl. Acad. Sci. U. S. A.*, 2021, **118**, e2102949118.
32. D. Yun, S. Y. Yoon, A. J. Park and Y. J. Park, The anticancer effect of natural plant alkaloid isoquinolines, *Int. J. Mol. Sci.*, 2021, **22**, 1653.
33. Z. Qing, Y. Xu, L. Yu, J. Liu, X. Huang, Z. Tang, P. Cheng and J. Zheng, Investigation of fragmentation behaviours of isoquinoline alkaloids by mass spectrometry combined with computational chemistry, *Sci. Rep.*, 2020, **10**, 733.
34. W. A. Kukula-Koch and J. Widelski, Alkaloids, in *Pharmacognosy*, Elsevier Inc., 2017, pp. 163–198.
35. A. R. Dos Santos and N. P. Vaz, Isoquinoline alkaloids and chemotaxonomy, in *Biodiversity and Chemotaxonomy*, ed. K. G. Ramawat, Sustainable Development and Biodiversity, Springer Nature Switzerland AG, 2019, vol. 24, pp. 167–193.
36. J. Lundstrom, Simple isoquinoline alkaloids, in *The Alkaloids*, Academic Press, 1983, **vol. XXI**, pp. 255–327.

37. J. M. Hagel and P. J. Facchini, Benzylisoquinoline alkaloid metabolism: a century of discovery and a brave new world, *Plant Cell Physiol.*, 2013, **54**, 647–672.
38. C. Weber and T. Opatz, Bisbenzylisoquinoline alkaloids, in *The Alkaloids Chem Biol.*, Elsevier Inc., 2019, vol. 81, pp. 1–114.
39. V. K. Nguyen and K. G. M. Kou, The biology and total syntheses of bisbenzylisoquinoline alkaloids, *Org. Biomol. Chem.*, 2021, **19**, 7535–7543.
40. W. C. Evans, *Trease and Evans Pharmacognosy*, Saunders Elseviers, 16th edn, 2009, pp. 353–413.
41. A. Croaker, G. J. King, J. H. Pyne, S. Anoopkumar-Dukie and L. Liu, *Sanguinaria canadensis*: Traditional medicine, phytochemical composition, biological activities and current uses, *Int. J. Mol. Sci.*, 2016, **17**, 1414.
42. A. Zhang, Y. Zhang, A. R. Branfman, R. J. Baldessarini and J. L. Neumeyer Advances in development of dopaminergic apophinoids, *J. Med. Chem.*, 2007, **50**, 171–181.
43. T. Meelaph, K. Kobtrakul, N. N. Chansilpa, Y. Han, D. Rani, W. De-Eknamkul and S. Vimolmangkang, Corregulation of biosynthetic genes and transcription factors for aporphine-type alkaloid production in wounded lotus provides insight into the biosynthetic pathway of nuciferine, *ACS Omega*, 2018, **3**, 8794–8802.
44. J. R. Jacobowitz and J. Weng, Exploring uncharted territories of plant specialized metabolism in the postgenomic era, *Annu. Rev. Plant Biol.*, 2020, **71**, 631–658.
45. E. Lee, J. M. Hagel and P. J. Facchini, Role of phloem in the biochemistry and ecophysiology of benzylisoquinoline alkaloid metabolism, *Front. Plant Sci.*, 2013, **4**, 182.
46. B. E. Cheong, T. Takemura, K. Yoshimatsu and F. Sato, Molecular cloning of an O-methyltransferase from adventitious roots of *Carapichea ipecacuanha*, *Biosci., Biotechnol., Biochem.*, 2011, **75**, 107–113.
47. A. Itoh, Y. Ikuta, T. Tanahashi and N. Nagakura, Two Alangium alkaloids from Alangium lamarckii, *J. Nat. Prod.*, 2000, **63**, 723–725.
48. I. Desgagne-Penix, Biosynthesis of alkaloids in Amaryllidaceae plants: a review, *Phytochem. Rev.*, 2021, **20**, 409–431.
49. S. Ka, M. Koirala, N. Merindol and I. Desgagne-Penix, Biosynthesis and biological activities of newly discovered Amaryllidaceae alkaloids, *Molecules*, 2020, **25**, 4901.
50. R. S. Nett, W. Lau and E. S. Sattely, Discovery and engineering of colchicine alkaloid biosynthesis, *Nature*, 2020, **584**, 148–153.
51. A. P. Santos and P. R. H. Moreno, Alkaloids derived from Histidine: Imidazole (Pilocarpine, Pilosine), in *Natural Products*, ed. K. G. Ramawat and J. M. Merillon, Springer-Verlag Berlin Heidelberg, 2013, pp. 861–882.
52. A. C. H. F. Sawaya, I. N. Abreu, N. L. Andreazza, M. N. Eberlin and P. Mazzafera, Pilocarpine and related alkaloids in *Pilocarpus* Vahl (Rutaceae), *Alkaloids: Properties, Applications and Pharmacological Effects*, ed. N. M. Cassiano, Nova Science Publishers, Inc., New York, 2010, pp. 63–80.

53. A. C. H. F. Sawaya, Y. D. Costa and P. Mazzafera, Unravelling the biosynthesis of Pilocarpine in *Pilocarpus microphyllus*, *Nat. Prod. Commun.*, 2015, **10**, 721–724.
54. T. M. Rocha, N. J. Machado, J. A. C. de Sousa, E. V. O. Araujo, M. A. Guimaraes, D. F. Lima, J. R. S. A. Leite and L. K. A. M. Leal, Imidazole alkaloids inhibit the pro-inflammatory mechanisms of human neutrophil and exhibit anti-inflammatory properties in vivo, *J. Pharm. Pharmacol.*, 2019, **71**, 849–859.
55. H. Kaur, S. Mukherjee, F. Baluska and S. C. Bhatla, Regulatory roles of serotonin and melatonin in abiotic stress tolerance in plants, *Plant Signaling Behav.*, 2015, **10**, e1049788.
56. R. Hardeland, Melatonin in plants – diversity of levels and multiplicity of functions, *Front. Plant Sci.*, 2016, **7**, 198.
57. D. Zhao, Y. Yu, Y. Shen, Q. Liu, Z. Zhao, R. Sharma and R. J. Reiter, Melatonin synthesis and function: evolutionary history in animals and plants, *Front. Endocrinol.*, 2019, **10**, 249.
58. H. R. Bravo, M. J. Iglesias, S. V. Copaja and V. H. Argandona, Phytotoxicity of indole alkaloids from cereals, *Rev. Latinoam. Quim.*, 2010, **38**, 123–129.
59. G. E. Zuniga and L. J. Corcuera, Effect of gramine in the resistance of barley seedlings to the aphid *Rhopalosiphum padi*, *Entomol. Exp. Appl.*, 1986, **40**, 259–262.
60. S. Lebecque, J. Crowet, L. Lins, B. M. Delroy, P. du Jardin, M. Fauconnier and M. Deleu, Interaction between the barley allelochemical compounds gramine and hordenine and artificial lipid bilayers mimicking the plant plasma membrane, *Sci. Rep.*, 2018, **8**, 9784.
61. M. Maver, B. Miras-Moreno, L. Lucini, M. Trevisan, Y. Pii, S. Cesco and T. Mimmo, New insights in the allelopathic traits of different barley genotypes: Middle Eastern and Tibetan wild-relative accessions vs. cultivated modern barley, *PLoS One*, 2020, **15**, e0231976.
62. J. M. Barnby and M. A. Mehta, Psilocybin and mental health – don't lose control, *Front. Psychiatry*, 2018, **9**, 293.
63. R. B. Kargbo, Psilocybin therapeutic research: The present and future paradigm, *ACS Med. Chem. Lett.*, 2020, **11**, 399–402.
64. D. E. Nichols, Psilocybin: from ancient magic to modern medicine, *J. Antibiot.*, 2020, **73**, 679–686.
65. T. Herraiz and J. Galisteo, Tetrahydro-β-carboline alkaloids occur in fruits and fruit juices. Activity as antioxidants and radical scavengers, *J. Agric. Food Chem.*, 2003, **51**, 7156–7161.
66. K. Pari, C. S. Sundari, S. Chandani and D. Balasubramanian, β-Carbolines that accumulate in human tissues may serve a protective role against oxidative stress, *J. Biol. Chem.*, 2000, **275**, 2455–2462.
67. R. Cao, W. Peng, Z. Wang and A. Xu, β-Carboline alkaloids: biochemical and pharmacological functions, *Curr. Med. Chem.*, 2007, **14**, 479–500.
68. N. S. R. S. Mota, M. R. Kviecinski, K. B. Felipe, V. M. A. S. Grinevicius, T. Siminski, G. A. M. Almeida, R. C. Zeferino, C. T. Pich, D. W. Filho and

R. C. Pedrosa, β-Carboline alkaloid harmine induces DNA damage and triggers apoptosis by mitochondrial pathway: study in silico, in vitro, and in vivo, *Int. J. Funct. Nutr.*, 2020, **1**, 1–12.

69. S. Li, Y. Wang, S. Qi, Y. Zhang, G. Deng, W. Ding, C. Ma, Q. Lin, H. Guan, W. Liu, X. Cheng and C. Wang, Analogous β-carboline alkaloids harmaline and harmine ameliorate scopolamine induced cognition dysfunction by attenuating acetylcholinesterase activity, oxidative stress, and inflammation in mice, *Front. Pharmacol.*, 2018, **9**, 346.
70. L. Zhang, D. Li and S. Yu, Pharmacological effects of harmine and its derivatives: a review, *Arch. Pharmacal Res.*, 2020, **43**, 1259–1275.
71. S. Ramani, N. Patil, S. Nimbalkar and C. Jayabaskaran, Alkaloids derived from tryptophan: Terpenoid indole alkaloids, in *Natural Products*, ed. K. G. Ramawat and J. M. Merillon, Springer-Verlag Berlin Heidelberg, 2013, pp. 575–603.
72. A. E. Mohammed, Z. H. Abdul-Hameed, M. O. Alotaibi, N. O. Bawakid, T. R. Sobahi, A. Abdel-Lateff and W. M. Alarif, Chemical diversity and bioactivities of monoterpene indole alkaloids (MIAs) from six Apocynaceae genera, *Molecules*, 2021, **26**, 488.
73. M. El-Sayed and R. Verpoorte, Catharanthus terpenoid indole alkaloids: biosynthesis and regulation, *Phytochem. Rev.*, 2007, **6**, 277–305.
74. B. St-Pierre, F. A. Vazquez-Flota and V. de Luca, Multicellular compartment of *Catharanthus roseus* alkaloid biosynthesis predicts intercellular translocation of a pathway intermediate, *Plant Cell*, 1999, **11**, 887–900.
75. J. Liu, J. Cai, R. Wang and S. Yang, Transcriptional regulation and transport of terpenoid indole alkaloid in *Catharanthus roseus*: Exploration of new research directions, *Int. J. Mol. Sci.*, 2017, **18**, 53.
76. N. Gerhards, L. Neubaur, P. Tudzynski and S. Li, Biosynthetic pathways of ergot alkaloids, *Toxins*, 2014, **6**, 3281–3295.
77. C. Crews, Analysis of ergot alkaloids, *Toxins*, 2015, **7**, 2024–2050.
78. D. Jakubczyk, J. Z. Cheng and S. E. O'Connor, Biosynthesis of ergot alkaloid, *Nat. Prod. Rep.*, 2014, **31**, 1328–1338.
79. D. G. Panaccione, Origins and significance of ergot alkaloid diversity in fungi, *FEMS Microbiol. Lett.*, 2005, **251**, 9–17.
80. P. L. Schiff, Ergot and its alkaloids, *Am. J. Pharm. Educ.*, 2006, **70**, 98.
81. E. Muller-Schweinltzer, P. Ellis and R. Ziegler, Vasoconstrictor responses to ergosine and ergosinine: evidence for the isomerization of ergosine, *Drug Dev. Res.*, 1992, **25**, 149–159.
82. U. Steiner and E. Leistner, Ergot alkaloids and their hallucinogenic potential of morning glories, *Planta Med.*, 2018, **84**, 751–758.
83. X. Shang, S. L. Morris-Natschke, Y. Liu, X. Guo, X. Xu, M. Goto, J. Li, G. Yang and K. Lee, Biologically active quinoline and quinazoline alkaloids part I, *Med. Res. Rev.*, 2018, **38**, 775–828.
84. H. Ashihara, T. Yokota and A. Crozier, Biosynthesis and catabolism of purine alkaloids, *Adv. Bot. Res.*, 2013, **68**, 111–138.

85. H. Ashihara, I. A. Ludwig and A. Crozier, *Plant Nucleotide Metabolism – Biosynthesis, Degradation, and Alkaloid Formation*, John Wiley & Sons Ltd, 1st edn, 2020, pp. 231–258.
86. H. Ashihara, H. Sano and A. Crozier, Caffeine and related purine alkaloids: Biosynthesis, catabolism, function and genetic engineering, *Phytochemistry*, 2008, **69**, 841–856.
87. Y. Sheng, J. Xiang, Z. Wang, J. Jin, Y. Wang, Q. Li, D. Li, Z. Fang, J. Lu, J. Ye, Y. Liang and X. Zheng, Theacrine from *Camellia kucha* and its health beneficial effects, *Front. Nutr.*, 2020, **7**, 596823.

CHAPTER 7

Essential Oils and Resins

7.1 Essential Oils

Essential oils (also known as ethereal oils or volatile oils) are odoriferous complex mixtures of **hydrocarbons** and their **oxygenated derivatives** produced by specialized cells present in different plant parts *e.g.*, flowers, fruits and peels, leaves, bark, roots, seeds *etc.*[1–4] in angiosperms (Table 7.1) and gymnosperms such as cedar leaf oil, cedar wood oil, juniper oil and pine oil.[3] The specialized cells (secretary structures) may be glandular hairs in Lamiaceae, modified parenchyma cells in Piperaceae, oil tubes called vittae in Apiaceae or lysigenous or schizogenous passages as in Pinaceae and Rutaceae. Essential oils may occur in all tissues as in rose petals.[2] The essential oil, obtained by mainly steam distillation or by other methods, is a **fragrant liquid** that is non-greasy, inflammable, and volatile in nature. Essential oils are obtained from more than 17 500 aromatic plants.[1] The oxygenated derivatives may include alcohols, ketones, aldehydes, ethers, oxides, esters *etc.* The essential oils are partially water-soluble, and soluble in ether, alcohol, and most organic solvents.

In general, essential oil constituents are terpenes (monoterpenes and sesquiterpenes) and phenylpropenes. Monoterpenes may be acyclic, monocyclic, and bicyclic compounds (details of structure and biosynthesis of monoterpenes are discussed in Chapter 3). Sesquiterpenes may exist in acyclic (linear) or mono-, bi-, or tricyclic forms (details of structure and biosynthesis of sesquiterpenes are discussed in Chapter 3). Phenylpropenes are plant volatiles derived from the amino acid phenylalanine (details of structure and biosynthesis of phenylpropenes have been discussed in Chapter 5). Diterpenes and triterpenes are present at very low concentrations. There are also some essential oils which are dominated by allyl polysulfides.

Chemical Diversity of Plant Specialized Metabolites: A Biosynthetic Approach
By Bratati De

Published by the Royal Society of Chemistry, www.rsc.org

Table 7.1 Some important angiosperm sources of volatile oils.

Leaves	**Flowers**
Lamiaceae	**Lamiaceae**
Mentha spicata (spearmint)	*Lavandula spica* (lavender)
Mentha piperita (peppermint)	**Oleaceae**
Ocimum basilicum (basil)	*Jasminum officinale* (jasmine)
Origanum vulgare (oregano)	**Rutaceae**
Pogostemon cablin (patchouli)	*Citrus* × *aurantium* (orange)
Salvia rosmarinus (rosemary)	**Myrtaceae**
Salvia officinalis (sage)	*Syzygium aromaticum* (clove)
Thymus vulgaris (thyme)	**Asteraceae**
Ericaceae	*Matricaria chamomilla* (chamomile)
Gaultheria procumbens (wintergreen)	*Matricaria recutita* (matricaria)
Rutaceae	**Rosaceae**
Citrus limon (Rutaceae) (lemon)	*Rosa* spp. (rose)
Lauraceae	**Poaceae**
Cinnamomum tamala (Indian bay leaf)	*Cymbopogon martini* (palmarosa)
Cinnamomum verum	**Fruits**
Myrtaceae	**Apiaceae**
Eucalyptus spp.	*Anethum graveolens* (dill)
Melaleuca leucadendron	*Carum carvi* (caraway)
Poaceae	*Coriandrum sativum* (coriander)
Cymbopogon citratus (lemon grass)	*Cuminum cyminum* (cumin)
Bark	*Foeniculum vulgare* (fennel)
Lauraceae	*Pimpinella anisum* (aniseed)
Cinnamomum zeylanicum (cinnamon)	**Schisandraceae**
Wood	*Illicium verum* (star anise)
Lauraceae	**Zingiberaceae**
Cinnamomum camphora (camphor)	*Elettaria cardamomum* (cardamon)
Santalaceae	**Fruit peel**
Santalum album (sandalwood)	**Rutaceae**
Rhizomes	*Citrus* × *aurantium* (orange)
Zingiberaceae	*Citrus limon* (lemon)
Curcuma longa (turmeric)	**Seed**
Zingiber officinale (ginger)	**Myristicaceae**
Roots	*Myristica fragrans* (nutmeg)
Poaceae	
Chrysopogon zizanioides (vetiver)	

Examples are diallyl sulfide, diallyl disulfide, diallyl trisulfide, allyl methyl trisulfide and allyl methyl disulfide detected in steam distilled essential oil of garlic (*Allium sativum*, Amaryllidaceae).[5] Dipropyl trisulfide was highest in *Allium cepa*[6] along with dimethyl trisulfide, dipropyl disulfide, allyl propyl disulfide, diethyl trisulfide *etc.* The essential oil of asafoetida (plant source *Ferula asafoetida*, Apiaceae) obtained by steam distillation[7] contained 2-butyl-1-propenyl disulfide, 1-(1-thiopropyl) 1-propenyl disulfide, and 2-butyl-3-methylthioallyl disulfide. Essential oil of mustard seeds (*Sinapis alba*, Brassicaceae) obtained by steam distillation contained 71% allyl isothiocyanate[8] (Figure 7.1). Important essential oil constituents obtained from different parts of plants are presented in Table 7.2.

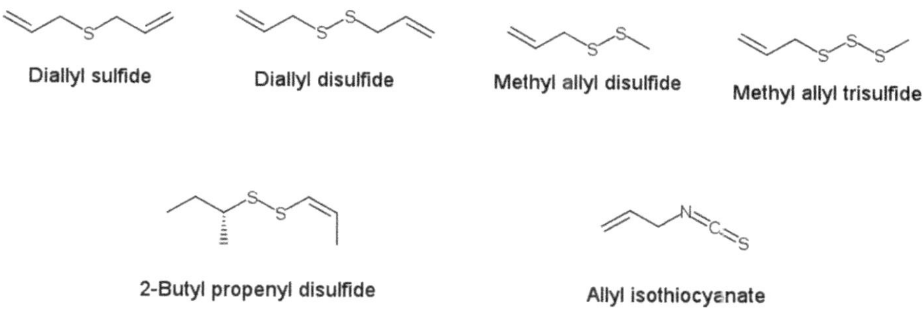

Figure 7.1 Examples of some sulfur containing volatile oil constituents.

Table 7.2 Major components of different essential oils.

- **Terpenes**
  - **Monoterpenes**
    - **Hydrocarbons**
      - **Acyclic:** Myrcene, *E*-β-Ocimene
      - **Monocyclic:** α-Terpinene, β-Phellandrene, Terpinolene, Limonene, *p*-Cymene
      - **Bicyclic:** α-Pinene, β-Pinene, 3-Carene, Camphene, α-Thujene, Sabinene
    - **Oxygenated monoterpenes**
      - **Alcohols:**
        - **Acyclic:** Geraniol, Nerol, Citronellol, Linalool
        - **Monocyclic:** Menthol, Pulegol, Thymol, Carvacrol, Terpineol
        - **Bicyclic:** Thujol, Sabinol, Myrtenol, Fenchol, Borneol
      - **Aldehydes:**
        - **Acyclic:** Geranial, Neral, Citronellal
        - **Monocyclic:** Perillaldehyde
        - **Bicyclic:** Myrtenal
      - **Ketones:**
        - **Monocyclic:** Menthone, Piperitone, Pulegone, Carvone
        - **Bicyclic:** Thujone, Pinocarvone, Fenchone, Camphor
      - **Esters:** Linalyl acetate
  - **Sesquiterpenes**
    - **Hydrocarbons**
      - **Acyclic:** Farnesane
      - **Monocyclic:** Zingiberene, ar-Curcumene, β-Bisabolene, Elemane, Germacrene-D, Humulene
      - **Bicyclic:** Cadinene, Eudesmene, Himachalene, β-Caryophyllene, Muurolene
      - **Tricyclic:** Patchoulene, Aromademdrene
    - **Oxygenated sesquiterpenes**
      - **Alcohols:** Spathulenol, α-Eudesmol, β-Eudesmol, Santalol
        - **Acyclic:** Farnesol, Nerolidol
        - **Monocyclic:** Bisabolol, Elemol, Humulol
        - **Bicyclic:** Eudesmol, Himachalol
        - **Tricyclic:** Patchouli alcohol
      - **Ketone:**
        - **Monocyclic:** Elemenone, Germacrone
        - **Tricyclic:** Patchoulenone
      - **Aldehyde acyclic:** Sinesal
      - **Esters:** Bornyl acetate
      - **Epoxide:** Humulene oxide, Caryophyllene oxide

(*continued*)

Table 7.2 (*Continued*)

Phenylpropenes	
	Eugenol, Isoeugenol, Methyleugenol, Methyl isoeugenol, Elemicin
	Anethole, *cis*-Asarone, *trans*-Asarone, Chavicol, Isochavicol, Estragole, Cinnamyl acetate, Cinnamaldehyde, Apiole, Dillapiole
Sulfur containing compounds	
	Diallyl sulfide, Diallyl disulfide, Diallyl trisulfide, Allyl methyl trisulfide, Allyl methyl disulfide, 2-Butyl-1-propenyl disulfide, 1-(1-Thiopropyl) 1-Propenyl disulfide, and 2-Butyl-3-methylthioallyl disulfide
	Allyl isothiocyanate

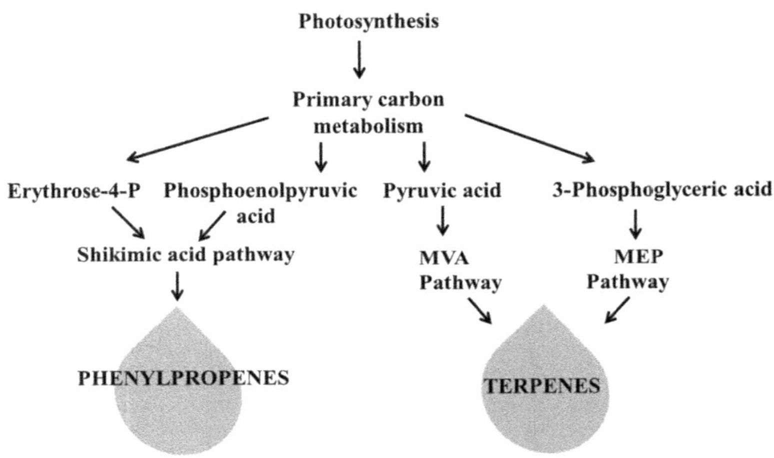

Figure 7.2 A general scheme for biosynthesis of essential oil constituents.

Biosynthetic pathways of different types of terpenes and phenylpropenes as essential oils have been discussed in detail under Chapters 3 and 5. A general scheme for biosynthesis of essential oil constituents is shown in Figure 7.2.

7.1.1 Role in Plants

Essential oils seem to have a role in plant defense against abiotic factors (light, temperature, drought, ozone levels) and biotic factors (pathogens, herbivores, harmful insects). They have beneficial effects in attracting animals for pollination and seed dispersal. Constitutive essential oils act on herbivores by repelling them or by inhibiting their feeding. Pathogens or herbivores also may induce formation of essential oils through the activation of defense signals and defense responses. Herbivore induced plant volatiles (HIPVs) not only communicate between infested plants and natural attacking insect enemies but also warn the neighbouring undamaged plants (inter-plant signaling) regarding forthcoming attacks. HIPVs also communicate within different parts of the same plant (intra-plant signaling). Such volatile oils are specific for the plant and insect/herbivore system. Plants exposed to HIVPs also alter their defense related metabolites. In addition to mono- and sesquiterpenes, other specialized metabolites such as *n*-hexenal, (*Z*)-3-hexenol, (*Z*)-3-hexen-1-yl acetate and isomers as well as phytohormones like ethylene, methyl salicylate, jasmonic acid *etc.* are produced within a few seconds of wounding caused by herbivore attack, particularly by insects. These metabolites also modulate the systemic acquired resistance (SAR) in the plant preparing them for future attacks by intra-plant signaling. Monoterpenes increase thermal tolerance to photosynthesis and protect photosynthetic apparatus under high temperature stress.[9–11]

The most famous example of allelopathic interaction between higher plants is the "*Salvia* phenomenon". A *Salvia leucophylla* (Lamiaceae) population is surrounded by a zone of bare soil. This pattern of vegetation is due to the monoterpenes, particularly 1,8-cineole and camphor. These volatile compounds released from the *Salvia leucophylla* leaves are absorbed in the soil and inhibit germination and seedling growth of annual herbs.[9,12]

7.1.2 Role in Human Use

Essential oils have been known as antimicrobial agents. Carvacrol, the major essential oil constituent of oregano oil and thyme oil, has antimicrobial activity against most gram-positive and gram-negative bacteria. This monoterpene alcohol disintegrates the outer membrane of gram-negative bacteria increasing permeability of the cytoplasmic membrane. Carvacrol also interacts with the membranes of gram-positive bacteria altering the permeability for cations like H^+ and K^+. This component has higher activity on gram-positive bacteria than on gram-negative bacteria. Essential oil of rosemary and the major monoterpene components α-pinene, β-pinene, myrcene, 1,8-cineole, borneol, camphor and verbenone show antimicrobial activity by disruption of integrity of the bacterial membrane.[4,13]

Many of the essential oil constituents are also reported to possess cytotoxic activity, anticancer activity, antifungal activity and antiparasitic activity against human protozoal pathogens. Several terpenoid essential oil constituents such as carvacrol, thymol and citronellol mimic plant phenols and possess antioxidant activity. However, other constituents like camphor, bornyl acetate and 1,8-cineole exhibit low antioxidant activity. Phenolic essential oil constituents *e.g.*, eugenol in clove are antioxidant in activity. However, paucity of clinical trials limits the effective utilization of essential oils as phytotherapeutic agents.[4,9] Essential oils protect foods against pathogenic and spoilage microorganisms. Film containing clove oil has been used for packaging fish and meat. It reduced gram-negative bacteria and H_2S-producing bacteria but did not inhibit growth of lactic acid bacteria.[4]

Aromatherapy is one of the complementary and alternative therapies where essential oils are utilized to treat different human ailments. Inhalations, baths, or local application are different methods by which essential oils penetrate the skin surface. Aromatherapy has been used in Egypt, China, and India for about 6000 years.[14]

Essential oils are used in the cosmetic industries for their pleasant aromas. The high valued essential oils used as fragrances are lavender, eucalyptus, citrus, rose oil and other floral oils. The aroma of linalool, geraniol, citronellol, limonene, and citral are appreciated by consumers in cosmetics.[3]

Eugenol, a phenylpropene obtained from clove, cinnamon, pepper, *Ocimum sanctum*, *Ocimum basilicum* (Lamiaceae) and *Piper betle* (Piperaceae) leaf, is a strong antimicrobial agent. Eugenol is also antioxidant and has effects against inflammation, hyperglycemia, cardiovascular diseases, neural disorders, and cancer. Eugenol is considered as safe by World Health Organization and is non-mutagenic. However, more studies are required for functional applications. Clove oil is an expectorant. Eugenol has long been used topically for toothache. Eugenol is used in dentistry in filling materials, endodontic sealers, dental cements and periodontal dressing materials mainly for its analgesic, and anti-inflammatory properties.[15,16]

Menthol, a monoterpene, is the major constituent of *Mentha piperita* and *Mentha canadensis* (Lamiaceae). Other constituents are menthone, isomenthone, pulegone, *etc.* Menthol occurs as different isomers (Figure 7.3). (−)-Menthol (L-menthol) is the major form of menthol and retains better

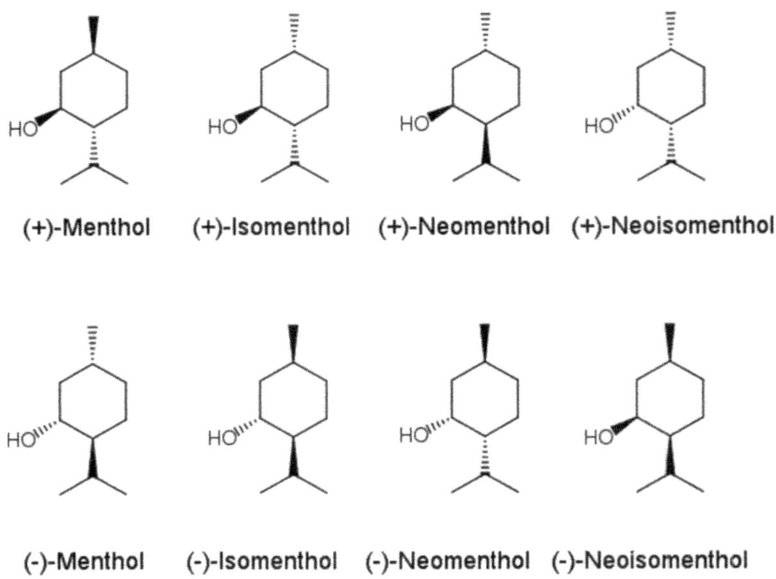

Figure 7.3 Stereoisomers of menthol.

cooling properties than the other isomers. Menthol causes a cooling sensation when inhaled or chewed. It is used in the tobacco and food industries as a flavour enhancer and preservative. Menthol is used in topical formulations with anti-inflammatory analgesics because of its cooling effect and penetration enhancing effect. Menthol is also used in pharmaceuticals, confectionary, oral hygiene products and cosmetics as a flavouring agent. It has also been used as pain reliever since ancient ages.[17–19]

Other important essential oils are described in Table 7.3 mentioning their sources, major constituents, and uses.[2–4,20–29]

7.2 Resins

Resins are amorphous products of plant tissues. These are chemically complex products. Based on the chemical components, Langenheim[30] has defined resin as a lipid soluble mixture of volatile or non-volatile terpenoid and/or phenolic compounds, usually secreted in specialized structures located internally or on the surface of the plant and has potential significance in ecological interactions. Terpenoid resins contain, predominantly, monoterpenes and sesquiterpenes as volatile components and diterpenes and triterpenes as non-volatile components. Phenolic resins contain phenylpropanoids or lipophilic flavonoids. Flavonoids exist as water soluble glycosides or lipophilic aglycones. **Balsams** consist of cinnamic acid and benzoic acid and are phenolic resins. However, some of them also contain terpenoids. Balsams are known from the fragrance of their volatile components. Balsams are soft and malleable. **Oleoresins** are fluid terpenoid resins. They have a higher ratio of volatile to non-volatile terpenes. Some resins are mixed with polysaccharides and are known as **gum resins**. Resins containing polysaccharides and volatile oils are known as **oleo-gum resins. Miscellaneous resins** do not fit into other types of resins. Resins often found in glycosidal combinations are known as **glycoresins**. **Amber** is fossilized plant resin. Conifers produce internally secreted terpenoid resins. Angiosperms produce both terpenoid and phenolic resins internally or on the surface of plants. Most resins are obtained by applying incisions to the bark of trees or shrubs. Fatty acids also occur in some resins. Resins have different applications (Table 7.4).[2,30–38]

Resins can provide defense producing toxic, deterrent effects to herbivores or inhibitory effects to pathogens. Different chemical components in resins enable them to interact with different organisms. The terpenes and phenolic compounds may attract beneficial parasitoids or predators of herbivores attacking the plant thus, indirectly providing defense to plants. Plant defense may break down due to development of resistance amongst herbivores and pathogens to the toxic resins. Constitutive resins are stored in the secretary structures whereas induced resins are produced at the site of injury.[30]

Table 7.3 Major constituents and uses of some commercially important essential oils.

Essential oil	Source	Major constituents	Uses
Aniseed oil	Dried ripe fruits of *Pimpinella anisum* (Apiaceae)	*trans*-Anethole, fenchone, estragole, linalool, anisaldehyde	Flavouring agent and carminative
Caraway oil	Dried ripe fruits of *Carum carvi* (Apiaceae)	Carvone, limonene, dihydrocarvone, carveol, β-myrcene	Flavouring agent and carminative, used in non-ulcer dyspepsia
Cardamon oil	Dried ripe fruits of *Elettaria cardamomum* (Zingiberaceae)	Terpinyl acetate, 1,8-cineole, α-terpineol, β-linalool, sabinene, nerolidol, *p*-cresol	Flavouring agent in curries, tea and cake; some medicinal properties
Coriander oil	Dried ripe fruits of *Coriandrum sativum* (Apiaceae)	(+)-Linalool, γ-terpinene, limonene, *p*-cymene, α-pinene	Flavouring agent and carminative
Cumin	Dried ripe fruits of *Cuminum cyminum* (Apiaceae)	Pinene, α-terpineol, cumin aldehyde (cuminic aldehyde/4-isopropylbenzaldehyde)	Cumin is a spice
Dill oil	Dried ripe fruits of *Anethum graveolens* (Apiaceae)	Carvone, dihydrocarvone, carveol, limonene, dihydrocarveol, α-terpinene, γ-terpinene, α-phellandrene	Flavouring agent and carminative, used in infant's gripe water.
Fennel	Dried ripe fruits of *Foeniculum vulgare* (Apiaceae)	*trans*-Anethole, anisaldehyde, fenchone, estragole, limonene, fenchol	Carminative and aromatic
Star anise oil	Fruits of *Illicium verum* (Schisandraceae)	*trans*-Anethole, estragole, anisaldehyde, small amount of safrole, chavicol methyl ether	Flavouring agent and carminative
Citronella oil	Leaf of *Cymbopogon nardus* (Poaceae)	Citronellal, geraniol, borneol	Mosquito and other insect repellent; perfume in manufacture of soaps, detergents, aerosol sprays, *etc.*
Lemon grass	Leaves of *Cymbopogon citratus* (Poaceae)	Citral, citronellol, geraniol, myrcene	Oil used in perfume, cosmetic, soap industries; leaves used for flavouring soups, currics, tea *etc.*

Eucalyptus oil	Fresh leaves of various species of *Eucalyptus e.g.*, *Eucalyptus globulus*, *Eucalyptus citriodora*, *Eucalyptus australiana*, *Eucalyptus smithii*, *Eucalyptus polybractea* (Myrtaceae)	1,8-Cineole, *o*-cresol, citronellol, citronellal, citronellyl acetate, *p*-cymene, limonene, linalool, aromadendrene, α-pinene, α-terpineol, γ-terpinene	For alleviation of nasopharyngeal infections, for treating coughs, as nasal decongestant, taken as inhalations, lozenges, externally as ointments; used in pharmaceuticals, perfumery, and industry (soaps, detergents *etc*); antibacterial, fungicidal, antiseptic; allelopathic
Geranium oil	Fresh leaves of *Pelargonium graveolens* (Geraniaceae)	Geraniol, citronellol, linalool, menthol, borneol	Flavouring agent in perfume, toothpastes, soap industry, ointments
Patchouli oil	Leaves of *Pogostemon patchouli* (Lamiaceae)	Patchoulol, eugenol, cadinene, cinnamaldehyde, β-patchoulene, elemene, caryophyllene	Insecticide, used for imparting odour to garments; ingredient of soap, cosmetics; finest fixatives for heavy perfumes
Peppermint oil	Dried leaves of *Mentha* × *piperita* (Lamiaceae)	Menthol, menthone, menthyl acetate, menthofuran, pulegone, limonene, cineole	Used in digestive disorders (irritable bowel syndrome, non-ulcer dyspepsia), topically in the treatment of tension headache. Shows adverse effect at higher dose
Spearmint oil	Dried leaves of *Mentha spicata* (Lamiaceae)	(−)-Carvone, (−)-limonene, phellandrene and esters	Used in flavouring of chewing gums, toothpastes, and other oral products, cosmetics, food industries
Sage oil	Several species of *Salvia e.g.*, *Salvia sclarea* (Lamiaceae)	α-Thujone, β-thujone, linalool, linalyl acetate, α-terpineol, germacrene D, sclareol	Perfumery and flavouring
Lavender oil	Dried flowers of *Lavandula angustifolia* (Lamiaceae)	Linalool, linalyl acetate, camphor, lavandulol, lavandulyl acetate, eucalyptol	Fragrant oil used in toiletry and perfumery industry, aromatherapy, occasionally in ointments; used as insect repellent for protection of linen garments; has a long history of medicinal use as an antidepressant, sedative, for epilepsy and migraine attack

(*continued*)

Table 7.3 (*Continued*)

Essential oil	Source	Major constituents	Uses
Clove oil	Dried flowers of *Syzygium aromaticum* (Myrtaceae)	Eugenol, acetyl eugenol, α-caryophyllene, β-caryophyllene	Flavouring agent, aromatic, antiseptic
Matricaria oil (chamomile oil)	Fresh or dried flowers or flowering tops of *Matricaria recutita*, (Asteraceae); common name chamomile	Rich in (−)-α-bisabolol or bisabolol oxides A and B, farnesene. Matricin is converted to chamazulene which is not present in the fresh flowers. It is produced during steam distillation and is responsible for the blue colour of the oil	Used as mild sedative and for digestion; used in perfumery, cosmetics, aromatherapy, food industry, toothpaste. Chamazulene and α-bisabolol are anti-inflammatory in nature
Jasmine oil	*Jasminum officinale* var. *grandiflorum* (Oleaceae)	Benzyl acetate, *cis*-jasmone, benzyl benzoate, geraniol, linalool, linalyl acetate, nerol, terpineol, cresol, methyl anthranilate *etc. cis*-Jasmone and methyl jasmonate contribute strongly to the jasmine fragrance	Perfumery, cosmetics, in air fresheners, deodorants
Oil of rose. Attar of rose	Fresh flowers of *Rosa* × *damascena*, *Rosa* × *gallica*, *Rosa* × *alba*, *Rosa centifolia* (Rosaceae)	Stearoptene, geraniol, citronellol, nerol, 2-phenylethanol	Perfumery
Rosemary oil	Flowering aerial part of *Rosmarinus officinalis* (Lamiaceae)	α-Pinene, β-pinene, camphene, borneol, camphor, bornyl acetate, cymene, eucalyptol, *trans*-caryophyllene	Used in aromatherapy to stimulate the central nervous system and blood circulation, to relieve muscular pain; used in perfumery as components of soaps, room fresheners, deodorants, perfumes, skin lotions, shampoo (can help to control dandruff and greasy hair)
Thyme oil	Fresh flowering aerial parts of *Thymus vulgaris*, *Thymus zygis* (Lamiaceae)	Thymol, carvacrol, linalool, β-myrcene, γ-terpinene, *p*-cymene, terpine-4-ol	Antibacterial and antioxidant properties reported
Juniper oil	Cones of *Juniperous communis* (Cupressaceae)	α-Pinene, β-pinene, β-myrcene, sabinene, limonene, terpinen-4-ol	Traditionally used as diuretic, carminative, anti-rheumatic

Orange oil	Fruit peel of *Citrus aurantium* (Rutaceae)	(+) Limonene, citral, citronellal, methyl anthranilate	Flavouring agent
Orange flower oil; oil of Neroli	Fresh flowers of *Citrus aurantium*	*trans*-Nerolidol, geranyl acetate, α-terpineol, linalyl acetate, linalool, limonene, β-pinene	Aromatherapy, perfumery in scenting soaps, shampoo, skin cream lotions
Lemon oil	Fruit peel of *Citrus limon* (Rutaceae)	(+)-Limonene, citronellal, geranyl acetate, geranial	Flavouring agent, in perfumery in beverages, cakes, pastries, desserts, ice creams; used in cosmetics
Nutmeg oil	Dried kernels of the seeds of *Myristica fragrans* (Myristicaceae)	α-Pinene, β-pinene, γ-terpinene, sabinine, myristicin, limonene, terpinen-4-ol, car-3-ene, safrole and minor constituents *e.g.*, eugenol, methyl eugenol, methoxyeugenol, methyl isoeugenol, isoeugenol, elemicin, isoelemicin	Flavouring agent
Camphor	Wood of *Cinnamomum camphora* (Lauraceae)	Camphor	Externally as counterirritant, rubefacient, antiseptic; a component of incense
Cinnamon oil	Dried stem bark of *Cinnamomum verum* (Lauraceae)	*trans*-Cinnamic aldehyde, eugenol, pinene, phellandrene, caryophyllene	Flavouring agent, mild astringent, carminative, germicide
Sandalwood oil	Heart wood of *Santalum album* (Santalaceae)	α-Santalol, β-santalol, α-santalenes, β-santalenes, bergamotene, γ- and β-curcumene, β-bisabolene, α-bisabolol	Perfumery
Tea-tree oil	Leaves and terminal branches of *Melaleuca alternifolia* (Myrtaceae)	Terpinen-4-ol, α-terpinene, γ-terpinene, *p*-cymene, cineole	Included in antiseptic creams
Vetiver	Roots and rhizomes of *Vetiveria zizanioides* (Poaceae)	Vetiverols, α- and β-vetivones	Perfumes, soaps, cosmetics

Table 7.4 Resin types.

Type	Source	Major components	Uses
Terpenoid resins			
Rosin or colophony	*Pinus palustris* (Pinaceae) and other species of *Pinus*	90% resin is resin acids that consists of abietic acid type (abietic acid, neoabietic acid, palustric acid, levopimaric acid, and dehydroabietic acid), and pimaric acid types (pimaric acid, isopimaric acid, and sandaracopimaric acid), 10% resin consists of corresponding esters, aldehydes, and many other terpenes	To make waxes, varnishes, for coating strings of bows to play musical instruments, in permanent colour markers, printing inks, printing paper surfaces, furniture polish, sealing wax, floor coverings; known for allergenic activity (main component being 15-hydroperoxyabietic acid, a contact allergen)
Mastic	Stem of shrubby tree *Pistacia lentiscus* var. *chia* (Anacardiaceae)	Triterpenoids moronic acid, oleanonic acid, masticadienolic acid, masticadienoic acid, oleanolic acid; dipterocarpol, lupeol, β-amyrin, β-amyrone, germanicol, 1,4-poly-β-myrcene	Used as food additive, in chewing gum, dental varnish to seal cavities, used as wound adhesive and on surgical tapes used for dermatologic surgery
Turpentine (Oleoresin)	*Pinus palustris* and other species of *Pinus* (Pinaceae)	α-Pinene, β-pinene	Externally as counterirritant
Ginger (Oleoresin)	Rhizome of *Zingiber officinale* (Zingiberaceae)	Gingerol (the main pungent ingredient), shogaols (major gingerol degradation product), gingeridione *etc.*; gingerol being the major compound of the oleoresin	Condiment, aromatic stimulant, carminative; used in different food items, soft drinks, and many medicinal substances
Copaiba (Oleoresin)	Trunk of the tree of different species of *Copaifera e.g.*, *Copaifera officinalis, Copaifera reticulata, Copaifera duckei etc.* (Fabaceae).	Main sesquiterpenes (β-caryophyllene, α-farnesene, β-bisabolene, α-humulene, α-copaene, δ-cadinene, α-cadinol, β-elemene, β-cubebene, α- and β-selinene, *trans*-α-bergamotene *etc.*), diterpenes (copalic acid, *ent*-agathic acid, kaurenoic acid, polyalthic acid, hardwickiic acid)	In folk medicine it is used for purposes such as wound healing, anti-inflammatory, anti-gastric ulcer, and other digestive diseases

Agarwood (oleoresin)	Trunks and branches of different species of *Aquilaria* and *Gyrinops* (Thymelaeaceae). Resins are formed by damaging insects or other organisms. Fungal infection promotes accumulation of volatile compounds.	Sesquiterpenes (guaiol. espatulenol, α-eudesmol, α-copaen-11-ol, agarospirol, 1,3,3-trimethyl-2-vinylcyclohexane); diterpenes [2-methylenecholestan-3-ol, 4,6-di-*tert*-butyl-*o*-cresol, (1-methoxy-4-methyl-3-pentenyl)benzene], monoterpenes (thymol). alkanes (hexadecane, 2,8-dimethylundecan, heptatriacontan-1-ol), *etc.*	Medicine, incense across Asia
Phenolic resins			
Myrrh (oleo-gum resin)	*Commiphora myrrha* (Burseraceae)	Volatile oils (pinene, limonene, cadinene, furanoeudesma-1,3-diene, curzarene), cinnamaldehyde, eugenol, non-volatile terpenoids commiphoric acid and amyrins, cell wall polysaccharides	Used in mouthwashes, toothpastes, gargles; employed as stimulant, stomachic
Balsams			
Tolu Balsam	*Myroxylon balsamum* var. *genuinum* (Fabaceae)	Toluresinotannol, cinnamic acid, benzoic acid, benzyl benzoate, benzyl cinnamate, free benzoic acid, volatile oil (eugenol, farnesol, geraniol, menthol, citronellal, linalool, safrole, cinnamaldehyde, nerolidol pinene), vanillin, ferulic acid, cinnamyl benzyl ferulate, flavonoids (7-hydroxy-4-methoxyisoflavone and some others) *etc.*	Expectorant, flavouring in medicinal syrups, confectionery, chewing gum, perfumery; sun block agent

(*continued*)

Table 7.4 (*Continued*)

Type	Source	Major components	Uses
Peru Balsam	Sap wood of *Myroxylon balsamum* var. *pereire*, (Fabaceae)	Peruresinotannol, benzoic acid, cinnamic acid, cinnamyl cinnamate (or styracin), benzyl cinnamate (or cinnamein), benzyl benzoate, terpenes (α-pinene, *p*-cymene, α- and β-bourbonene, caryophyllene, α- and β-muurolene)	Used for aroma in ointments, cosmetics, perfumes, and soaps; also used in some regions to treat asthma, rheumatism, hemorrhoidal pain, antiseptic dressing for ulcerated surfaces; found to be allergenic agent
Storax	**Strorax** is obtained from the trunk of *Liquidambar orientalis* (known as Levant storax) or *Liquidambar Styraciflua* (known as American storax) (Altingiaceae)	**Storax (from** *L. orientalis*) contains volatile oil (styrene, α- and β-pinene), cinnamyl alcohol, cinnamic acid, isomers of 2,3-butanediol, 3-phenylpropanol, oleanolic acid, 3-epioleanolic acid. **Storax (from** *L. styraciflua*) contains cinnamic acid, cinnamyl cinnamate (styracin), phenylpropyl cinnamate, vanillin, borneol, bornyl acetate	**Storax (from** *L. orientalis*): Pharmaceutic aid for benzoin tincture, stimulant, expectorant, antiseptic. **Storax (from** *L. styraciflua*) used in tobacco industry to flavour cigarettes
Benzoin	The balsamic resin obtained from *Styrax tonkinensis* (Styracaceae) is known as **Siam benzoin**. Benzoin obtained from *Styrax benzoin* and *Styrax paralleloneurus* is known as **Sumatra benzoin**	Siam benzoin consists of coniferyl benzoate, benzoic acid, siaresinol; Sumatra benzoin consists of coniferyl cinnamate, cinnamyl cinnamate (styracin), coniferyl benzoate, cinnamic acid, benzoic acid, vanillin	Siam benzoin is used as perfumery in alcoholic beverages, baked goods, candies, chewing gum, ice cream, puddings, and as components in cosmetics, lotions, perfumes, toiletries. Sumatra benzoin possesses antiseptic stimulant, expectorant, carminative, and diuretic properties; also reported to be a contact allergen
Canada balsam	Bark of *Abies balsamea* (Pinaceae)	20% monoterpenes and sesquiterpenes, 80% non-volatile diterpenes	'Cooked' Canada balsam is used for cementing achromatic lenses for telescopes and microscopes and mounting specimens for microscopical examination

Miscellaneous resins			
Asafoetida	*Ferula assa-foetida* (Apiaceae)	Ferulic acid, umbelliferone, asaresinotannols, coumarins, sesquiterpene farnesiferols, α- and β-pinene, vanillin, gums composed of galactose-arabinose, rhamnose, glucuronic acid. The pungent odour is due to the presence of disulfides	Used in cooking as a condiment, perfumery, for getting rid of gas and bloating
Jalap	Dried tuberous roots of *Ipomoea purga* (Convolvulaceae)	Resin glycosides (complex mixtures of glycolipids): hexasaccharides of jalapinolic acids, convolvulinic acid, purgic acids A, B and C, jalapinoside	Hydragogue cathartic
Hashish and marijuana	Trichomes on floral bracts of *Cannabis sativa* (Cannabinaceae) in female plants	About 30 cannabinoids *e.g.*, cannabinoid, Δ^1-tetrahydrocannabinol (Δ^1-THC), cannabinol, cannabidiol, cannabidiolic acid, cannabichromene, cannabigerol; monoterpenes (limonene, β-phellandrene, α- and β-pinene); sesquiterpenes (β-caryophyllene, *trans*-bergamotene, β-farnesene, α- and β-selinene)	Narcotic, possesses euphoric activity (due to tetrahydrocannabinol), sedative (cannabichromene, cannabidiolic acid); tetrahydrocannabinol available for control of nausea in cancer chemotherapy in terminally ill patients
Podophylloresin	Root and rhizome of *Podophyllum peltatum*, *Podophyllum emodi* (synonym: *Podophyllum hexandrum*) (Berberidaceae)	Podophyllotoxin	Podophyllotoxins are the sources of semisynthetic anticancer compounds etoposide and teniposide; podophyllins have purgative properties
Guaiacum resin	Heart wood of *Guaiacum officinale* and *Guaiacum sanctum* (Zygophyllaceae)	Phenolic resin consists of different lignan resin acids (guaiaretic acid, guaiacic acid, dihydro-guaiaretic acid), terpene (guaiagutin), vanillin	The resin is used as a food additive, to alleviate skin disease, rheumatism, treat venereal diseases, for sore throat in lozenges

References

1. H. Eslahi, N. Fahimi and A. R. Sardarian, Chemical composition of essential oils, in *Essential Oils in Food Processing: Chemistry, Safety and Applications*, ed. S. M. B. Hashemi, A. M. Khaneghah and A. de S. Sant'Ana, John Wiley & Sons Ltd, 2018, pp. 119–171.
2. V. E. Tyler, L. R. Brady and J. E. Robbers, *Pharmacognosy*, Lea & Febiger, 600 Washington Square, Philadelphia, USA, 1988, pp. 139–155.
3. J. B. Sharmeen, F. M. Mahomoodally, G. Zengin and F. Maggi, Essential oils as natural sources of fragrance compounds for cosmetics and cosmeceuticals, *Molecules*, 2021, **26**, 666.
4. P. Tonguanchan and S. Benjakul, Essential oils: Extraction, Bioactivities, and their uses for food preservation, *J. Food Sci.*, 2014, **79**, R1231.
5. P. Satyal, J. D. Craft, N. S. Dosoky and W. N. Setzer, The chemical compositions of the volatile oils of garlic (*Allium sativum*) and wild garlic (*Allium vineale*), *Foods*, 2017, **6**, 63.
6. L. Zhang, Z. Xue, L. Ni and D. Ma, Extraction and quality analysis of volatiles from onions by coupling pilot and laboratory equipment based on multi-rectification, *Sep. Purif. Technol.*, 2014, **137**, 36–42.
7. B. Rajanikanth, B. Ravindranathan and M. L. Shankaranarayana, Volatile polysulphides of asafoetida, *Phytochemistry*, 1984, **23**, 899–900.
8. C. Peng, S. Zhao, J. Zhang, G. Huang, L. Chen and F. Zhao, Chemical composition, antimicrobial property and microencapsulation of mustard (*Sinapis alba*) seed essential oil by complex coacervation, *Food Chem.*, 2014, **165**, 560–568.
9. J. Sharifi-Rad, A. Sureda, G. C. Tenore, M. Daglia, M. Sharifi-Rad, M. Valussi, R. Tundis, M. Sharifi-Rad, M. R. Loizzo, A. O. Ademiluyi, R. Sharifi-Rad, S. A. Ayatollahi and M. Iriti, Biological activities of essential oils: from plant chemoecology to traditional healing systems, *Molecules*, 2017, **22**, 70.
10. L. J. Cseke, P. B. Kaufman and A. Kirakosyan, The biology of essential oils in the pollination of flowers, *Nat. Prod. Commun.*, 2007, **12**, 1317–1336.
11. A. R. War, H. C. Sharma, M. G. Paulraj, M. Y. War and S. Ignacimuthu, Herbivore induced plant volatiles: their role in plant defense for pest management, *Plant Signaling Behav.*, 2011, **12**, 1973–1978.
12. A. Sakai and H. Yoshimura, Monoterpenes of Salvia leucophylla, *Curr. Bioact. Compd.*, 2012, **8**, 90–100.
13. M. Friedman, Chemistry and multibeneficial bioactivities of carvacrol (4-isopropyl-2-methylphenol), a component of essential oils produced by aromatic plants and spices, *J. Agric. Food Chem.*, 2014, **62**, 7652–7670.
14. B. Ali, N. A. Al-Wabel, S. Shams, A. Ahmad, S. A. Khan and F. Anwar, Essential oils used in aromatherapy: A systemic review, *Asian Pac. J. Trop. Biomed.*, 2015, **5**, 601–611.
15. A. A. Khalil, U. Rahman, M. R. Khan, A. Sahar, T. Mehmood and M. Khan, Essential oil eugenol: sources, extraction techniques and nutraceutical perspectives, *RSC Adv.*, 2017, 7, 32669.

16. G. Chung and S. B. Oh, Eugenol as local anesthetic, in *Natural Products*, ed. K. G. Ramawat and J. M. Merillon, Springer-Verlag Berlin Heidelberg, 2013, pp. 4001–4015.
17. M. Oz, E. G. E. Nebrisi, K. S. Yang, F. C. Howarth and L. T. A. Kury, Cellular and molecular targets of menthol actions, *Front. Pharmacol.*, 2017, **8**, 472.
18. J. V. Pergolizzi, R. Taylor, J.-A. LeQuang and R. B. Raffa, The role and mechanism of action of menthol in topical analgesic products, *J. Clin. Pharm. Ther.*, 2018, **43**, 313–319.
19. G. P. P. Kamatou, I. Vermaak, A. M. Viljoen and B. M. Lawrence, Menthol: A simple monoterpene with remarkable biological properties, *Phytochemistry*, 2013, **96**, 15–25.
20. W. C. Evans, Volatile oils and resins, in *Trease and Evans' Pharmacognosy*, Elsevier Limited, 2009, pp. 263–303.
21. B. Sasikumar, Rosemary, in *Handbook of Herbs and Spices*, ed. K. V. Peter, Woodhead Publishing, 2nd edn, 2012, vol. 1, pp. 452–468.
22. S. L. Kochhar, Essential oil yielding Plants, in *Economic Botany: A Comprehensive Study*, Cambridge University Press, 2018, pp. 576–614.
23. J. K. Srivastava, E. Shankar and S. Gupta, Chamomile: A herbal medicine of the past with bright future, *Mol. Med. Report.*, 2010, 3, 895–901.
24. B. Kiligler and S. Chaudhury, Peppermint oil, *Am. Fam. Physician*, 2007, **75**, 1027–1030.
25. R. Scherer, M. F. Lemos, M. F. Lemos, G. C. Martinelli, J. D. L. Martins and A. G. da Silva, Antioxidant and antibacterial activities and composition of Brazilian spearmint (*Mentha spicata* L.), *Ind. Crops Prod.*, 2013, **50**, 408–413.
26. R. Sharma, R. Rao, S. Kumar, S. Mahant and S. Khatkar, Therapeutic potential of Citronella essential oil: A review, *Curr. Drug Discovery Technol.*, 2019, **16**, 330–339.
27. P. H. Koulivand, K. Ghadiri and A. Gorji, Lavender and the nervous system, *J. Evidence-Based Complementary Altern. Med.*, 2013, 681304.
28. D. R. Batis, H. P. Singh, R. K. Kohli, S. Kaur, Eucalyptus essential oil as a natural pesticide, *For. Ecol. Manage.*, 2008, **256**, 2166–2174.
29. K. Ashokkumar, S. Vellaikumar, M. Murugan, M. K. Dhanya, G. Ariharasutharsan, S. Aiswarya, M. Akilan, T. D. Warkentin and A. Karthikeyan, Essential oil profile diversity in cardamom accessions from Southern India, *Front. Sustainable Food Syst.*, 2021, **5**, 639619.
30. J. H. Langenheim, *Plant Resins: Chemistry, Evolution, Ecology, Ethnobotany*, Timber Press, Portland, Cambridge, 2003.
31. P. S. Murthy, R. Gautam and P. Naik, Ginger oleoresin chemical composition, bioactivity and application as bio-preservatives, *J. Food Process. Preserv.*, 2015, **39**, 1905–1912.
32. J. Liu, X. Zhang, J. Yang, J. Zhou, Y. Yuan, C. Jiang, X. Chi and L. Huang, Agarwood locations provide insight into the association between fungal diversity and volatile compounds in *Aquilaria sinensis*, *R. Soc. Open Sci.*, 2019, **6**, 190211.

33. S. Sadhra, I. S. Foulds, C. N. Gray, D. Koh and K. Gardiner, Colophony – uses, health effects, airborne measurement and analysis, *Ann. Occup. Hyg.*, 1994, **38**, 385–396.
34. A.-T. Karlberg, Colophony, in *Handbook of Occupational Dermatology*, ed. L. Kanerva, *et al.*, Springer-Verlag Berlin Heidelberg, 2000, pp. 509–516.
35. A. C. Goren, G. Bilsel, A. H. Ozturk and G. Topcu, Chemical composition of natural colophony from Pinus brutia and comparison with synthetic colophony, *Nat. Prod. Commun.*, 2010, **5**, 1729–1732.
36. R. Pereda-Miranda, M. Fragoso-Serrano, E. Escalante-Sanchez and B. Hernandez-Carlos, Profiling of the resin glycosides content of Mexican Jalap roots with purgative activity, *J. Nat. Prod.*, 2006, **69**, 1460–1466.
37. E. Bautista, M. Fragoso-Serrano and R. Pereda-Miranda, Jalapinoside, a macrocyclic bisdesmoside from the resin glycosides of *Ipomea purga*, as a modulator of multidrug resistance in human cancer cells, *J. Nat. Prod.*, 2015, **78**, 168–172.
38. D. L. Custodio and V. F. Veiga-Junior, True and common balsams, *Rev. Bras. Farmacogn.*, 2012, **22**, 1372–1383.

Subject Index